MW00830038

Material Encounters and Indigenous Transformations in the Early Colonial Americas

The Early Americas: History and Culture

VOLUME 9

The titles published in this series are listed at *brill.com/eahc*

Material Encounters and Indigenous Transformations in the Early Colonial Americas

Archaeological Case Studies

Edited by

Corinne L. Hofman
Floris W.M. Keehnen

BRILL

LEIDEN | BOSTON

Cover illustration: Precolonial and early colonial ceramics from the Cibao Valley, Hispaniola. Chicoid anthropozoomorphic *adornos* characterizing twelfth- to fifteenth-century indigenous ceramics and mixed Amerindian, African, and European ceramics typical of early colonial Spanish towns in the Americas (Photos by Marlieke Ernst and Menno L.P. Hoogland, image by Menno L.P. Hoogland).

The Library of Congress Cataloging-in-Publication Data is available online at http://catalog.loc.gov

Typeface for the Latin, Greek, and Cyrillic scripts: "Brill". See and download: brill.com/brill-typeface.

ISSN 1875-3264
ISBN 978-90-04-39245-8 (hardback)
ISBN 978-90-04-27368-9 (e-book)

This book is printed on acid-free paper and produced in a sustainable manner.

Contents

Preface: What's in a Name?

Charles R. Cobb

In a notable article on the perils of typological nomenclature as applied to the European colonial era, Stephen Silliman (2005) took me and other North American scholars to task for our reliance on the expression "contact" – as in contact-period archaeology. As he observed, by using the term contact rather than colonialism, archaeologists were in danger of emphasizing short-term encounters over entanglements, eliding the relations of power and inequality more directly implied by the concept of colonialism, and focusing on predefined traits over processes of hybridization and creolization.

There clearly is good merit to Silliman's concerns over the conflation of the notions of contact and colonialism. What perhaps got overlooked in his argument, however, is that North American archaeologists primarily use contact as short-hand for the first century or so of interactions between Native American and European cultures. It is not usually meant as a brief encounter shorn of history, domination, and resistance.

Certainly, concepts like "contact" or "protohistoric" – and "conquest period" as frequently seen in this volume – do carry considerable semantic freight and we must be wary in how we use them. Nevertheless, I still maintain that it is extremely useful to delineate distinctive chronological and spatial pulses in the colonial experience. And I am gratified to say that I think the contributions to this pathbreaking volume support my argument. The "Early Colonial Americas" alluded to in the title – more or less spanning the late 1400s to early 1600s CE in the larger circum-Caribbean region – embody relations between Native Americans and Europeans that are wholly different than for other times and places.

The authors demonstrate that the earliest phase of European colonialism in the Americas was manifested as a highly uneven event. The event can be characterized as a large wave of *conquistadores*, clergy, and colonists – largely from Iberia – rapidly roiling into a newfound world in the decades following Columbus' first landing in 1492. Moving down from a panoramic to a historically granular viewpoint, though, it is obvious that any hope of regimenting this westward surge by the Spanish or Portuguese Crowns was thwarted by the complex personalities and objectives that comprised the many expeditions and colonies. Equally important, the immense variety of indigenous societies that awaited guaranteed that each European landing on the American mainland and the Caribbean islands would spark a unique trajectory of cultural interactions. In the material world, the outcome was a form of normative flexibility, where attempts to meld the old and the new by a variety of Native American and European cultures led to complex bends and folds of similarity and difference.

A case in point is the Spanish gift kit, described in several of the chapters herein. In his original definition of the gift kit, Brain (1975) noted that several types of objects and materials seemed to be consistently favored by a wide range of Native American societies in southeastern North America, leading him to conclude that Spanish explorers had arrived at a fairly standardized suite of gifting items. These included such things as small, copper alloy bells and glass beads.

In a subsequent study, DePratter and Smith (1980) relied on an extensive analysis of written accounts to demonstrate that, rather than relying on a uniform package of gifts, European expeditions maintained a more flexible inventory with the idea that they could address variation in taste and consumption patterns between Native American leaders who they hoped to draw to their cause. In turn, indigenous groups used and modified these objects in incredibly diverse ways. As displayed through the gift kit, neither Europeans nor Native Americans represented homogeneous groups with similar beliefs, strategies, and practices.

So, yes, we must be cautious in our use of nomenclature that refers to European colonialism. Nevertheless, as documented in the case studies in this volume, there is an argument to be made for focusing on a horizon of initial colonialism dominated by Iberians. Here we see different European expeditions or settlements hailing from the same broad background, pursuing somewhat similar objectives, and carrying the same general types of material culture. Yet as soon as they made contact with new peoples and new places, there ensued an incredible ramifying of entanglement, resistance, innovation, and hybridization. This represents a vital form of baseline for comparative study.

Importantly, the patterns established during the first century or so of early colonialism defined a variety of regional traditions that built upon the histories preceding contact, and which continued to shape colonialism in the following centuries. As the authors of Chapter 16 emphasize, this very critical period was to set the stage for the globalized world as we know it today.

References

Brain, Jeffrey P. 1975. "Artifacts of the Adelantado." In *Conference on Historic Site Archaeology Paper* 8, edited by Stanley South, 129–138. Columbia: South Carolina Institute for Archaeology and Anthropology.

DePratter, Chester B. and Marvin T. Smith. 1980. "Sixteenth Century European Trade in the Southeastern United States: Evidence from the Juan Pardo Expeditions (1566–1568)." In *Spanish Colonial Frontier Research*, edited by Henry F. Dobyns, 67–77. Albuquerque: Center for Anthropological Studies.
Silliman, Stephen W. 2005. "Culture Contact or Colonialism? Challenges in the Archaeology of Native North America." *American Antiquity* 70 (1): 55–74.

Acknowledgments

The present volume is the outcome of the session entitled 'Material Encounters and Indigenous Transformations in the Early Colonial Americas' organized at the 82nd Annual Meeting of the Society for American Archaeology (SAA) in Vancouver, British Columbia, Canada, from March 29–April 2, 2017. The session was organized in the context of the Synergy project NEXUS1492: New World Encounters in a Globalizing World financially supported by the European Union's Seventh Framework Programme (FP7/2007-2013) / ERC-NEXUS1492 (grant agreement n° 319209) and directed by Prof. Dr. Corinne L. Hofman, and the NWO PhDs in the Humanities project Values and Valuables (PGW-13-02) carried out by Floris W.M. Keehnen. The aim of the session was to bring together scholars working with the topic of early colonial encounters in the Americas from a material culture perspective, particularly based on novel field data, and to contribute to global dialogues about the archaeology of colonialism.

We want to thank all the participants to the SAA session in Vancouver and the additional contributors to this volume, as well as session discussants Craig N. Cipolla and Neal Ferris. We are grateful for the thoughtful comments and useful suggestions made by two anonymous reviewers of this volume, as well as the support and guidance offered to us by Brill Publishing in realizing this project. The final preparation stages of this work were supported by a Fellowship from "The Netherlands Institute for Advanced Study in the Humanities and Social Sciences (NIAS)" awarded to Corinne L. Hofman. Finally, we would like to thank Konrad Antczak, Valeria Corona, Andrzej Antczak, and Arie Boomert for their help with the Spanish-English translations and Melissa Riesen for proofreading the English texts. Menno Hoogland is acknowledged for his help with many of the figures and great thanks to Emma de Mooij for her excellent help with the editorial work.

Corinne L. Hofman and Floris W.M. Keehnen
Leiden, September 2018

Illustrations

Figures

Tables

Notes on Contributors

Andrzej T. Antczak
(PhD University College London) is currently Associate Professor in Caribbean Archaeology and Chair of the Department of World Archaeology, Faculty of Archaeology, Leiden University, The Netherlands, and curator of the archaeological collections at the Archaeology Research Unit at Universidad Simón Bolívar, Caracas, Venezuela. His interests include island archaeology, indigenous ontologies, colonial encounters, zooarchaeology, and community archaeology. Since 1982, together with Ma.M. Antczak, he has carried out pioneering archaeological research on the off-shore islands of the Venezuelan Caribbean.

Maria Magdalena Antczak
(PhD University College London 2000) is currently a Researcher at the Faculty of Archaeology, Leiden University, The Netherlands, and Associate Professor Archaeology Research Unit, Universidad Simón Bolívar, Caracas, Venezuela. She is founder and curator together with A.T. Antczak of the Archaeology Research Unit at the Simón Bolívar University, Caracas, Venezuela. Since 1982, Antczak has been co-director (also with A.T. Antczak) of pioneering archaeological investigations on the off-shore islands of the Venezuelan Caribbean. Her interests include the (re)construction of social past in pre-Hispanic north-central Venezuela, Amerindian ontologies, and the theory and method of signifying practices applied to the study of indigenous imagery.

Oliver Antczak
is a Researcher at the Archaeology Research Unit, Universidad Simón Bolívar, Caracas, Venezuela. Holding a Liberal Arts and Sciences BA from Leiden University College The Hague, and an MPhil from the University of Cambridge he thus far has specialized in the intersections between diverse disciplines in the social-sciences/humanities. He currently focuses on heritage research, interested in how archaeological practice meets with the interests of modern populations, especially with relation to the constitution of identity, and the interrelations between identities, material culture, politics, and colonialism. He has conducted research mainly in the Venezuelan Caribbean.

Jaime J. Awe
is Associate Professor of Anthropology at Northern Arizona University, Director of the Belize Valley Archaeological Reconnaissance Project, and Emeritus member of the Belize Institute of Archaeology. He received his Bachelor's and Master's Degree in Anthropology at Trent University in Ontario, Canada, and

his PhD from the Institute of Archaeology at University College London. Between 2003–2014, he served as Director of the Belize Institute of Archaeology. During his extensive career in archaeology, Dr. Awe has conducted important research and conservation at several major sites in Belize, and his research focuses on diverse topics on the Maya, ranging from the preceramic to the colonial periods.

Martijn van den Bel

After many archaeological wanderings in the Lesser Antilles, French Guiana, and The Netherlands between 1995 and 2004 Martijn (Haarlem 1971) now works as project leader for Inrap (the French National Institute for Compliance Archaeology) in the French Antilles and French Guiana, where he lives. Next to the archaeology of the latter regions, for which he earned a PhD title at Leiden University, The Netherlands, in 2015, he is also interested in the Dutch occupation and colonization of this particular area during the sixteenth and seventeenth centuries. Recently, he published two books on this matter: *The Voyages of Adriaan van Berkel to Guiana* (with Lodewijk Hulsman and Lodewijk Wagenaar; Sidestone Press, 2014) and *Entre deux mondes. Amérindiens et Européens sur les côtes de Guyane, avant la colonie (1560–1627)* (with Gérard Collomb; CTHS, 2014).

Mary Jane Berman

(PhD SUNY-Binghamton) is co-director of the Lucayan Ecological Archaeology Project, an interdisciplinary project studying the indigenous inhabitants of the Bahama archipelago. Berman has published extensively on the archaeology of the Lucayans, focusing on Lucayan agriculture, plant use, ceramics, lithics, and mobility and exchange. She is an Associate Professor of Anthropology at Miami University, Ohio, having rejoined the Faculty, after 15 years as director of the University's Center for American and World Cultures.

Arie Boomert

studied cultural anthropology and cultural prehistory at the University of Amsterdam and Leiden University, The Netherlands. He worked as an archaeologist at the Surinaams Museum, Paramaribo, Suriname, Leiden University, the University of Amsterdam, and the University of the West Indies, St. Augustine, Trinidad. In 2011 he retired as an Assistant Professor and Senior Researcher from the Faculty of Archaeology, Leiden University, The Netherlands. He is a Curatorial Affiliate in the Division of Anthropology, Yale University, New Haven, USA. In 2005 the International Association for Caribbean Archaeology awarded him a plaque 'in recognition of years of dedicated service and commitment to the promotion and development of the Archaeology of Trinidad and Tobago.'

Jeb J. Card
is Visiting Assistant Professor and Assistant for Special Projects in the Department of Anthropology of Miami University, Ohio. He received his PhD from Tulane University in 2007. He specializes in historical archaeology, early colonialism, material culture hybridity, ethnogenesis, ceramic analysis, and pre-Hispanic Maya political history, working in Mesoamerica, chiefly in El Salvador. At the University of Miami, he also works on three-dimensional documentation and analysis of artifacts. He is editor of *The Archaeology of Hybrid Material Culture* (Southern Illinois University Press, 2013), co-editor of *Lost City, Found Pyramid* (with David S. Anderson; University of Alabama Press, 2016), and author of *Spooky Archaeology: Myth and the Science of the Past* (University of New Mexico Press, 2018).

Charles R. Cobb
is Curator and Lockwood Professor of Historical Archaeology at the Florida Museum of Natural History, University of Florida. His research focuses on Native American engagements with European colonialism in southeastern North America. He has been involved in a sustained study of Indian towns and English forts on the Carolina frontier. In a collaborative project with the Chickasaw Nation, he is also exploring the complex interactions between the Chickasaw, English, Spanish, and French in Mississippi. With funding from the National Endowment for the Humanities, Cobb is also developing an online archaeological database for the Franciscan missions of La Florida.

Gérard Collomb
is an anthropologist and Associated Researcher at the Institut Interdisciplinaire d'Anthropologie du Contemporain (EHESS/CNRS) and at the Centre Enseignement et recherche en ethnologie amérindienne (Université Paris Nanterre – CNRS) in Paris. His research interests include ethnohistory and political anthropology of Amerindians in French Guiana and Surinam during the colonial and post-colonial periods, focusing on the building processes of a multiethnic society in these two countries. He published *Les Indiens de la Sinnamary. Journal du père Jean de la Mousse en Guyane (1684–1691)* (Chandeigne, 2006) and *Entre deux mondes. Amérindiens et Européens sur les côtes de Guyane, avant la colonie (1560–1627)* (with Martijn van den Bel; CTHS, 2014).

Shannon Dugan Iverson
is an archaeologist specializing in the Aztec-to-colonial religious transition in central Mexico. She received her PhD in Anthropology from the University of Texas, Austin, in 2015. From 2016–2017 she served as a Mellon Postdoctoral Fellow at Rice University. She currently works as a Senior User Experience Researcher at Daito Design in Austin, Texas.

Marlieke Ernst

is a PhD Candidate at the Faculty of Archaeology, Leiden University, The Netherlands. Within the NEXUS1492 project, she focuses on ceramic material transformations. She investigates transcultural processes within intercultural communications at the islands of Hispaniola and Cubagua. The material reflection of this multicultural society and the agency of the enslaved and colonized are studied through the continuities and changes in the manufacture between precolonial and colonial non-European ceramics present at colonial sites. Both Amerindian (local and non-local), Spanish, and African presences are studied within the ceramic assemblages. Her study assesses the extent to which indigenous pottery traditions disappeared and the degree to which new techniques and forms appeared.

William R. Fowler

is an Associate Professor in the Department of Anthropology, Vanderbilt University. He received a PhD in Archaeology from the University of Calgary in 1981. His principal research interests include historical archaeology, architecture, landscapes, colonialism, urbanism, migrations, and the Nahua cultures of Mesoamerica. He is founder and editor-in-chief of *Ancient Mesoamerica* (Cambridge University Press). He has directed the Ciudad Vieja Archaeological Project in El Salvador since 1996.

Perry L. Gnivecki

(PhD SUNY-Binghamton) is an Assistant Teaching Professor in the Department of Social and Behavioral Sciences, Miami University, Hamilton, and the Department of Anthropology, Miami University, Oxford, Ohio. He is co-director of the Lucayan Archaeological Ecological Project. He has conducted excavations on numerous islands in the Bahamas and, for the last 15 years, has led the Department of Anthropology's archaeology field school on San Salvador and Eleuthera islands. He has also conducted archaeological research in the USA and Middle East. His research has focused on colonial encounters and site formation processes. He is a member of the Register of Professional Archaeologists.

Christophe Helmke

is Associate Professor of American Indian Languages and Cultures at the Institute of Cross-Cultural and Regional Studies, University of Copenhagen, Denmark. He teaches undergraduate and graduate courses on the archaeology, epigraphy, iconography, and languages of Mesoamerica. Besides Maya archaeology and epigraphy, other research interests include the pre-Columbian use of caves, Mesoamerican writing systems, as well as rock art and comparative Amerindian mythology.

Shea Henry
(PhD Simon Fraser University 2018) recently completed her PhD researching the precontact and the contact-era transition in Jamaica. She has studied and worked in the field of historical and contact-era archaeology for the past 12 years, focusing on zooarchaeological evidence of the change in diet over time and the intersection of diet and culture. She is currently focusing on the realm of heritage and public education as the curator of the Maple Ridge Museum.

Gilda Hernández Sánchez
was Adjunct Researcher at the Department of Anthropology, Leiden University, The Netherlands, from 2000 to 2010. Her previous work focused on the analysis of pictographic decoration on ancient and present-day ceramic vessels found in Mexico. She has published in various journals such as Journal de la Société des Américanistes, Mexicon, and Latin American Antiquity. At present she teaches Spanish and History in Germany.

Corinne L. Hofman
is Professor of Caribbean Archaeology at the Faculty of Archaeology, Leiden University, The Netherlands. Hofman has conducted fieldwork throughout the Caribbean over the past 30 years (together with Menno Hoogland). Her research and publications are highly multi-disciplinary and major themes of interest center around mobility and exchange, colonial encounters, intercultural dynamics, settlement archaeology, artifact analyses, and provenance studies. Hofman's projects are designed to contribute to the historical awareness, valorization of archaeological heritage, and knowledge exchange in the Caribbean. Since 1998, Hofman has obtained numerous research grants and prizes, including the ERC-Synergy Grant for the NEXUS1492 project in 2012. Her recent publications include *Managing our Past into the Future: Archaeological Heritage Management in the Dutch Caribbean* (with Jay B. Haviser; Sidestone Press, 2015) and *The Caribbean Before Columbus* (with William F. Keegan; Oxford University Press, 2017).

Menno L.P. Hoogland
is an Associate Professor at the Faculty of Archaeology, Leiden University, The Netherlands. Hoogland studied cultural anthropology in Leiden with a focus on prehistory and physical anthropology and wrote his PhD thesis on settlement patterns of the Amerindian population of Saba, Netherlands Antilles. He is an expert in archaeothanatology and Caribbean archaeology. Hoogland's research focuses on the funerary practices of precolonial and early colonial Amerindian societies in the Caribbean and the application of taphonomical methods for the reconstruction of funerary behaviour. He was PI of the NWO

project Houses for the Living and the Dead. Currently he is a Senior Researcher in the ERC-Synergy project NEXUS1492 at Leiden University.

Rosemary A. Joyce
is Professor of Anthropology at the University of California, Berkeley. She received her PhD from the University of Illinois-Urbana in 1985. As curator and faculty member at Harvard University from 1985 to 1994, she moved to Berkeley in 1994, and served as Director of the Hearst Museum of Anthropology until 1999. She began participating in archaeological fieldwork in Honduras as an undergraduate in 1977, and co-directed projects on early village life, the Classic period, and the colonial and Republican periods. While collaborating in research in the western Maya area with Mexican colleagues, she continues research on Honduran collections in museums.

Floris W.M. Keehnen
is a PhD Candidate at the Faculty of Archaeology, Leiden University, The Netherlands. Funded by a grant from the Netherlands Organisation for Scientific Research (NWO), he investigates indigenous Caribbean attitudes towards European-introduced material culture in early colonial times (AD 1492–1550). His research interests include the archaeology and ethnohistory of the Caribbean, indigenous value systems, colonial encounters, and trade and exchange.

Luis A. Lemoine Buffet
is President of the ARCA-Arqueología del Caribe Foundation and Researcher at Unidad de Estudios Arqueológicos (USB) in Caracas, Venezuela. He has a degree in Archaeology from the University of Leicester, UK (2005). Lemoine Buffet has more than 15 years of field archaeology experience in Venezuela and abroad. In 2008, he discovered several Archaic Age sites on Margarita Island where, since then, he has been conducting fieldwork together with a team from the Archaeology Research Unit (USB) and Instituto Venezolano de Investigaciones Científicas (IVIC), Venezuela. He specializes in osteoarchaeology of human remains including isotopic analysis, aDNA studies, and musculoskeletal biomechanics analysis.

John Angus Martin
is an Archivist, Researcher, and Historian, specializing in Grenadian history, and European colonization and slavery in the Caribbean. He is currently pursuing a PhD in Heritage Management at Leiden University, The Netherlands. He holds Master's degrees in History and Agricultural and Applied Economics

from Clemson University, South Carolina. He is the author of *A-Z of Grenada Heritage* (2007) and *Island Caribs and French Settlers in Grenada, 1498–1763* (2013). He also co-authored *The Temne Nation of Carriacou: Sierra Leone's Lost Family in the Caribbean* (with Joseph Opala and Cynthia Schmidt; Polyphemus Press, 2016) and co-edited *Perspectives on the Grenada Revolution, 1979–83* (with Nicole Phillip-Dowe; Cambridge Scholars Publishing, 2017).

Clay Mathers
is the Executive Director of the Coronado Institute, a registered non-profit in Albuquerque, New Mexico, focused on investigating and preserving early historic sites in the Western US. Mathers is a professional archaeologist with a PhD in Iberian Prehistory (University of Sheffield, UK), a MPhil in GIS and Remote Sensing (University of Cambridge, UK), and a BA in Anthropology (University of Pennsylvania). He has worked on numerous Vázquez de Coronado sites in New Mexico and has published widely on early *entradas* in the US Southwest, including two recent edited volumes: *Native and Spanish New Worlds* (with Jeffrey M. Mitchem and Charles M. Haecker; University of Arizona Press, 2013), and *The Destiny of Their Manifests* (University Press of Florida, forthcoming).

Maxine Oland
is a Lecturer in Anthropology at the University of Massachusetts-Amherst. Her archaeological research is focused on the Maya people of Belize from the fifteenth through the twentieth centuries, their interactions with European and other indigenous groups, and their eventual incorporation into the global economy. Her work on the early colonial Maya has been published in Antiquity, the International Journal of Historic Archaeology, Lithic Technology, and several edited volumes. She is co-editor of *Decolonizing Indigenous Histories: Exploring Prehistoric/ Colonial Transitions in Archaeology* (with Siobhan M. Hart and Liam Frink; University of Arizona Press, 2012).

Alberto Sarcina
is an archaeologist, graduated from the Sapienza University of Rome, Italy, with postgraduate studies at the same University. He currently is a PhD Candidate at the Faculty of Archaeology, Leiden University, The Netherlands. He has twenty years of experience as a field archaeologist in Italy and is specialized in stratigraphy and archaeological graphical documentation. He teaches Methods and Techniques of Archaeological Research at the Faculty of Heritage Studies at the Externado University, Bogotá. He is Researcher of the Colombian Institute of Anthropology and History (ICANH) and Coordinator of the Archaeological and Historical Park of Santa María de la Antigua del Darién. Since

2013, he has directed the archaeological works in Santa María de la Antigua del Darién (Colombia), which led to the recognition and delimitation of the city and the declaration of it as a Heritage of Cultural Interest of the Nation.

Russell N. Sheptak
is Research Associate at the Archaeological Research Facility of the University of California, Berkeley. He received his PhD from Leiden University, The Netherlands, in 2013. His dissertation was a study of indigenous continuity in Honduras from the sixteenth to nineteenth centuries. He conducted his first archaeological fieldwork in Meadowcroft rock shelter and has contributed to Honduran archaeology since 1980, working in archaeological sites that date from the earliest settled villages through to the twentieth century. His main research interest is in indigenous and Afro-descendent peoples in colonial-period Honduras.

Roberto Valcárcel Rojas
is a Postdoctoral Researcher at Leiden University (ERC-Synergy NEXUS1492) and Visiting Professor at the Santo Domingo Institute of Technology, Dominican Republic. His main research interests are cultural interaction, indigenous social organization in the Caribbean, and archaeology and history of the early colonial times in the Americas. He specifically focuses on the study of the Indian as a colonial category. Dr. Valcárcel Rojas is author of several books and articles about Cuban and Caribbean precolonial and colonial archaeology including *Archaeology of Early Colonial Interaction at El Chorro de Maíta, Cuba* (University Press of Florida, 2016).

Robyn Woodward
(PhD Simon Fraser University 2006) is an Adjunct Professor of archaeology at Simon Fraser University. She is also a trustee and governor of the Vancouver Maritime Museum and the vice-president of the Institute of Nautical Archaeology (INA). She has worked as an underwater archaeologist at the pirate city of Port Royal, Jamaica, and as an archaeologist and conservator on various INA projects in Turkey and the Yukon Territories, Canada. She has directed the excavation program at the site of Sevilla la Nueva, in Jamaica since 2001.

CHAPTER 1

Material Encounters and Indigenous Transformations in the Early Colonial Americas

Floris W.M. Keehnen, Corinne L. Hofman, and Andrzej T. Antczak

1 Introduction[1]

Contributions of indigenous peoples to colonial encounters in the Americas were profound, varied, and dynamic. Instead of mere respondents, let alone passive bystanders, indigenous peoples were active agents in processes of colonialism, vital in the negotiation and recreation of new colonial realities. Paradoxically, they have long been some of the most invisible craftsmen of today's societies. However, recent archaeological scholarship continues to provide material evidence that suggests that notwithstanding the severe and enduring impacts of intruding colonial powers, indigenous peoples continued to make choices that would benefit them. Among the many strategies they chose were alliance making, intermarriage, cooperation, negotiation, trading, escape, resistance, rebellion, and armed conflict. Engagement in this range of (flexible) friendly and antagonistic social and material relationships was not restricted to two-sided indigenous-European affairs. Quite the contrary, colonial processes resulted as much in shifting relations and identities among indigenous peoples, Africans, and Europeans themselves, as well as between indigenous peoples and Africans, and Europeans and Africans.

Over the past few decades, the study of colonial contact and interaction has progressed significantly with the adoption of new and revised theoretical paradigms, innovative research approaches, and multiscalar perspectives. Since the late 1990s, a conceptual framework has come to fruition that highlights colonialism's entangled and transformative nature on the premise that all parties contributed to and were impacted by the process of interactions through negotiation, creativity, and innovation. Focusing on these and related aspects including local agency, power, and resistance, as well as social constructs such as gender, race, class, and identity, archaeologists have advanced considerably in reconstructing indigenous lives in colonial settings (e.g., Anderson-Córdova

1 This introductory chapter is largely based on the introduction chapter of the PhD dissertation Values and Valuables by Floris Keehnen (forthcoming).

 | DOI:10.1163/9789004273689_002

2017; Cipolla and Hayes 2015; Deagan 2003, 2004; Dietler 2010; Ferris 2009; Ferris et al. 2014; Funari and Senatore 2015; Given 2004; Graham 2011; Liebmann and Murphy 2011; Loren 2008, 2010; Murray 2004; Oland et al. 2012; Rodríguez-Alegría 2016; Rothschild 2003, 2006; Rubertone 2000; Scheiber and Mitchell 2010; Silliman 2005, 2009; Stein 2005; Torrence and Clarke 2000; Valcárcel Rojas 2012, 2016; Van Buren 2010; Voss and Casella 2012; Whitehead 2011).

Despite addressing these important nuances, European colonialism was an unmistakably painful process, the effects of which – many still felt today – cannot and should not be minimized. Many groups were severely restricted in their self-determination, some of whom never were able to stand up to their oppressor or make choices for themselves. The current trend of decolonizing indigenous histories aims to investigate this interplay of individual experiences amidst (hostile) colonial realities (see also Atalay 2006; Bruchac et al. 2010; Jansen and Raffa 2015; McNiven 2016; Silliman 2012; Smith and Wobst 2005). To date, efforts to disrupt the many and long persistent "Grand Narratives" of European colonialism (Voss 2015) and to decolonize archaeology by recognizing long-term indigenous trajectories in fine-grained views of history (Oland et al. 2012) continue to be successful.

The developments outlined above have also promoted the understanding of the roles of material culture in processes of colonialism (e.g., Card 2013; Cipolla 2017; Cobb 2003; Cusick 1998; Funari and Senatore 2015; Gosden 2004; Liebmann 2015; Lightfoot et al. 1998; Lyons and Papadopoulos 2002; Maran and Stockhammer 2012; Richard 2015; Rodríguez-Alegría 2008, 2010; Rogers 1990; Rothschild 2003, 2006; Scaramelli and Tarble de Scaramelli 2005; Silliman 2010, 2016; Thomas 1991; Van Dommelen 2006). With respect to the study of colonial encounters in the Americas, an interest was raised in exploring the indigenous appropriation of European material culture through gifts, trade, or imitation. It resulted in the rethinking of ideas about the indigenous adoption or resistance of foreign objects, and how such differential choices not only altered indigenous material assemblages, but also affected existing social, political, and economic structures. Nowadays, our understanding of material encounters in the colonial Americas largely comes from case studies in North America, where over the past decades updated theories on, for example, consumption, hybridity, and entanglement, have already been successfully applied. Building upon these efforts, this volume will specifically target the previously underrepresented Caribbean and its surrounding mainland(s), thereby focusing on the period of Spanish/European colonialism from AD 1492 to 1800. Working from a critical understanding of indigenous long-term historical trajectories, the authors will discuss how foreign goods were differentially employed across time, space, and scale; how these were considered within indigenous ontologies and value systems; what implications their adoption had for larger indigenous society; and which theoretical approaches and methodologies better help us understand indigenous material practices.

2 Towards a Material Perspective of Colonial Encounters in the Americas

In the Americas, including the broader Caribbean region defined above, the efforts of many scholars have greatly contributed to the creation of a more balanced and better-informed understanding of early colonial dynamics. Yet, whereas social and cultural issues of the colonial period have been important topics of inquiry, only a few studies on indigenous histories in this part of the Americas have taken a more profound interest into material aspects of the colonial encounter (e.g., Boomert 2002; Cooper et al. 2008; Crosby 1972; Deagan 2004; Ernst and Hofman 2015; Funari and Senatore 2015; Graham 2011; Hofman et al. 2014; Ibarra Rojas 2003; Keehnen 2011, 2012; Mol 2008; Morsink 2015; Oland 2014; Oliver 2009; Ortiz 1995; Ostapkowicz 2013; Pugh 2009; Rodríguez-Alegría 2008, 2010; Scaramelli and Tarble de Scaramelli 2005; Valcárcel Rojas 2012, 2016; Valcárcel Rojas et al. 2010, 2013; Vega 1979, 1987). In fact, the central importance of (foreign) objects has never been fully examined for early colonial settings in this region. Of course, transformations of indigenous material culture repertoires resulting from European contact have been discussed earlier. However, these realizations often followed occasional findings of European materials in indigenous assemblages or the labeling of recovered ceramics as "transcultural." An integrated, synthetic approach on the topic, however, has been lacking so far; a missing part of a much larger issue, which involves the extremely limited work done on the archaeology of the Spanish/European colonial period. Although new insights over the last few decades have promoted the study of transculturation and indigenous and African responses to colonialism, the field seems to have remained an offshoot of pre-Columbian research with a predisposition towards European perspectives (Deagan 2004; Ewen 2001). As a result, indigenous experiences amidst new colonial realities are still poorly understood and merit further investigation if we aim to reach a more inclusive understanding of this turbulent period in the history of the region (Valcárcel Rojas 2012, 2016).

Recently, numerous studies of colonialism worldwide have shown that the exchange and adoption ("consumption") of material elements of the 'other' were vital for the establishment and structuring of interethnic relationships (e.g., Cusick 1998; Van Dommelen 2006; Gosden 2004; Lightfoot et al. 1998;

Lyons and Papadopoulos 2002; Maran and Stockhammer 2012; Rogers 1990; Rubertone 2000; Silliman 2010; Thomas 1991). In addition, objects embodied and directed transformations in social, cultural, and material domains for all of those involved. Especially in the Caribbean – the nexus of first interactions between groups from Europe, Africa, and the Americas – it is most interesting to unravel the entanglement of widely divergent material culture repertoires (Hofman 2019; Hofman et al. 2012; Keehnen and Mol 2018). Here, with no previous contact, indigenous communities and Spanish/European explorers and colonists used objects to negotiate a mutual base of understanding. Studying the material components of colonial encounters provides insights into respective value systems and reveals the various social mechanisms that were at play. Furthermore, it is through the analysis of practices including gift giving or the voluntary adoption of previously unknown types of objects, that it is possible to retrace aspects of indigenous agency – particularly important here or in similar situations in which these tend to become obscured.

From reading late fifteenth- and early sixteenth-century (ethno)historical sources on the first encounters in the Caribbean islands, the basic material constituents of these interactions are known: on the one hand, a European assembly of "trinkets" (beads, bells, and other shiny items) previously already used to successfully entice native peoples living in West Africa and the Canary Islands (cf. Fernández-Armesto 1987; Graeber 1996); and, on the other hand, indigenous Caribbean gold, pearls, and foodstuffs, and especially valued lustrous objects (Hofman et al. 2018; Keehnen 2011, 2012; Oliver 2000; Saunders 1999).[2] However, we know very little about the materiality of these things, their interconnections, and the underlying social and cognate mechanisms facilitating the flow of objects within and between the different cultural groups involved in the encounters. With the present volume, it is our aim that this contribution fills part of this lacuna by focusing on the materiality of things in early colonial encounters in the Caribbean and adjacent mainland areas. We specifically focus on the nature of indigenous and Spanish/European object realms, the types of objects that enabled both parties to connect upon contact, and the underlying systems of value allowing for the adoption of new types of material culture. We want to explore how objects transcended cultural boundaries, how 'cultural others' dealt with such articles after their 'foreign' adoption, and what changes we can observe in the material interrelations, practices, and valuations of indigenous

---

2 What is more, in due time Europeans started to import indigenous material culture from other parts of the circum-Caribbean area into the islands to use in their interactions with the island inhabitants (Valcárcel Rojas 2012, 2016).

societies. Finally, we evaluate how these material dynamics contribute to our broader understanding of early colonial encounters in the first regions of the Americas that were impacted by the European colonization.

3 A Matter of Concept(s)

Early colonial patterns and connections have not always been appreciated for their entangled and transformative character. To provide a contextual framing of previous studies and ideas, we will discuss a selection of the most important past scholarly advances in the field and their influence on the development of Spanish/European colonial archaeology in the broader Caribbean region, including the adjacent parts of the continental coasts (for summaries, see also Deagan 1998; Hernández Mora 2011; Valcárcel Rojas et al. 2013; Van Buren 2010).

The theme of culture contact and change has been ingrained in the disciplines of anthropology and archaeology since their very beginnings, particularly in the Americas, given the profound impact of post-1500 European colonialism in the region. Starting in the 1930s, and firmly established in the 1950s and 1960s, interactions between European and non-European peoples were generally explained in terms of the "acculturation" model (Beals 1953; Herskovits 1938; Kroeber 1948, 425–437; Redfield et al. 1936). The theory's key premise was that the dominant colonizing "donor" culture transforms the more passive "recipient" culture, resulting in the loss and eventual disappearance of traditional lifeways, materialities, and even entire cultures. Analyses of contact situations thus focused on determining the degree of change indigenous "subordinate" cultures had undergone because of Western contact. These changes (classified into types or stages) were considered unidirectional, given the idea that power resided exclusively with the "culture bearing" (and, by extension, "civilization bearing") Europeans (but see Foster 1960; Spicer 1961). In an influential study, George Quimby and Alexander Spoehr (1951) proposed a methodology to assess the rate of indigenous acculturation in colonial artifact assemblages. Their classification scheme distinguished between (modified) traditional, introduced, and mixed objects based on artifact form, material, use, and manufacture. The model long remained popular and has been adopted and amplified by archaeologists in different ways (for relatively recent examples, see Farnsworth 1989, 1992; cf. Rogers 1990, 1993). In this context, European technologies and materials were generally regarded as superior (i.e., more efficient and sophisticated) to indigenous ones, leading to ideas about the inevitability of cultural and technological progress to promote a theory of immediate replacement (e.g., Elliott 2002; Foster 1960; Pasztory 2005; Trigger 1991).

For the ensuing decades, the acculturation concept remained the dominant model for studies of culture contact. Ironically, it was the search for the origins of modern capitalism during the 1970s – the basis of Wallerstein's world-systems theory (1974, 1980) – that resulted in an increased scholarly awareness of the position of indigenous communities. In the establishment of the colonial network, these indigenous – or "pre-state" – societies were the "peripheries" of the nation-state "cores," distant nodes of exploitation vital for the emergence of Europe's global mercantilist and imperialist structures. Within the world system, both entities were crucial for the larger interlinked and interdependent whole. As a result of their intimate connection, changes on either side were understood to directly affect the other, although it was the nation-state dictating this relation, a structured inequality that eventually hindered further development of the colonies (see, e.g., Frank 1966). For one thing, it was realized that a core-periphery entanglement was required to study both parts in tandem, thereby stretching the unit of analysis to encompass both (cf. Braudel 1981).

A turning point in the anthropological perception of indigenous peoples contacted by Europeans was reached with the publication of Eric Wolf's (1982) influential work *Europe and the People Without History*, which for the first time advocated a "bottom-up" understanding of colonial situations. Although world-systems theory proponents had urged scholars to include peripheral colonies, the approach contributed little to understand the active participation of these areas in the creation of the larger commercial system, nor as (precontact) entities on their own, with local historically defined dynamics and particularities. To restore this imbalance, anthropologists were to take up responsibility for the study of indigenous historical trajectories, even to look for similar phenomena among past societies on a worldwide scale (e.g., Abu-Lughod 1989; Algaze 1993; Blanton and Feinman 1984; Schneider 1977; for a review, see Hall and Chase-Dunn 1993). Moreover, many influential publications sprouted from the burgeoning interest in issues of European colonialism and indigenous responses (Fitzhugh 1985; Ramenofsky 1987; Rogers 1990; Rogers and Wilson 1993; Thomas, ed., 1989, 1990, 1991; Wylie 1992).

Meanwhile, as post-processual and postcolonial theories became more widespread, both the acculturation model and the world-systems theory were discarded for being passive, unidirectional, and ethnocentric (see essays in Cusick 1998). The emphasis on concepts such as agency, practice, and identity, building upon the foundational works of Sahlins (1976), Bourdieu (1977), and Giddens (1979), as well as the reconsideration of Western cultural representations of "the other," represented by Edward Said's *Orientalism* (1978), called upon a more inclusive and multidimensional approach to colonial interaction in pluralistic social settings (Lightfoot et al. 1998). These theoretical advances resulted in the formulation of a range of new and updated concepts about cultural mixture, including bricolage (Comaroff 1985), creolization (Dawdy 2000; Deagan 1996; Deetz 1996; Delle 2000; Ewen 2000; Ferguson 1992; Hannerz 1987; Mintz and Price 1992), ethnogenesis (Deagan 1998; Hill 1996; Moore 1994), hybridity (Bhabha 1994; Hall 1990; Silliman 2015; Young 1995), *mestizaje* (Deagan 1974, 1983), syncretism (Palmié 1995; Stewart and Shaw 1994), and transculturation (Deagan 1998; Domínguez 1978; Ortiz 1995; Romero 1981). But, as part of this ongoing reconfiguration of historical anthropology, it was realized that to understand cultural continuities and changes following European contact, it is critical to perceive of these as grounded in a precolonial past. Hence, the categorical separation between a "prehistoric" or "precontact" period and a "historical" or "postcontact" period became heavily criticized for being an artificial and Eurocentric construct (e.g., Lightfoot 1995; Rubertone 1996, 2000; Schneiber and Mitchell 2010; Williamson 2004). Such a "historical divide" frames European contact as a defining moment in the lives of indigenous peoples, overlooking indigenous long-term histories as well as their presence and performances in early colonial times and up until the present (Boomert 2016; Hofman et al. 2012; Oland et al. 2012, 1).

In anticipation of the Columbian Quincentenary in 1992 (the 500-year "anniversary" of Columbus' fortuitous arrival in the Caribbean), scholars all over the Americas increasingly embraced the topic of European colonization and "first contact" situations (e.g., Bray 1993; Deagan 1985, 1987, 1988; Fitzhugh 1985; Greenblatt 1991, 1993; Hulme 1986; Pagden 1993; Ramenofsky 1987; Rogers 1990; Rogers and Wilson 1993; Wilson 1990; Wylie 1992; see also Axtell 1995). In the insular Caribbean, it was only then that archaeological research of the "contact period" started to boom.[3] Before this time, Spanish colonial archaeology – a nascent specialization during the 1940s and 1950s – focused mostly on the part of the colonizer, searching and digging for the remains of first European settlement (e.g., Cruxent 1955, 1972; Goodwin 1946; de Hostos 1938; Palm 1945, 1952, 1955; Wing 1961) and drawing up descriptions and classifications of European artifacts (e.g., Goggin 1960, 1968; Lister and Lister 1974; Mendoza 1957; Rouse 1942). Paradoxically, indigenous-European interaction was observed

3 Historical developments in the field of Spanish colonial archaeology have followed a parallel pattern throughout the broader Caribbean region, including the adjacent parts of the continental coasts. Here, however, we take the insular Caribbean as our point of reference, as this is the area we work in and we know best.

mainly through the haphazard recovery of Spanish artifacts and their purported indigenous modifications or imitations in sites previously identified as pre-Columbian. Such objects were not contemplated much other than being indications for barter or gift exchange (García Castañeda 1949; Morales Patiño and Pérez de Acevedo 1945; Rouse 1942). The methodology Quimby and Spoehr (1951) had used to measure culture change did not become as popular in the Caribbean as it had been in North America. Instead, the works that gained ground in the former region were those of anthropologists Fernando Ortiz (1995) and, later, George M. Foster (1960).

Ortiz' seminal work *Cuban Counterpoint* (published in Spanish in 1940, and in English in 1947 [1995]) was an effort to explain the creation of Cuban popular culture based on the historical roles of sugar and tobacco. In this study, he coined the term "transculturation" in direct opposition to the concept of acculturation. Changes resulting from culture contact, he argued, must be seen as the two-way adoption of elements from "the other." Beyond a mere acquisition, this process entails both the loss of traditional cultural expressions and the creation of new (blended) ones. Cuban archaeologists Oswaldo Morales Patiño and Roberto Pérez de Acevedo (1945) and José A. García Castañeda (1949) were the first to apply Ortiz' idea of transculturation as an analytical construct for the study of indigenous-European interaction. Although they did not appreciate the truly processual, transformative, and creative nature of the original concept (Valcárcel Rojas 2012, 32), their pioneering work served as an example for others in subsequent decades (e.g., García Arévalo 1978a; Guanche 1983; Rey Betancourt 1972; Romero 1981; see also Deagan 2010).

Foster's (1960) contribution can be found in his idea of "conquest culture," a concept he developed to understand culture change and colonial interaction in Mexico. The "donor" culture, he argued, only transfers to the "recipient" culture a selection of its culture traits and complexes, called "conquest culture," from which the latter adopts or rejects elements. This "stripped down" version comes about through two "screening processes," an initial "formal" one by the colonial power (directed policy) and a second "informal" one determined locally by the contact situation itself (founder's effect). In Foster's view, this filtering led to "cultural crystallization," when the "recipient" culture internalizes the new traits, with the acculturated society as final creation. Much of this process depends on the (differential) amount of power exercised by the dominant culture. Spanish colonial archaeologists widely adopted the model after 1970, especially since Kathleen A. Deagan (1974, 1983, 1990) first applied it for her interpretation of archaeological patterns observed at St. Augustine, Florida (e.g., Ewen 1991; García Arévalo 1990; Smith 1995; see South 1978).

In 1978, Cuban archaeologist Lourdes S. Domínguez synthesized prevailing ideas to come up with a methodology relating the distribution of Spanish artifacts in indigenous sites to patterns of interaction. To this end, she distinguished between "contact sites" and "transculturation sites" (Domínguez 1978, 37). A "contact site" points to short or indirect interaction, expressed by few and unmodified European objects scattered across the surface. A "transculturation site," in contrast, is identified through larger quantities of introduced materials, modified or reused for indigenous purposes, and occurring at deeper site levels, indicating more prolonged interaction and cultural exchange. The classification was important for offering an analytical tool for the archaeological understanding of indigenous-European interactions. During this time, also some of the first studies contemplating the indigenous use and valuation of Spanish artifacts appeared (García Arévalo 1978a; Vega 1987). Archaeological investigations into the period of indigenous-Spanish interaction increased over the years, although a strong focus on Spanish colonial spaces remained (for some examples from the Greater Antilles, see, e.g., Agorsah 1991; Curtin 1994; Domínguez 1984, 1991; Luna Calderón 1992; Ortega 1982; Ortega and Fondeur 1978; Pantel et al. 1988; but cf. Ortega 1988). These localities yielded various "transcultural" ceramics or "colonowares," referring to non-European and locally manufactured household vessels that combine both European and non-European stylistic and compositional elements (Domínguez 1980; García Arévalo 1978b, 1990, 1991; Ortega 1980; Ortega and Fondeur 1978; Woodward 2006). While these are typically blends of indigenous and European traditions, African-European or Creole ceramics have also been documented (Ernst 2015; Smith 1995; Solis Magaña 1999). Thus, observed changes in burial, culinary, and material practices created an interest in further exploring the transformations set in motion by the early colonial interactions. Although it was acknowledged, particularly in Cuba (Valcárcel Rojas 1997), that better-informed analyses needed new methodologies and research strategies, it was not until the 1990s that these started to come to fruition. It was only then that Spanish colonial-era archaeology experienced a shift in orientation, discarding its ideology of dominance and unidirectionality. Largely in response to the emerging post-modern theories within the broader discipline, a more nuanced, inclusive, and local view of the early colonial period was sought (Keegan 1996; Patterson 1991; Wilson 1993; see also Van Buren 2010). Nowadays, the entangled and transformative nature of indigenous-European encounters has been widely acknowledged, and is considered the foundational bricks of the later colonial period and the postcolonial present. It is within this context that this volume is placed and aims to contribute.

4 Time, Space, and Scale

In late 1950s, Cruxent and Rouse (1958, I, 415) used the very first radiocarbon dates to define five arbitrary chronological periods for the history of the Southern Caribbean and the Venezuelan mainland; "Period 5" in this chronology lasted from 1500 onwards. A few years later, Rouse and Cruxent (1963, 22) defined four subsequent "epochs" connecting technology with subsistence strategies in Caribbean culture history, where the "Indo-Hispanic epoch" comprised the period from AD 1500 to the present. Later, Rouse (1972, 136–138) redefined the epochs in purely technological terms to "ages" where the "Historic Age" lasted from AD 1500 until the present. However, the "Indo-Hispanic" or "Historic Age" resulted in broad historical-cultural conceptualizations, which had to be challenged by future researchers whose aim was attaining more socially-tuned and spatially localized levels of understanding of the colonial past. In the insular Caribbean, the very early colonial period became equated with a so-called "Columbian era" or "contact period," a time frame of roughly two or three decades of indigenous-Spanish interaction following the arrival of Columbus (Deagan 1988; García Arévalo 1978a). Historically, the end of this period corresponds to the Spanish shift of attention to the Central American mainland. Moreover, it was wrongly believed that at this moment indigenous sociopolitical structures had disintegrated in the insular Caribbean and their numbers reduced to a negligible minimum. Nowadays we know, acknowledging the different colonial temporalities within and between islands, that this period lasted until the 1520s or 1530s in the Bahamas, Jamaica, and Puerto Rico, and the mid-sixteenth century for Cuba and Hispaniola (Valcárcel Rojas 2016, 11–12; see also Valcárcel Rojas 2012, 2016 for his conceptual application of distinguishing contact versus colonial situations in the Antillean context). Indigenous slavery and resistance, European colonization, and the influx of African slaves beginning in the sixteenth century led to the mixing of biological ancestries and the formation of new identities as well as social and material worlds (Deagan 2003; Jiménez 1986; Mira Caballos 1997). Indigenous peoples and indigenous traditions survived in this area until today, which ultimately contributed to the formation of present-day, multi-ethnic Caribbean society on these islands (García Molina et al. 2007; Guitar et al. 2006; Hofman 2019; Hofman et al. 2012; Pesoutova and Hofman 2016).

The indigenous presence clearly continues beyond the above-mentioned fifteenth- to eighteenth-century time slot. For example, by 1800, the Carib presence in the Lesser Antilles was dramatically reduced, but the colonial encounters had resulted in new and unique social formations influenced by indigenous, European, and African cultural elements. Carib communities absorbed large numbers of escaped African slaves, leading to the formation of a Black Carib ethnic identity, alongside traditional *Kalinago* communities. After several wars with the British, many Black Carib were deported to Central America in 1797, where they are known as the *Garifuna*. Descendants of the *Kalinago* and *Garifuna* live throughout the Lesser Antilles where they reclaim their Amerindian roots as an integral part of their Caribbean identity. Based on this and many other similar examples known from across the Caribbean, the archeological sites and materials discussed in this volume rely on data that embody the wide range of fifteenth- through eighteenth-century processes of negotiation, exchange, hybridization, resignification and transformation, and resistance and struggle between the colonized and the colonizer. The case studies cover the circum-Caribbean macroregion and include the Bahamas, the Dominican Republic and Haiti (previously La Hispaniola), Jamaica, the southern Lesser Antilles, the Guianas, Venezuela, Colombia, San Salvador, Honduras, Belize, Guatemala, Mexico, and the American Southwest. Although this selection inevitably leaves out other parts and islands of our area of focus, we believe the case studies assembled here are representative of the colonial processes, initial European settlement, and Amerindian-European-African intercultural dynamics, we wish to explore with this contribution (Fig. 1.1).

5 Outline of the Volume

It is important to delineate the place of this volume in the recently burgeoning archaeological scholarship on colonial encounters in the Americas. For centuries, a bricolage of facts and fiction have permeated the reconstruction of early colonial encounters and their later colonial trajectories in the Caribbean. Scholarly data has been freely incorporated into popular lore and contributed to the creation of grand narratives, which nourished literature and cinematography, and became an important cog in the creation of national identities throughout the entire region. However, these narratives continue the perpetuation of incomplete and unidirectional drifts in which, (1) Amerindian and African social actors are essentialized (portrayed distortedly) or invisible (lacking their place in history); (2) the materiality of colonial encounters has often been reduced to the mere description of indigenous versus European artifacts; and (3) the role of non-urban spaces in the re-organization of early colonial territories in the Caribbean has barely been investigated. Clearly, the role of archaeology in this scenario cannot lay only in nourishing the existing narratives. Because of new interdisciplinary field investigations, novel theoretical approaches, and cutting-edge methods and techniques, archaeology

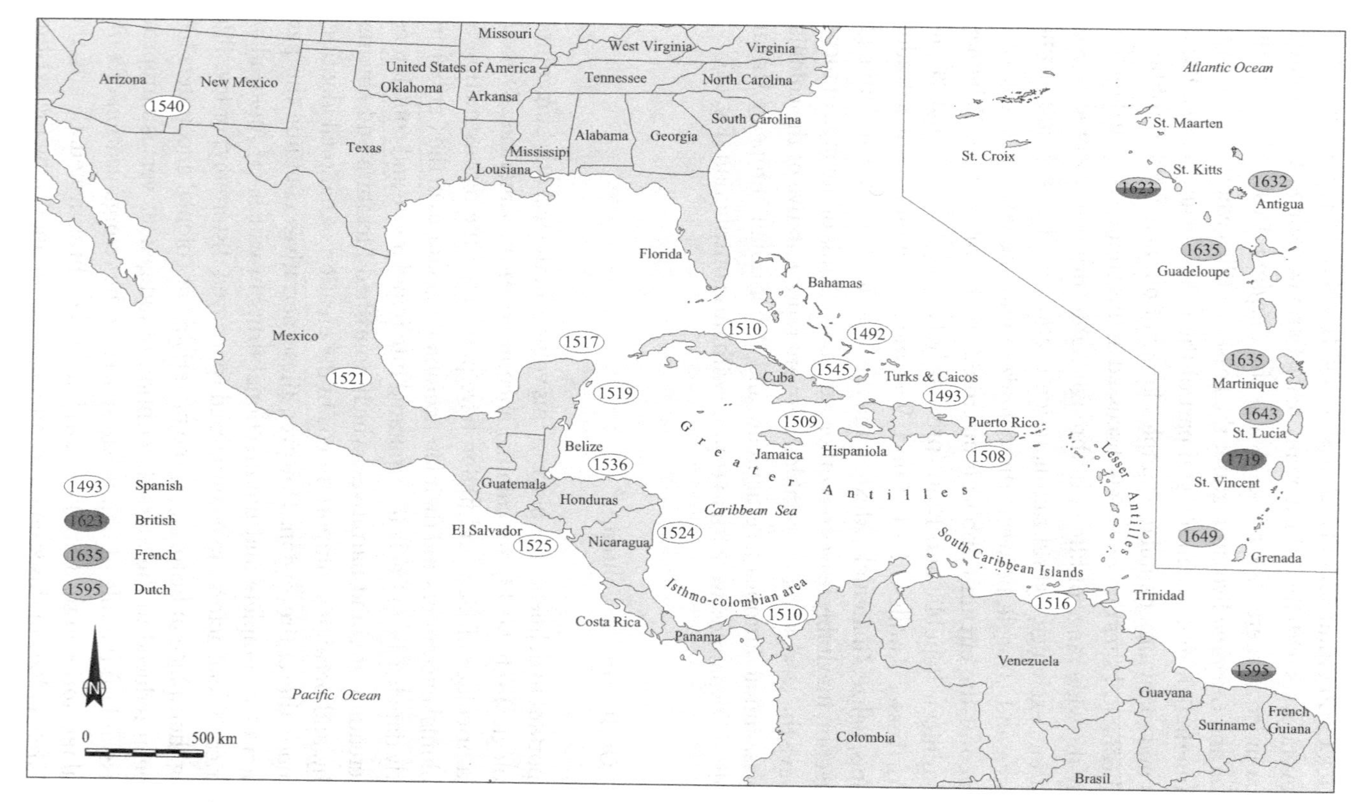

FIGURE 1.1 Map of the circum-Caribbean and adjacent areas included in this volume with dates of initial European settlement
MAP BY MENNO L.P. HOOGLAND

may amend and complement some of the narratives. But most importantly, it should critically evaluate, contest or replace them in constant interaction with, and sensitive to, the tensions of the globalizing world in and beyond the walls of academia. We are confident that this volume will contribute to this task by exploring early colonial encounters in the broader Caribbean region from a material culture perspective, one especially based on new field data.

In Chapter 2, Mary Jane Berman and Perry Gnivecki explore the first indigenous-Spanish encounters in the Bahama archipelago, a colonial frontier impacted by Spanish policies and practices during the late fifteenth and early sixteenth centuries. They argue that European objects, which made their way here through numerous pathways, were easily incorporated into the indigenous Lucayan economic system due to the precedents set by the Lucayan's familiarity with non-local items and peoples through trade, exchange, and raids. Additionally, the Lucayans found European objects to be analogous to materials they knew and understood and so they were easily assimilated into the local system. Berman and Gnivecki conclude that in the Bahamas the indigenous alteration of European objects was minimal because of the absence of direct colonial control, the short period of time during which the Lucayans were part of the Spanish empire, the sporadic and intermittent duration of direct contact experiences with the Spanish, and the way the Lucayans were removed from their homeland.

In Chapter 3 of this volume, Floris Keehnen discusses the cultural implications of European materials recovered from early colonial indigenous spaces on the island of Hispaniola, where during the late fifteenth- and early sixteenth-century indigenous and Spanish material worlds increasingly entangled. Reviewing more than fifty years of archaeological work on the island, Keehnen presents an up-to-date overview of indigenous sites yielding European or European-influenced objects. An examination of the recovered materials illustrates that these are found in a variety of contexts, range from singular finds to direct associations to indigenous valuables, and occasionally appear in reworked, repurposed, or copied forms. Through several case-study sites, Keehnen explores how the various ways in which indigenous peoples handled these European-introduced objects relates to the differential impacts of colonial power on Hispaniola.

Shea Henry and Robyn Woodward (Chapter 4) present novel data on their excavations at the indigenous village of Maima in Jamaica, which they compare with material culture and faunal remains from nearby Spanish Sevilla la Nueva. In June 1503, Columbus and his crewmembers spent a year marooned in the sheltered harbor of St. Ann's Bay, 1.4 kilometers from Maima. In 1509, the Spanish returned to find the Jamaican colonial capital of Sevilla la Nueva.

By the time Sevilla la Nueva was abandoned in 1534, Maima was deserted. Historical records kept by the colonists indicate that the villagers were brought to the colony and made into laborers and wives. The material culture and dietary practices at Sevilla la Nueva reflect this through the presence of colonoware and indigenous-adapted European goods. Henry and Woodward observe that at Maima very few European goods and domesticate animals were found. They argue that this, and the presence of traditional material culture and diet throughout the site, indicates continuity of a traditional way of life until their final act of resistance, abandoning Maima.

In Chapter 5, Roberto Valcárcel Rojas focuses on northeastern Cuba, particularly the modern-day province of Holguín, which is one of the areas of the Caribbean with the largest number of indigenous sites yielding European objects. In the sixteenth century, most of these sites maintained direct or indirect links with Europeans, while others were transformed into permanent colonial spaces by the Spaniards. The study of European objects found at these sites suggests that some of these items were acquired through exchange or as gifts. However, as Valcárcel Rojas shows, the largest collections of objects appear to have originally functioned as tools or other items used by both Europeans and indigenous peoples for mining and agricultural labor. He proposes that this pattern was established as a result of a process of conquest and colonization specific to Cuba, during which European colonizers rapidly managed to control the local population, thus limiting the indigenous capacity for negotiation.

Marlieke Ernst and Corinne Hofman take us back to early colonial Hispaniola in Chapter 6, where they contrast the incorporation of European earthenwares in the indigenous sites of El Cabo and Playa Grande with the presence of indigenous ceramics and new manufacturing traditions in the early Spanish colonial sites of Cotuí and Concepción de la Vega. Using theories of gift giving, appropriation, and imitation, combined with archaeological and ethnoarchaeological studies of the operational sequence (*chaîne opératoire*) of ceramic manufacture, Ernst and Hofman specifically assess transformation processes in ceramic repertoires, providing new insights into the dynamics of indigenous-European-African interactions, mutual influences, and resilience.

In Chapter 7, Andrzej Antczak and colleagues delve into the early sixteenth century on the islands of Margarita, Coche, and Cubagua, which lay at the core of the so-called 'Coast of Pearls' in northeastern Venezuela. Cubagua has hosted Nueva Cádiz, one of the earliest Spanish towns in South America, since 1528. Despite such precocious credentials, the understanding of early colonial realities on these islands has almost entirely relied on documentary sources with only a small contribution from archaeology. By analyzing the ecology of the pearl oyster, documentary information, and archaeological data obtained in recent surveys on Margarita and Coche, from museum collections of materials recovered in the 1950s, and from the database of the Venezuelan Islands Archaeology project, Antczak and colleagues shed new light on the earliest colonial non-urban settlements or *rancherías* established on these islands and on the nature of the intercultural dynamics that took place there. They discuss specific sites on Margarita and Coche, which yielded abundant sixteenth-century European ceramics, indigenous pottery, and possibly intercultural hybrids. Their findings allow Antczak and colleagues to provide new insights into the beginnings of the early Spanish town of Nueva Cádiz and underscore the role of pearls and other material culture in the early colonial endeavors in northeastern Venezuela.

Alberto Sarcina (Chapter 8) investigates the relationships between the indigenous peoples of the western region of the Gulf of Urabá, Colombia, and the Spaniards in the early years of the conquest. He focuses on what happened in Santa María de la Antigua del Darién, the first European city founded on the American mainland, in the course of its short history, and immediately after its abandonment (1510–1524). Sarcina offers new reflections on these questions, based on historical sources (Oviedo and the reports of the travels of Julian Gutiérrez) and archaeological data obtained during excavations carried out by the author between 2014 and 2016, which include ritual caches dated to the phase of the city's abandonment; "contact" pottery from the Basurero Norte area; and, "excavation unit F," a possible house inhabited by indigenous servants (*naborías*).

In the following Chapter 9, William Fowler and Jeb Card introduce the conquest-period and early colonial site of Ciudad Vieja, the ruins of the first *villa* of San Salvador, El Salvador, settled from 1525 to about 1550/60. They evaluate subsequent developments from investigations of the indigenous town of Caluco, in the Izalcos region of western El Salvador, during the last quarter of the sixteenth century. A theoretical framework inspired by Bourdieu's structural theory of practice allows them to intepret differing strategies of practice during the early Spanish colonial period. Fowler and Card argue that the early "Spanish" community of San Salvador, potentially located on an already ancient Mesoamerican ritual site, was an incubator of experimentation and transformation of Mesoamerican roles and identities. They illustrate that by the time of Caluco's colonial community in the late sixteenth century, practices and structures found in later Latin American communities, built on tensions between indigenous communities and state extraction, were increasingly apparent.

In Chapter 10, Russell Sheptak and Rosemary Joyce emphasize the novel construction of defensive walls at Ticamaya, a pre-Columbian settlement in Caribbean Honduras that continued to be occupied into the nineteenth century, and at allied sites along the coast of the Gulf of Honduras as likely material traces of innovations mediated by Spanish knowledge mobilized for indigenous resistance to Spanish colonization. Archaeological excavations at Ticamaya, described in sixteenth-century Spanish documents as the seat of a leader of indigenous resistance, identified confirmed deposits from the period covering initial conflict with the Spanish, from roughly 1520 to 1536. Yet, as Sheptak and Joyce demonstrate, these excavations produced no use of European goods until the late eighteenth century. Contemporary with Ticamaya, the site of Naco to the west hosted troops sent by Hernán Cortés, and at least one majolica vessel was discarded there. The contrast could lead to the conclusion that Ticamaya was unaffected by the Spanish encounter until it was incorporated into the colony. By considering apparently indigenous things as outcomes of tactical coping with Spanish invasion, Sheptak and Joyce seek to blur seemingly firm lines between native and foreign materialities and define a third option of hybrid cultures.

Jaime Awe and Christophe Helmke in Chapter 11 focus on the early Maya-Spanish interactions in Belize. They note that researchers who have focused attention on the Belize colonial frontier describe Maya-Spanish relationships as anything but amicable. Because of this bellicose relationship, some authors suggest that few material goods of European origin were traded or integrated into frontier settlements. They also contend that while ethnohistoric reports describing the missionizing efforts of Spanish priests provide us with important data on Maya life during the early colonial period, the Spanish *entradas* provide precious little information about the material goods they gifted to the Maya, and even less about how the Maya utilized these foreign goods. In this chapter, Awe and Helmke discuss how the ethnohistoric record offers us considerable information concerning the consumption of European objects by the Maya, and that archaeological discoveries in Belize, Guatemala, and Yucatan provide increasing evidence to suggest that a variety of objects of European origin were integrated into Maya material culture. The archaeological record also indicates that, as Awe and Helmke show, objects of European origin were used as status symbols by the Maya elite, that they sometimes served mundane purposes, or were deposited in caches and offerings in sacred places where they were ritually decommissioned.

In Chapter 12, Shannon Dugan Iverson discusses archaeological assemblages from two early colonial religious sites at Tula, Hidalgo, Mexico. These assemblages are nearly indistinguishable from pre-Columbian ones at the same sites, indicating that, as Iverson argues, colonial changes in material culture were much more gradual than expected, and driven to a surprising degree by indigenous traditions and aesthetic choices. Taking these data into account, Iverson reconsiders various models of social change that would adequately account for the observations of material culture at Spanish religious sites. While documentary sources inform us that the colonial encounter was not an equal exchange of ideas, models of top-down power alone could not account for the data in Tula. Conversely, models that posited cultural continuity – an indigenous "core" with a Spanish colonial "veneer" – seemed inadequate to account for genuine indigenous relationships with the Church. Iverson uses the case of Tula to explore the legacies and problems of several models, including acculturation and syncretism, before positing Judith Butler's concept of resignification as an appropriate model of colonial power and religious change.

In Chapter 13, Gilda Hernández Sánchez presents insights into the process of indigenous cultural continuity and change by focusing on pottery technology in the region of central Mexico during the early colonial period (AD 1521–1650). Her analysis is based on the integration of previous research on ceramics, as well as on the consulting of several archaeological collections of early colonial ceramics from many contexts in the Valley of Mexico. Hernández-Sánchez shows that the collapse of the Aztec empire, the emergence of a new colonial society, and the introduction of Spanish ceramic traditions (e.g., potters' wheel, glazing, and majolica ware) differentially impacted the native production of ceramics. While clay recipes, method of forming, and firing technology were maintained without change, surface finishing and decoration evidenced great creativity, providing proof of indigenous peoples' varied responses and adaptations to the changing circumstances Spanish colonization had set in motion.

Clay Mathers argues in Chapter 14 that, although conflict and *conquista* campaigns characterized many of the earliest encounters between indigenous and European groups in New Spain and La Florida, the transformation of objects, communities, and strategic policies in these areas was locally variable and changed dramatically by the close of the sixteenth century. Mathers points out that materials characteristic of these changes and variegated responses are found widely in the archaeological record of the American Southwest, but have seldom been explored for the insights they provide into broader anthropological themes such as resistance, exchange, and agency. While Mathers focuses on the fine-grained, contextual analysis of objects, the broader goal of his contribution is to compare cultural trajectories at the regional and interregional scales, particularly the congruence and contrasts between the American Southwest and Southeast in the first century of New-Old World contact. Both areas transitioned from initial imperial strategies of acquisition and conflict, to policies of settlement and missionization by the end of the 1600s, and in both areas a similar suite of European objects was available. Nevertheless, as Mathers illustrates, the way these objects were employed by indigenous peoples and Europeans varies significantly and in ways that reveal to us important aspects of the earliest colonial encounters in North America.

In Chapter 15, Martijn van den Bel and Gérard Collomb argue that during the sixteenth century, the indigenous population of the Guianas was already aware of and in contact with the Spanish settlement at Margarita Island. The *Aruacas*, the privileged allies of the Spanish, relied on their large sociopolitical (trade) network to obtain victuals and commercial goods from the Guianas, but also raided *Caribe* villages to assure indigenous slaves for the Spanish plantations and mines in the insular Caribbean. Van den Bel and Collomb explain that although the indigenous peoples of the eastern Guianas feared and fled the Spanish and the *Aruacas*, they did engage in encounters with the English, Dutch, and French, in whom they found allies to wage war against their local and Spanish enemies. These encounters with Europeans, they point out, took place mainly in the embouchures of rivers along the Guiana Coast, establishing a '*zone franche*' or socio-economical free zone. In this chapter, Van den Bel and Collomb focus upon the policies and alliances of the Yao of the Oyapock estuary, who, through their access of the interior, managed to control this coastal area.

Corinne Hofman and co-authors (Chapter 16) focus on the impacts of colonial encounters on indigenous Island Carib/*Kalinago* societies in the southern Lesser Antilles by studying transformations in settlement pattern and organization, material culture, and network strategies using historical information and new archaeological data. They present their results of the recent excavations at the early colonial sites of Argyle, St. Vincent, and La Poterie, Grenada, which have revealed the remains of indigenous villages and a set of material culture evidencing the first indigenous, European, and African interactions in this area. Hofman and colleagues advance novel perspectives on intercultural dynamics in colonial encounter situations and contribute to discussions of indigenous resistance, cultural transformations, and cultural diversity in an ever-globalizing world.

In the epilogue, Maxine Oland brings closure to the volume by commenting on the contributions it brings to the understanding of the material encounters and indigenous transformations in the early colonial Americas and delineating the avenues for future research.

Considered together, these individual contributions are illustrative of the diversity, plurality, and complexity of early colonial situations in the Americas. With novel theoretical insights and fresh, interdisciplinary, and fine-grained views of local histories, the chapters offer new perspectives on materiality and indigenous agency in colonial encounters and entanglements. This volume highlights the importance of studying these issues in the Caribbean and surrounding mainland given the early dates of indigenous-European interactions and their foundational impact for the subsequent unfolding of colonial processes in the wider Americas. With this unique combination of geographical scope and approach, this volume contributes to the further decolonization of (indigenous) colonial histories and to global dialogues about the archaeology of colonialism.

References

Abu-Lughod, Janet L. 1989. *Before European Hegemony: The World System A.D. 1250–1350*. Oxford: Oxford University Press.

Agorsah, E. Kofi. 1991. "Recent Developments in Archaeological Research in Jamaica." In *Proceedings of the 14th Congress of the International Association for Caribbean Archaeology*, edited by Alissandra Cummins and Philippa King, 416–424. Barbados: The Barbados Museum and Historical Society.

Algaze, Guillermo. 1993. *The Uruk World System: The Dynamics of Expansion of Early Mesopotamian Civilization*, 2nd ed. Chicago: University of Chicago Press.

Anderson-Córdova, Karen F. 2017. *Surviving Spanish Conquest: Indian Fight, Flight, and Cultural Transformation in Hispaniola and Puerto Rico*. Tuscaloosa: University of Alabama Press.

Atalay, Sonya. 2006. "Indigenous Archaeology as Decolonizing Practice." *American Indian Quarterly* 30 (3/4): 280–310.

Axtell, James. 1995. "Columbian Encounters: 1992–1995." *The William and Mary Quarterly* 52 (4): 649–696.

Beals, Ralph. 1953. "Acculturation." In *Anthropology Today: An Encyclopedic Inventory*, edited by Alfred L. Kroeber, 621–641. Chicago: University of Chicago Press.

Bhabha, Homi K. 1994. *The Location of Culture*. London: Routledge.

Blanton, Richard and Gary Feinman. 1984. "The Mesoamerican World System." *American Anthropologist* 86 (3): 673–682.

Boomert, Arie. 2002. "Amerindian-European Encounters on and around Tobago (1498–Ca. 1810)." *Antropológica* 97/98: 71–207.

Boomert, Arie. 2016. *The Indigenous Peoples of Trinidad and Tobago: From the First Settlers until Today*. Leiden: Sidestone Press.

Bourdieu, Pierre. 1977. *Outline of a Theory of Practice*. Cambridge: Cambridge University Press.

Braudel, Fernand. 1981. *Civilization and Capitalism 15th–18th Century, Vol. 1: The Structures of Everyday Life*. New York: Harper & Row.

Bray, Warwick, ed. 1993. *The Meeting of Two Worlds: Europe and the Americas, 1492–1650*. Proceedings of the British Academy, Volume 81. Oxford: Oxford University Press.

Bruchac, Margaret M., Siobhan M. Hart, and H. Martin Wobst. 2010. *Indigenous Archaeologies: A Reader on Decolonization*. Walnut Creek: Left Coast Press.

Card, Jeb J. 2013. *The Archaeology of Hybrid Material Culture*. Center for Archaeological Investigations Occasional Paper No. 39. Carbondale: Southern Illinois University Press.

Cipolla, Craig N., ed. 2017. *Foreign Objects: Rethinking Indigenous Consumption in American Archaeology*. Tucson: University of Arizona Press.

Cipolla, Craig N. and Katherine Howlett Hayes, eds. 2015. *Rethinking Colonialism: Comparative Archaeological Approaches*. Gainesville: University Press of Florida.

Cobb, Charles R., ed. 2003. *Stone Tool Traditions in the Contact Era*. Tuscaloosa: University of Alabama Press.

Comaroff, Jean. 1985. *Body of Power, Spirit of Resistance: The Culture and History of a South African People*. Chicago: University of Chicago Press.

Cooper, Jago, Marcos Martinón-Torres, and Roberto Valcárcel Rojas. 2008. "American Gold and European Brass: Metal Objects and Indigenous Values in the Cemetery of El Chorro de Maíta, Cuba." In *Crossing the Borders: New Methods and Techniques in the Study of Archaeological Materials from the Caribbean*, edited by Corinne L. Hofman, Menno L.P. Hoogland, and Annelou L. van Gijn, 34–42. Tuscaloosa: University of Alabama Press.

Crosby, Alfred W. 1972. *The Columbian Exchange: Biological and Cultural Consequences of 1492*. Westport: Greenwood.

Cruxent, José M. 1955. "Nueva Cádiz: Testimonio de Piedra." *El Farol* 159 (17): 2–5.

Cruxent, José M. 1972. "Algunas Noticias Sobre Nueva Cádiz (Isla de Cubagua), Venezuela." In *Memorias VI Conferencia Geológica Del Caribe, Isla de Margarita, 6–14 July*, 33–35. Caracas.

Cruxent, José M. and Irving B. Rouse. 1958. *An Archaeological Chronology of Venezuela, Vol. 1 and 2*. Social Sciences Monographs 6. Washington, DC: Pan American Union.

Curtin, Marguerite. 1994. "Carvings from New Seville." *Jamaica Journal* 25 (2): 19–23.

Cusick, James G., ed. 1998. *Studies in Culture Contact: Interaction, Culture Change, and Archaeology*. Occasional Paper No. 25. Carbondale: Center for Archaeological Investigations, Southern Illinois University.

Dawdy, Shannon Lee. 2000. "Understanding Cultural Change through the Vernacular: Creolization in Louisiana." *Historical Archaeology* 34 (3): 107–123.

Deagan, Kathleen A. 1974. "Sex, Status, and Role in the Mestizaje of Spanish Colonial Florida." PhD dissertation, University of Florida, Gainesville.

Deagan, Kathleen A. 1983. *Spanish St. Augustine: The Archaeology of a Colonial Creole Community*. New York: Academic Press.

Deagan, Kathleen A. 1985. "Spanish-Indian Interaction in Sixteenth-Century Florida and Hispaniola." In *The European Impact on Native Cultural Institutions in Eastern North America, A.D. 1000–1800*, edited by William W. Fitzhugh, 281–318. Washington, DC: Smithsonian Institution Press.

Deagan, Kathleen A. 1987. "Initial Encounters: Arawak Responses to European Contact at the En Bas Saline Site, Haiti." In *Proceedings of the 1st San Salvador Conference: Columbus and His World*, edited by Donald T. Gerace, 341–359. Fort Lauderdale: College Center of the Finger Lakes and San Salvador: Bahamian Field Station.

Deagan, Kathleen A. 1988. "The Archaeology of the Spanish Contact Period in the Caribbean." *Journal of World Prehistory* 2 (2): 187–233.

Deagan, Kathleen A. 1990. "Sixteenth-Century Spanish-American Colonization in the Southeastern United States and the Caribbean." In *Columbian Consequences, Vol. 2: Archaeological and Historical Perspectives on the Spanish Borderlands East*, edited by David Hurst Thomas, 225–250. Washington, DC: Smithsonian Institution Press.

Deagan, Kathleen A. 1996. "Colonial Transformation: Euro-American Cultural Genesis in the Early Spanish-American Colonies." *Journal of Anthropological Research* 52 (2): 135–160.

Deagan, Kathleen A. 1998. "Transculturation and Spanish American Ethnogenesis: The Archaeological Legacy of the Quincentenary." In *Studies in Culture Contact: Interaction, Culture Change, and Archaeology*, edited by James G. Cusick, 23–43. Occasional Paper No. 25. Carbondale: Center for Archaeological Investigations, Southern Illinois University.

Deagan, Kathleen A. 2003. "Colonial Origins and Colonial Transformations in Spanish America." *Historical Archaeology* 37 (4): 3–13.

Deagan, Kathleen A. 2004. "Reconsidering Taíno Social Dynamics after Spanish Conquest: Gender and Class in Culture Contact Studies." *American Antiquity* 69 (4): 597–626.

Deagan, Kathleen A. 2010. "Cuba and Florida: Entwined Histories of Historical Archaeologies." In *Beyond the Blockade: New Currents in Cuban Archaeology*, edited by Susan Kepecs, L. Antonio Curet, and Gabino La Rosa Corzo, 16–25. Tuscaloosa: University of Alabama Press.

Deetz, James. 1996. *In Small Things Forgotten: An Archaeology of Early American Life*. Expanded and revised edition. Garden City: Doubleday.

Delle, James A. 2000. "The Material and Cognitive Dimensions of Creolization in Nineteenth-Century Jamaica." *Historical Archaeology* 34 (3): 56–72.

Dietler, Michael. 2010. *Archaeologies of Colonialism: Consumption, Entanglement, and Violence in Ancient Mediterranean France*. Berkeley: University of California Press.

Domínguez, Lourdes S. 1978. "La Transculturación En Cuba (S. XVI–XVII)." *Cuba Arqueológica* 1: 33–50.

Domínguez, Lourdes S. 1980. "Cerámica Transcultural En El Sitio Colonial Casa de La Obrapía." In *Cuba Arqueológica II*, edited by Manuel Rivero de la Calle, 15–26. Santiago de Cuba: Editorial Oriente.

Domínguez, Lourdes S. 1984. *Arqueología Colonial Cubana: Dos Estudios*. Havana: Editorial de Ciencias Sociales.

Domínguez, Lourdes S. 1991. *Arqueología Del Centro-Sur de Cuba*. Havana: Editorial Academia.

Elliott, John H. 2002. *Imperial Spain, 1469–1716*. London: Penguin.

Ernst, Marlieke. 2015. "(Ex)Changing the Potter's Process: Continuity and Change in the Non-European Ceramics of Cotuí, the First Colonial Mine in Hispaniola, after 1505." Research Master thesis, Leiden University.

Ernst, Marlieke and Corinne L. Hofman. 2015. "Shifting Values: A Study of Early European Trade Wares in the Amerindian Site of El Cabo, Eastern Dominican Republic." In *GlobalPottery 1: Historical Archaeology and Archaeometry for Societies in Contact*, edited by Jaume Buxeda i Garrigós, Marisol Madrid i Fernández, and Javier Garcia Iñañez, 195–204. Oxford: British Archaeological Reports, International Series 2761.

Ewen, Charles R. 1991. *From Spaniard to Creole: The Archaeology of Cultural Formation at Puerto Real, Haiti*. Tuscaloosa: University of Alabama Press.

Ewen, Charles R. 2000. "From Colonists to Creole: Archaeological Patterns of Spanish Colonization in the New World." *Historical Archaeology* 34 (3): 36–45.

Ewen, Charles R. 2001. "Historical Archaeology in the Colonial Spanish Caribbean." In *Island Lives: Historical Archaeologies of the Caribbean*, edited by Paul Farnsworth, 3–20. Tuscaloosa: University of Alabama Press.

Farnsworth, Paul. 1989. "The Economics of Acculturation in the Spanish Missions of Alta California." *Research in Economic Anthropology* 11: 217–249.

Farnsworth, Paul. 1992. "Missions, Indians, and Cultural Continuity." *Historical Archaeology* 26 (1): 22–36.

Ferguson, Leland. 1992. *Uncommon Ground: Archaeology and Early African America, 1650–1800*. Washington D.C.: Smithsonian Institution Press.

Fernández-Armesto, Felipe. 1987. *Before Columbus: Exploration and Colonisation from the Mediterranean to the Atlantic, 1229–1492*. London: Macmillan.

Ferris, Neal. 2009. *The Archaeology of Native-Lived Colonialism: Challenging History in the Great Lakes*. Tucson: University of Arizona Press.

Ferris, Neal, Rodney Harrison, and Michael V. Wilcox. 2014. *Rethinking Colonial Pasts through Archaeology*. Oxford: Oxford University Press.

Fitzhugh, William W. 1985. "Early Contacts North of Newfoundland before A.D. 1600: A Review." In *Cultures in Contact: The Impact of European Contacts on Native Cultural Institutions in Eastern North America, A.D. 1000–1800*, edited by William W. Fitzhugh, 23–43. Washington, DC: Smithsonian Institution Press.

Foster, George M. 1960. *Culture and Conquest: America's Spanish Heritage*. Viking Fund Publications in Anthropology 27. New York: Wenner-Gren Foundation.
Frank, André Gunder. 1966. "The Development of Underdevelopment." *Monthly Review* 18: 17–31.
Funari, Pedro P.A. and Maria X. Senatore, eds. 2015. *Archaeology of Culture Contact and Colonialism in Spanish and Portuguese America*. Cham: Springer.
García Arévalo, Manuel A. 1978a. "La Arqueología Indo-Hispana En Santo Domingo." In *Unidad Y Variedad: Ensayos Antropológicos En Homenaje a José M. Cruxent*, edited by Erika Wagner and Alberta Zucchi, 77–127. Caracas: Instituto Venezolano de Investigaciones Científicas.
García Arévalo, Manuel A. 1978b. "Influencias de La Dieta Indo-Hispanica En La Cerámica Taína." In *Proceedings of the 7th International Congress for the Study of the Pre-Columbian Cultures of the Lesser Antilles*, edited by Ripley P. Bullen, 263–277. Montreal: Centre de Recherches Caraïbes, Université de Montréal.
García Arévalo, Manuel A. 1990. "Transculturation in Contact Period and Contemporary Hispaniola." In *Columbian Consequences, Vol. 2: Archaeological and Historical Perspectives on the Spanish Borderlands East*, edited by David Hurst Thomas, 269–280. Washington, DC: Smithsonian Institution Press.
García Arévalo, Manuel A. 1991. "Influencias Hispánicas En La Alfarería Taína." In *Proceedings of the 13th Congress of the International Association for Caribbean Archaeology Part 1*, edited by Edwin N. Ayubi and Jay B. Haviser, 363–383. Willemstad, Curaçao: Archaeological-Anthropological Institute of the Netherlands Antilles.
García Castañeda, José A. 1949. "La Transculturación Indo-Española En Holguín." *Revista de Arqueología Y Etnología* 8/9: 195–205.
García Molina, José A., Mercedes Garrido Mazorra, and Daisy Fariñas Gutiérrez. 2007. *Huellas Vivas Del Indocubano*. Havana: Editorial de Ciencias Sociales.
Giddens, Anthony. 1979. *Central Problems in Social Theory: Action, Structure, and Contradiction in Social Analysis*. London: Palgrave Macmillan.
Given, Michael. 2004. *The Archaeology of the Colonized*. London: Routledge.
Goggin, John M. 1960. *The Spanish Olive Jar: An Introductory Study*. New Haven: Department of Anthropology, Yale University.
Goggin, John M. 1968. *Spanish Majolica in the New World: Types of the Sixteenth to Eighteenth Centuries*. New Haven: Department of Anthropology, Yale University.
Goodwin, William B. 1946. *Spanish and English Ruins in Jamaica*. Boston: Meador.
Gosden, Chris. 2004. *Archaeology and Colonialism: Cultural Contact from 5000 BC to the Present*. Cambridge: Cambridge University Press.
Graeber, David. 1996. "Beads and Money: Notes Toward a Theory of Wealth and Power." *American Ethnologist* 23 (1): 4–24.
Graham, Elizabeth A. 2011. *Maya Christians and Their Churches in Sixteenth-Century Belize*. Gainesville: University Press of Florida.

Greenblatt, Stephen. 1991. *Marvelous Possessions: The Wonder of the New World.* Chicago: University of Chicago Press.

Greenblatt, Stephen, ed. 1993. *New World Encounters.* Berkeley: University of California Press.

Guanche, Juan. 1983. *Procesos Étnoculturales En Cuba.* Havana: Ediciones Letros de Cubanos.

Guitar, Lynne, Pedro Ferbel-Azcarate, and Jorge Estevez. 2006. "Ocama-Daca Taíno (Hear Me, I Am Taíno): Taíno Survival on Hispaniola, Focusing on the Dominican Republic." In *Indigenous Resurgence in the Contemporary Caribbean: Amerindian Survival and Revival*, edited by Maximilian C. Forte, 41–67. New York: Peter Lang.

Hall, Stuart. 1990. "Cultural Identity and Diaspora." In *Identity: Community, Culture, Difference*, edited by Jonathan Rutherford, 222–237. London: Lawrence and Wishart.

Hall, Thomas D. and Christopher Chase-Dunn. 1993. "The World-Systems Perspective and Archaeology: Forward into the Past." *Journal of Archaeological Research* 1 (2): 121–43.

Hannerz, Ulf. 1987. "The World in Creolisation." *Africa* 57 (4): 546–559.

Hernández Mora, Iosvany. 2011. "La Arqueología Del Período Colonial En Cuba: Apuntes Teóricos de Sus Primeros Cincuenta Años (Parte I)." *El Caribe Arqueológico* 12: 3–14.

Herskovits, Melville J. 1938. *Acculturation: The Study of Culture Contact.* New York: Augustin.

Hill, Jonathan D. 1996. *History, Power, and Identity: Ethnogenesis in the Americas, 1492–1992.* Iowa City: University of Iowa Press.

Hofman, Corinne L. 2019. "Indigenous Caribbean Networks in a Globalizing World." In *Power, Political Economy, and Historical Landscapes of the Modern World: Interdisciplinary Perspectives*, edited by Christopher R. DeCorse, 55–80. Fernand Braudel Center studies in historical social science, SUNY press.

Hofman, Corinne L., Gareth R. Davies, Ulrik Brandes, and Willem J.H. Willems. 2012. "ERC Synergy Grant 2012, Research Proposal. NEXUS1492: New World Encounters in a Globalising World." Leiden: Leiden University.

Hofman, Corinne L., Angus A.A. Mol, Menno L.P. Hoogland, and Roberto Valcárcel Rojas. 2014. "Stage of Encounters: Migration, Mobility and Interaction in the Pre-Colonial and Early Colonial Caribbean." *World Archaeology* 46 (4): 590–609.

Hofman, Corinne L., Jorge Ulloa Hung, Eduardo Herrera Malatesta, Joseph Sony Jean, and Menno L.P. Hoogland. 2018. "Indigenous Caribbean Perspectives: Archaeologies and Legacies of the First Colonized Region in the New World." *Antiquity* 92 (361): 200–216.

Hostos, Adolfo de. 1938. *Investigaciones Históricas.* San Juan: Gobierno de Puerto Rico, Oficina del Historiador.

Hulme, Peter. 1986. *Colonial Encounters: Europe and the Native Caribbean, 1492–1797.* Cambridge: Cambridge University Press.

Ibarra Rojas, Eugenia. 2003. "Gold in the Everyday Lives of Indigenous Peoples of Sixteenth-Century Southern Central America." In *Gold and Power in Ancient Costa Rica, Panama, and Colombia*, edited by Jeffrey Quilter and John W. Hoopes, 383–419. Washington, DC: Dumbarton Oaks.

Jansen, Maarten E.R.G.N. and Valentina Raffa, eds. 2015. *Tiempo Y Comunidad: Herencias E Interacciones Socioculturales En Mesoamérica Y Occidente*. Archaeological Studies Leiden University 29. Leiden: Leiden University Press.

Jiménez, G., A. Morella 1986. *La Esclavitud Indígena En Venezuela (Siglo XVI)*. Caracas: Academia Nacional de la Historia.

Keegan, William F. 1996. "West Indian Archaeology. 2. After Columbus." *Journal of Archaeological Research* 4 (4): 265–294.

Keehnen, Floris W.M. 2011. "Conflicting Cosmologies: The Exchange of Brilliant Objects between the Taíno of Hispaniola and the Spanish." In *Communities in Contact: Essays in Archaeology, Ethnohistory & Ethnography of the Amerindian Circum-Caribbean*, edited by Corinne L. Hofman and Anne Van Duijvenbode, 253–268. Leiden: Sidestone Press.

Keehnen, Floris W.M. 2012. "Trinkets (f)or Treasure? The Role of European Material Culture in Intercultural Contacts in Hispaniola during Early Colonial Times." Research Master thesis, Leiden University.

Keehnen, Floris W.M. and Angus A.A. Mol. 2018. "The Roots of the Columbian Exchange: An Entanglement and Network Approach to Early Caribbean Encounter Transactions." Unpublished manuscript, Faculty of Archaeology, Leiden University.

Kroeber, Alfred L. 1948. *Anthropology*. New York: Harcourt, Brace & Company.

Liebmann, Matthew. 2015. "The Mickey Mouse Kachina and Other 'Double Objects': Hybridity in the Material Culture of Colonial Encounters." *Journal of Social Archaeology* 15 (3): 319–341.

Liebmann, Matthew and Melissa S. Murphy, eds. 2011. *Enduring Conquests: Rethinking the Archaeology of Resistance to Spanish Colonialism in the Americas*. Santa Fe: School for Advanced Research Press.

Lightfoot, Kent G. 1995. "Culture Contact Studies: Redefining the Relationship between Prehistoric and Historical Archaeology." *American Antiquity* 60 (2): 199–217.

Lightfoot, Kent G., Antoinette Martinez, and Ann M. Schiff. 1998. "Daily Practice and Material Culture in Pluralistic Social Settings: An Archaeological Study of Culture Change and Persistence from Fort Ross, California." *American Antiquity* 63 (2): 199–222.

Lister, Florence C. and Robert H. Lister. 1974. "Maiolica in Colonial Spanish America." *Historical Archaeology* 8 (1): 17–52.

Loren, Diana DiPaolo. 2008. *In Contact: Bodies and Spaces in the Sixteenth- and Seventeenth-Century Eastern Woodlands*. Lanham: AltaMira Press.

Loren, Diana DiPaolo. 2010. "The Exotic in Daily Life: Trade and Exchange in Historical Archaeology." In *Trade and Exchange: Archaeological Studies from History and*

Prehistory, edited by Carolyn D. Dillian and Carolyn L. White, 195–204. New York: Springer.

Luna Calderón, Fernando. 1992. *La Isabela. Primer Cementerio Indohispano En El Nuevo Mundo*. Manuscript on file, Museo del Hombre Dominicano, Santo Domingo.

Lyons, Claire L. and John K. Papadopoulos. 2002. *The Archaeology of Colonialism*. Los Angeles: Getty Research Institute.

Maran, Joseph and Philipp W. Stockhammer, eds. 2012. *Materiality and Social Practice: Transformative Capacities of Intercultural Encounters*. Oxford: Oxbow Books.

McNiven, Ian J. 2016. "Theoretical Challenges of Indigenous Archaeology: Setting an Agenda." *American Antiquity* 81 (1): 27–41.

Mendoza, Lourdes C. 1957. "Cerámica de Las Ruinas de La Vega Vieja." *Casas Reales* 11: 101–113.

Mintz, Sidney W. and Richard Price. 1992. *The Birth of African American Culture*. Boston: Beacon.

Mira Caballos, Esteban. 1997. *El Indio Antillano: Repartimiento, Encomienda Y Esclavitud (1492–1542)*. Seville: Muñoz Moya.

Mol, Angus A.A. 2008. "Universos Socio-Cósmicos En Colisión: Descripciones Etnohistóricas de Situaciones de Intercambio En Las Antillas Mayores Durante El Período de Proto-Contacto." *El Caribe Arqueológico* 10: 13–22.

Moore, John H. 1994. "Ethnogenetic Theory." *Research and Exploration* 10 (1): 10–23.

Morales Patiño, Oswaldo, and Roberto Pérez de Acevedo. 1945. "El Período de Transculturación Indo-Hispánica." Contribuciones Del Grupo Guama. *Revista de Arqueología Y Etnología* 4/6: 5–36.

Morsink, Joost. 2015. "Spanish-Lucayan Interaction: Continuity of Native Economies in Early Historic Times." *Journal of Caribbean Archaeology* 15: 102–119.

Murray, Tim, ed. 2004. *The Archaeology of Contact in Settler Societies*. Cambridge: Cambridge University Press.

Oland, Maxine. 2014. "'With the Gifts and Good Treatment That He Gave Them': Elite Maya Adoption of Spanish Material Culture at Progresso Lagoon, Belize." *International Journal of Historical Archaeology* 18 (4): 643–667.

Oland, Maxine, Siobhan M. Hart, and Liam Frink, eds. 2012. *Decolonizing Indigenous Histories: Exploring Prehistoric/Colonial Transitions in Archaeology*. Tucson: University of Arizona Press.

Oliver, José R. 2000. "Gold Symbolism among Caribbean Chiefdoms: Of Feathers, Çibas and Guanín Power among Taíno Elites." In *Precolumbian Gold: Technology, Style and Iconography*, edited by Colin McEwan, 196–219. London: British Museum Press.

Oliver, José R. 2009. *Caciques and Cemí Idols: The Web Spun by Taíno Rulers Between Hispaniola and Puerto Rico*. Tuscaloosa: University of Alabama Press.

Ortega, Elpidio J. 1980. *Introducción a La Loza Común O Alfarería En El Período Colonial de Santo Domingo*. Santo Domingo: Fundación Ortega Álvarez.

Ortega, Elpidio J. 1982. *Arqueología Colonial En Santo Domingo*. Santo Domingo: Taller.

Ortega, Elpidio J. 1988. *La Isabela Y La Arqueología En La Ruta de Colón*. San Pedro de Macorís: Universidad Central del Este.

Ortega, Elpidio J. and Carmen G. Fondeur. 1978. *Estudio de La Cerámica Del Período Indo-Hispana de La Antigua Concepción de La Vega*. Santo Domingo: Taller.

Ortiz, Fernando. 1995. *Cuban Counterpoint: Tobacco and Sugar*. Durham: Duke University Press.

Ostapkowicz, Joanna. 2013. "'Made … With Admirable Artistry': The Context, Manufacture and History of a Taíno Belt." *The Antiquaries Journal* 93: 287–317.

Pagden, Anthony. 1993. *European Encounters with the New World: From Renaissance to Romanticism*. New Haven: Yale University Press.

Palm, Edward W. 1945. "Excavations at La Isabela, White Man's First Town in the Americas." *Acta Americana* 3: 298–303.

Palm, Edward W. 1952. "La Fortaleza de La Concepción de La Vega." In *Memoria Del V Congreso Histórico Municipal Interamericano*, 2: 115–118. Santo Domingo: Editora del Caribe.

Palm, Edward W. 1955. *Los Monumentos Arquitectónicos de La Española*. Vol. 1 & 2. Santo Domingo: Universidad de Santo Domingo.

Palmié, Stephan. 1995. "Against Syncretism: 'Africanizing' and 'Cubanizing' Discourses in North American Òrìsà Worship." In *Counterworks: Managing of the Diversity of Knowledge*, edited by Richard Fardon, 73–104. London: Routledge.

Pantel, A. Gus, Jalil Sued Badillo, Anibal Sepúlveda, and Beatriz del Cueto de Pantel. 1988. *Archaeological, Architectural and Historical Investigations of the First Spanish Settlement in Puerto Rico: Caparra*. Manuscript on file. The Foundation of Archaeology, Anthropology and History of Puerto Rico, San Juan.

Pasztory, Esther. 2005. *Thinking with Things: Toward a New Vision of Art*. Austin: University of Texas Press.

Patterson, Thomas C. 1991. "Early Colonial Encounters and Identities in the Caribbean: A Review of Some Recent Works and Their Implications." *Dialectical Anthropology* 16 (1): 1–13.

Pesoutova, Jana and Corinne L. Hofman. 2016. "La Contribución Indígena a La Biografía Del Paisaje Cultural de La República Dominicana: Una Revisión Preliminar." In *Indígenas E Indios En El Caribe: Presencia, Legado Y Estudio*, edited by Jorge Ulloa Hung and Roberto Valcárcel Rojas, 115–150. Santo Domingo: Instituto Tecnológico de Santo Domingo.

Pugh, Timothy W. 2009. "Contagion and Alterity: Kowoj Maya Appropriations of European Objects." *American Anthropologist* 111 (3): 373–386.

Quimby, George I., and Alexander Spoehr. 1951. "Acculturation and Material Culture-1." *Fieldiana – Anthropology* 36 (6): 107–147.

Ramenofsky, Ann F. 1987. *Vectors of Death: The Archaeology of European Contact*. Albuquerque: University of New Mexico Press.

Redfield, Robert, Ralph Linton, and Melville J. Herskovits. 1936. "Memorandum for the Study of Acculturation." *American Anthropologist* 38 (1): 149–152.

Rey Betancourt, Estrella. 1972. *La Transculturación Indohispánica En Cuba*. Serie Histórica 3. Havana: Academia de Ciencias de Cuba.

Richard, François G., ed. 2015. *Materializing Colonial Encounters: Archaeologies of African Experience*. New York: Springer.

Rodríguez-Alegría, Enrique. 2008. "Narratives of Conquest, Colonialism, and Cutting-Edge Technology." *American Anthropologist* 110 (1): 33–43.

Rodríguez-Alegría, Enrique. 2010. "Incumbents and Challengers: Indigenous Politics and the Adoption of Spanish Material Culture in Colonial Xaltocan, Mexico." *Historical Archaeology* 44 (2): 51–71.

Rodríguez-Alegría, Enrique. 2016. *The Archaeology and History of Colonial Mexico: Mixing Epistemologies*. Cambridge: Cambridge University Press.

Rogers, J. Daniel. 1990. *Objects of Change: The Archaeology and History of Arikara Contact with Europeans*. Washington, DC: Smithsonian Institution Press.

Rogers, J. Daniel. 1993. "The Social and Material Implications of Contact on the Northern Plains." In *Ethnohistory and Archaeology: Approaches to Postcontact Change in the Americas*, edited by J. Daniel Rogers and Samuel M. Wilson, 73–88. New York: Plenum Press.

Rogers, J. Daniel and Samuel M. Wilson, eds. 1993. *Ethnohistory and Archaeology: Approaches to Postcontact Change in the Americas*. New York: Plenum Press.

Romero, Leandro E. 1981. "Sobre Las Evidencias Arqueológicas de Contacto Y Transculturación En El Ámbito Cubano." *Santiago* 44: 77–108.

Rothschild, Nan A. 2003. *Colonial Encounters in a Native American Landscape: The Spanish and the Dutch in North America*. Washington, DC: Smithsonian Institution Press.

Rothschild, Nan A. 2006. "Colonialism, Material Culture, and Identity in the Rio Grande and Hudson River Valleys." *International Journal of Historical Archaeology* 10 (1): 73–108.

Rouse, Irving B. 1942. *Archaeology of the Maniabón Hills, Cuba*. Yale University Publications in Anthropology 26. New Haven: Yale University Press.

Rouse, Irving B. 1972. *An Introduction to Prehistory: A Systematic Approach*. New York: McGraw-Hill.

Rouse, Irving B. and José M. Cruxent. 1963. *Venezuelan Archaeology*. New Haven: Yale University Press.

Rubertone, Patricia E. 1996. "Matters of Inclusion: Historical Archaeology and Native Americans." *World Archaeological Bulletin* 7: 77–86.

Rubertone, Patricia E. 2000. "The Historical Archaeology of Native Americans." *Annual Review of Anthropology* 29 (1): 425–446.

Sahlins, Marshall D. 1976. *Culture and Practical Reason*. Chicago: University of Chicago Press.

Said, Edward W. 1978. *Orientalism*. New York: Pantheon Books.

Saunders, Nicholas J. 1999. "Biographies of Brilliance: Pearls, Transformations of Matter and Being, C. AD 1492." *World Archaeology* 31 (2): 243–257.

Scaramelli, Franz and Kay Tarble de Scaramelli. 2005. "The Roles of Material Culture in the Colonization of the Orinoco, Venezuela." *Journal of Social Archaeology* 5 (1): 135–168.

Scheiber, Laura L. and Mark D. Mitchell, eds. 2010. *Across a Great Divide: Continuity and Change in Native North American Societies, 1400–1900*. Tucson: University of Arizona Press.

Schneider, Jane. 1977. "Was There a Pre-Capitalist World System?" *Peasant Studies* 6 (1): 20–29.

Silliman, Stephen W. 2005. "Culture Contact or Colonialism? Challenges in the Archaeology of Native North America." *American Antiquity* 70 (1): 55–74.

Silliman, Stephen W. 2009. "Change and Continuity, Practice and Memory: Native American Persistence in Colonial New England." *American Antiquity* 74 (2): 211–230.

Silliman, Stephen W. 2010. "Indigenous Traces in Colonial Spaces: Archaeologies of Ambiguity, Origin, and Practice." *Journal of Social Archaeology* 10 (1): 28–58.

Silliman, Stephen W. 2012. "Between the Longue Durée and the Short Pureé: Postcolonial Archaeologies of Indigenous History in Colonial North America." In *Decolonizing Indigenous Histories: Exploring Prehistoric/Colonial Transitions in Archaeology*, edited by Maxine Oland, Siobhan M. Hart, and Liam Frink, 113–132. Tucson: University of Arizona Press.

Silliman, Stephen W. 2015. "A Requiem for Hybridity? The Problem with Frankensteins, Purées, and Mules." *Journal of Social Archaeology* 15 (3): 277–298.

Silliman, Stephen W. 2016. "Disentangling the Archaeology of Colonialism and Indigeneity." In *Archaeology of Entanglement*, edited by Lindsay Der and Francesca Fernandini, 31–48. Walnut Creek: Left Coast Press.

Smith, Claire and H. Martin Wobst, eds. 2005. *Indigenous Archaeologies: Decolonizing Theory and Practice*. London: Routledge.

Smith, Greg C. 1995. "Indians and Africans at Puerto Real: The Ceramic Evidence." In *Puerto Real: The Archaeology of a Sixteenth-Century Spanish Town in Hispaniola*, edited by Kathleen A. Deagan, 335–374. Gainesville: University of Florida Press.

Solis Magaña, Carlos. 1999. "Criollo Pottery from San Juan de Puerto Rico." In *African Sites Archaeology in the Caribbean*, edited by Jay Haviser, 131–141. Princeton: Markus Weiner.

South, Stanley. 1978. "Pattern Recognition in Historical Archaeology." *American Antiquity* 43 (2): 223–230.

Spicer, Edward H., ed. 1961. *Perspectives in American Indian Culture Change*. Chicago: University of Chicago Press.

Stein, Gil J., ed. 2005. *The Archaeology of Colonial Encounters: Comparative Perspectives*. Santa Fe: School of American Research Press.

Stewart, Charles and Rosalind Shaw, eds. 1994. *Syncretism/Anti-Syncretism: The Politics of Religious Synthesis*. London: Routledge.

Thomas, David Hurst, ed. 1989. *Columbian Consequences, Vol. 1: Archaeological and Historical Perspectives on the Spanish Borderlands West*. Washington, DC: Smithsonian Institution Press.

Thomas, David Hurst, ed. 1990. *Columbian Consequences, Vol. 2: Archaeological and Historical Perspectives on the Spanish Borderlands East*. Washington, DC: Smithsonian Institution Press.

Thomas, David Hurst, ed. 1991. *Columbian Consequences, Vol. 3: The Spanish Borderlands in Pan-American Perspective*. Washington, DC: Smithsonian Institution Press.

Thomas, Nicholas. 1991. *Entangled Objects: Exchange, Material Culture and Colonialism in the Pacific*. Cambridge: Harvard University Press.

Torrence, Robin and Anne Clarke, eds. 2000. *The Archaeology of Difference: Negotiating Cross-Cultural Engagements in Oceania*. London: Routledge.

Trigger, Bruce G. 1991. "Early Native North American Response to European Contact: Romantic versus Rationalistic Interpretations." *The Journal of American History* 77 (4): 1195–1215.

Valcárcel Rojas, Roberto. 1997. "Introducción a La Arqueología Del Contacto Indo-Hispánico En La Provincia de Holguín, Cuba." *El Caribe Arqueológico* 2: 64–77.

Valcárcel Rojas, Roberto. 2012. "Interacción Colonial en Un Pueblo de Indios Encomendados: El Chorro de Maíta, Cuba." PhD diss., Leiden University.

Valcárcel Rojas, Roberto. 2016. *Archaeology of Early Colonial Interaction at El Chorro de Maíta, Cuba*. Gainesville: University of Florida Press.

Valcárcel Rojas, Roberto, Marcos Martinón-Torres, Jago Cooper, and Thilo Rehren. 2010. "Turey Treasure in the Caribbean: Brass and Indo-Hispanic Contact at El Chorro de Maíta, Cuba." In *Beyond the Blockade: New Currents in Cuban Archaeology*, edited by Susan Kepecs, L. Antonio Curet, and Gabino La Rosa Corzo, 106–125. Tuscaloosa: University of Alabama Press.

Valcárcel Rojas, Roberto, Alice V.M. Samson, and Menno L.P. Hoogland. 2013. "Indo-Hispanic Dynamics: From Contact to Colonial Interaction in the Greater Antilles." *International Journal of Historical Archaeology* 17 (1): 18–39.

Van Buren, Mary. 2010. "The Archaeological Study of Spanish Colonialism in the Americas." *Journal of Archaeological Research* 18: 151–201.

Van Dommelen, Peter. 2006. "Colonial Matters: Material Culture and Postcolonial Theory in Colonial Situations." In *Handbook of Material Culture*, edited by Chris Tilley, Webb Keane, Susanne Küchler, Michael Rowlands, and Patricia Spyer, 104–124. London: Sage Publications.

Vega, Bernardo. 1979. *Los Metales Y Los Aborígenes de La Hispaniola*. Santo Domingo: Museo del Hombre Dominicano.

Vega, Bernardo. 1987. *Santos, Shamanes Y Zemíes*. Santo Domingo: Fundación Cultural Dominicana.

Voss, Barbara L. 2015. "Narratives of Colonialism, Grand and Not So Grand: A Critical Reflection on the Archaeology of the Spanish and Portuguese Americas." In *Archaeology of Culture Contact and Colonialism in Spanish and Portuguese America*, edited by Pedro P.A. Funari and Maria X. Senatore, 353–361. Cham: Springer.

Voss, Barbara L., and Eleanor Conlin Casella. 2012. *The Archaeology of Colonialism: Intimate Encounters and Sexual Effects*. Cambridge: Cambridge University Press.

Wallerstein, Immanuel M. 1974. *The Modern World-System, Vol. I: Capitalist Agriculture and the Origins of the European World-Economy in the Sixteenth Century*. New York: Academic Press.

Wallerstein, Immanuel M. 1980. *The Modern World-System, Vol. II: Mercantilism and the Consolidation of the European World-Economy, 1600–1750*. New York: Academic Press.

Whitehead, Neil L. 2011. "Native Americans and Europeans: Early Encounters in the Caribbean and along the Atlantic Coast." In *The Oxford Handbook of the Atlantic World: 1450–1850*, edited by Nicholas Canny and Philip Morgan, 55–70. Oxford: Oxford University Press.

Williamson, Christine. 2004. "Contact Archaeology and the Writing of Aboriginal History." In *The Archaeology of Contact in Settler Societies*, edited by Tim Murray, 176–199. Cambridge: Cambridge University Press.

Wilson, Samuel M. 1990. *Hispaniola: Caribbean Chiefdoms in the Age of Columbus*. Tuscaloosa: University of Alabama Press.

Wilson, Samuel M. 1993. "Structure and History: Combining Archaeology and Ethnohistory in the Contact Period Caribbean." In *Ethnohistory and Archaeology: Approaches to Postcontact Change in the Americas*, edited by J. Daniel Rogers and Samuel M. Wilson, 19–30. New York: Plenum Press.

Wing, Elizabeth S. 1961. "Animal Remains Excavated at the Spanish Site of Nueva Cádiz on Cubagua Island, Venezuela." *Nieuwe West-Indische Gids* 41 (2): 162–165.

Wolf, Eric R. 1982. *Europe and the People without History*. Berkeley: University of California Press.

Woodward, Robyn P. 2006. "Medieval Legacies: The Industrial Archaeology of an Early Sixteenth-Century Sugar Mill at Sevilla La Nueva, Jamaica." PhD diss., Simon Fraser University.

Wylie, Alison. 1992. "Rethinking the Quincentennial: Consequences for Past and Present." *American Antiquity* 57 (4): 591–594.

Young, Robert J.C. 1995. *Colonial Desire: Hybridity in Theory, Culture, and Race*. London: Routledge.

CHAPTER 2

Colonial Encounters in Lucayan Contexts

Mary Jane Berman and Perry L. Gnivecki

1 Introduction

Consumption patterns are informed by context, so, when studying indigenous consumption of European items, it is necessary to consider how colonial contexts varied (Dietler 1998; Lightfoot and Simmons 1998; Oland 2014, 646). Much of what has been written about the indigenous consumption of European artifacts during the early period of Spanish colonialism of the Caribbean has focused on the patterns observed at colonial settler sites such as La Isabela (Deagan 1988; Deagan and Cruxent 2000a,b), Puerto Real (Deagan 1995), and Concepción de la Vega (Ortega and Fondeur 1978) on Hispaniola or *encomienda* sites such as El Chorro de Maíta on Cuba (Valcárcel Rojas 2016). At these sites, the indigenous occupants and the Spanish lived and worked in close proximity under colonial scrutiny in mines, workshops, fields, and households (Kulstad-González 2015; Valcárcel Rojas et al. 2013).

Indigenous sites located on the geographic and political frontiers of colonial settlements also offer insights about local consumption of European goods during the early period of Spanish colonization. These communities were not subject to the same level of regulation as those in the colonial centers and their autonomy offered different opportunities for indigenous agency (Lightfoot and Martinez 1995; Oland 2014). In some cases, local peoples did not have direct contact with the Spanish, but acquired European objects by way of down-the-line trade. Such was the case of El Cabo, an indigenous site in southeastern Dominican Republic, which did not experience direct colonial control during the early years of colonization of Hispaniola (Hofman et al. 2014; Samson 2010; Valcárcel Rojas et al. 2013).

The Bahama archipelago, home to the Lucayans, the indigenous occupants of these islands, offers another opportunity to view native consumption of European goods during the early period of Spanish colonization. The Spanish regarded the Bahama Islands as useless, referring to them as *islas inútiles* (Anderson-Córdova 2017, 131) and did not establish settler communities or institute the *encomienda* system here as they did elsewhere in the Antilles. With the decimation of many of its indigenous inhabitants and the formal establishment of the *encomienda* system on Hispaniola in 1503, the Spanish

DOI:10.1163/9789004273689_003

turned to the Bahamas to secure labor, and the Lucayans were brought to the Antilles as slaves and *naborías* (Anderson-Córdova 2017; Sauer 1966). Since we view colonialism as the "process by which a city or nation-state exerts control over people – termed indigenous – and territories outside of its geographical boundaries" (Silliman 2005, 58), it can be said that the Bahamas functioned as a colonial "space," or frontier impacted by Spanish policies and practices. In this chapter, we examine the historical and cultural contexts in which Lucayan consumption of European objects occurred and the processes by which they were transformed into indigenous objects.

Studies have shown that European objects were integrated differently and for a variety of different reasons into indigenous systems of use. In areas falling under Spanish colonial rule, indigenous people were selective in what European objects they incorporated, accepting them at varying rates or not at all (Charlton 1968; Cobb 2003; Rodríguez-Alegría 2008; Rodríguez-Alegría et al. 2015). The scarcity of Spanish objects at En Bas Saline, a Taíno site located adjacent to the early colonial site of Puerto Real, for example, has been attributed to native "indifference to and rejection" of Spanish culture (Deagan 2004).

In colonial contexts, indigenous and European items were often transformed physically and given new meanings and uses as they crossed cultural borders (*sensu* Kopytoff 1986; Thomas 1991). Thomas (1991) and Cipolla (2017, 18) note, however, that such physical modification of European objects in indigenous contexts occurred infrequently. In his study of the Brothertown Indians of New England, Cipolla (2013, 2017, 18) observed no physical alterations of European-made artifacts. In the Caribbean, indigenous objects were changed occasionally to resemble European items, while some European objects were modified to look like native objects (Deagan 1988; Hofman et al. 2014; Rouse 1942). In short, there is limited evidence for such modifications. Cusick (1991, 452) has suggested that we begin to see significant "Europeanization" of or other changes in indigenous material culture only with the relocation of the Taíno (indigenous populations) to towns or *encomiendas*, contexts where cultures experienced major changes in social structure and cultural coherence. For example, native pottery did not show any evidence for the incorporation of Spanish elements at En Bas Saline (Cusick 1991). In contrast, the indigenous pottery at Puerto Real exhibited European attributes with the creation of the Spanish settlement (Deagan 1988, 214). Such hybrid forms retained traditional uses or were given new meanings or new uses. Keehnen's chapter (this volume) adds to the database of these kinds of objects in Hispaniola. Other items remained unchanged physically, but given new purposes, and assigned new meanings, as Valcárcel Rojas (2016) observed at a number of sites in Cuba, such as El Chorro de Maíta. Finally, European items, such as glass beads, mirrors, and brass ornaments were occasionally incorporated into Taíno material culture, such as belts and sculptures (Ostapkowicz 2018).

In addition to studying context, we are also interested in the length of time it took for European objects to be modified or assimilated into indigenous contexts in early colonial situations. In our study, we examine a time frame of less than 40 years (1492–1530). In this example, contact and interaction were brief and sporadic and did not result in the establishment of settlements where interaction was prolonged or continuous. By reframing our questions, reconsidering our assumptions, and reexamining the evidence, we present a new narrative of Lucayan acquisition and consumption of European items.

2 Setting the Stage for Consumption

There were numerous documented ways that the Lucayans interacted with the Spanish and acquired and consumed European objects (Dunn and Kelley 1989). These included direct trade, exchange, and theft. Other possible means include securing items from shipwrecks or vessels that stopped on the islands for repair or careening. Intra- and inter-island distribution of European objects was likely to have occurred through indirect means such as down-the-line exchange (Gnivecki 1995, 2011; Keegan 1992).

The first known direct engagement between the Spanish and the Lucayans occurred in 1492 when Columbus and his men made landfall on the island of Guanahaní, known today as San Salvador (Dunn and Kelley 1989; Morison 1942). Here they participated in trade and exchange and the Spanish took six men captive. This was followed by visits to several other Lucayan islands where the Spanish traded and exchanged objects with the local peoples. Soon after, explorers, traders, and enslavers passed through the archipelago (Table 2.1).

TABLE 2.1 Spanish activity in the Bahama archipelago (1492–1530)

Date	Individual(s) Involved	References
1492	Cristoforo Colón	Dunn and Kelley (1989, 57–117)
1499–1500	Juan de la Cosa	Parry and Keith (1984:II, 147)
1499–1500	Alonso de Hojeda	Sauer (1966, 112, 159)
1499–1500	Vincente Yáñez Pinzón	Burns (1965, 90–91); Quinn (1979:I, 234–235, 237–238)
1499–1500	Amerigo Vespucci	Parry and Keith (1984:II, 163–164)

Date	Individual(s) Involved	References
1508	Lucas Vásquez de Allyón	Anderson-Córdova (2017, 134)
1508–1509	Nicolás de Ovando	Sauer (1966, 158–159)
1508–1509	Alonso de Hojeda	Anderson-Córdova (2017, 131)
1508–1509	Diego de Nicuesa	Anderson-Córdova (2017, 131)
1513	Juan Ponce de León	Kelley (1991, 41–42, Footnote 22, 52); Weddle (1985, 40)
1513	Diego Miruelo	Weddle (1985, 46)
1513	Juan Bono de Quejo	Weddle (1985, 40)
1513–1514	Antón de Alaminos	Quinn (1979:I, 237–238); Weddle (1985, 40)
1513–1514	Diego Bermúdez	Quinn (1979 I, 237–238); Weddle (1985, 40)
1513–1514	Juan Pérez de Ortubia	Weddle (1985, 40)
1514	Diego Velázquez	Anderson-Córdova (2017, 141)
1514–1516	Pedro de Salazar	Hoffman (1990, 6)
1514–1517	Francisco Gordillo	Hoffman (1990, 5)
1514–1517	Toribio de Villafranca	Hoffman (1990, 5)
1515–1516	Diego Velázquez	Weddle (1985, 55)
1517	Diego Velázquez	Anderson-Córdova (2017, 141)
1519	Francisco de Barrionuevo	Parry and Keith (1984:II, 390)
1521	Lucas Vásquez de Allyón	Hoffman (1990); Quinn (1979:I, 248, 255, 257)
1521	Francisco Gordillo	Hoffman (1990, 6–7)
1521	Pedro de Quijos	Hoffman (1990, 6–7); Quinn (1979:I, 257–258)
1521	Alonso Fernández Sotil	Hoffman (1990, 7)
1521	Juan Ponce de León	Ober (1908, 197)
1525	Pedro de Quijos	Hoffman (1990, 36–37)
1526	Lucas Vásquez de Allyón	Hoffman (1990, 44, 55)
1521–1526	Bahamas depopulated	Quinn (1979:I, 258, 265)

Note: It is believed that the Highborne Key and Molasses Reef wrecks in Bahamian waters date to this period, but they have not been securely dated (Keith et al. 1984; Smith et al. 1985). Thus, they are excluded from this table. For the sake of brevity, we have mentioned only royal decrees of enslavement authorization. See Anderson-Córdova (2017) for the details of royal and local authorization of enslavement beyond Hispaniola and Puerto Rico.

Subsequent visits brought the Lucayans and Spanish into direct contact. In spite of the number of visits, however, such interactions were limited to short, irregular encounters.

Shipwrecks may have been significant sources of European goods (Turnbaugh 1993, 136) and in the Bahamas, as well as other coastal contexts, shipwrecked sailors offered opportunities for Spanish items to be traded, exchanged, or given as gifts, and grounded ships presented chances for items to be pilfered and scavenged (Gnivecki 1995, 2011). The numerous reefs and shoals of the archipelago and the destructive forces of hurricanes posed great challenges for fifteenth- and sixteenth-century seafaring, resulting in numerous shipwrecks (Sauer 1966). Although only two ships from that period, the Molasses Reef wreck (Keith et al. 1984) in 1499 and the Highbourn Cay wreck, one of the ships lost by Pinzón in 1500 (Smith et al. 1985) have been found, it is probable that many more went aground.

The acquisition of wood and water from the islands to provision ships offered occasions for Lucayan-Spanish exchanges. Stopping at islands for ship maintenance and repairs must have also brought Lucayans and Spanish in direct contact. Keegan (1992, 213) suggests that during late 1499 and early 1500 Vespucci may have established short-term encampments on several islands, in order to careen his vessel during his journey through the islands. (The trip included taking of Lucayan slaves, see below). These would have provided occasions for the introduction of European items. Spanish ships from this and other expeditions may also have moored close to shore as they passed through the islands and the Lucayans may have swum out to the boats to procure goods, as they did during Columbus' voyage (Dunn and Kelley 1989). The ship captains may have sent their crew in small boats to shore, as well, to engage in trade and exchange.

Spanish maps from the early and middle 1500s suggest that considerable shipping traffic took place through the archipelago during this period (Granberry 1979, 1980, 1981). Numerous secret missions by unnamed European powers also likely occurred (Harisse 1961; Keegan 1992, 207). Items may have been lost, discarded, or traded during passage through the islands, as the sailors associated with these activities stopped to secure fresh water, make repairs, investigate the landscape, or take captives (Keegan 1992, 203). The appropriation of items from shipwrecks and/or shipwrecked sailors, the procurement of items that washed up on shore from shipwrecks, the seizure of items from careened ships, and trade and exchange with sailors on the many expeditions that passed through the Bahamas during this period are all means by which European items found their way into Lucayan hands.

Spanish removal of the Lucayans began in the early days of colonization, although we have no written records to that effect until 1499. There is documentary evidence of at least 232 Lucayans having being taken by force to Spain by Amerigo Vespucci during his four month passage through the Bahamas (Keegan 1992, 212; Sauer 1966, 112, 159). During the early 1500s, Lucayans were brought to Hispaniola as *naborías*, but functioned as slaves and were sold illegally (Anderson-Córdova, 2017, 145). In 1508, raiding in the Bahamas was recognized legally and the following year the Crown decreed that those who resisted capture were to be designated as slaves (Anderson-Córdova 2017, 131). Even though the greatest period of enslavement occurred during 1509–1515 (Sauer 1966), such activity continued for at least another decade. In 1517, privately financed slave ships, commissioned by Governor Diego Velazquez of Cuba, set out from Santiago de Cuba, to secure Lucayan slaves to work on Cuba (Sauer 1966); that year 300 Indians from Bimini and Florida were taken to Puerto Rico (Anderson-Córdova 2017, 141). In 1518, privately financed slave ships originating from Puerto Plata and Santiago brought shiploads of Lucayans, as well as individuals from other islands, to Hispaniola (Anderson-Córdova 2017, 137; Sauer 1966). In 1520 Lucayans who had been taken to Hispaniola were sent to Nueva Cádiz to work in the pearl beds (Granberry 1979). The exact number of Lucayans brought to the Antilles varies between 20,000–40,000 people; of those captured, anywhere from 800 to 5000 people survived transport (Anderson Córdova 2017, 136–137). Lucayans may have been lured to ships through the enticement of European goods, although, according to Las Casas [in Granberry (1979, 15)], actual enslavement was achieved with "sword and lance." As an example, Pohl (1966, 87) in Keegan (1992, 212) notes that Vespucci took captives "by force." While no record exists as to how many such expeditions took place, Keegan (1992, 221) has projected that at least for the purposes of enslavement, 320 vessels may have traveled through the islands. The archipelago was ultimately depopulated by 1520–1530, primarily through these actions (Gnivecki 1995; Granberry 1981, 18).

3 The Long Bay Site

The Long Bay site, located on the western side of San Salvador (Figure 2.1), has yielded the greatest number of Spanish artifacts of any site in the archipelago. Three other excavated sites, the Three Dog site on San Salvador (Berman and Gnivecki 1995), MC-6 on Middle Caicos (Sullivan 1981), and CC-6 on Cotton Cay (Sinelli 2010), have each produced one or two Spanish objects. Surface finds of earthenware sherds have been found on Long Island, Little Exuma, Acklins Island, Conception Island, and Samana Cay (Keegan 1992), and Middle Caicos (Sinelli 2010). Some of these finds are associated with archaeological sites; others are isolated finds (Keegan 1992). Two shipwreck sites (Keith et al.

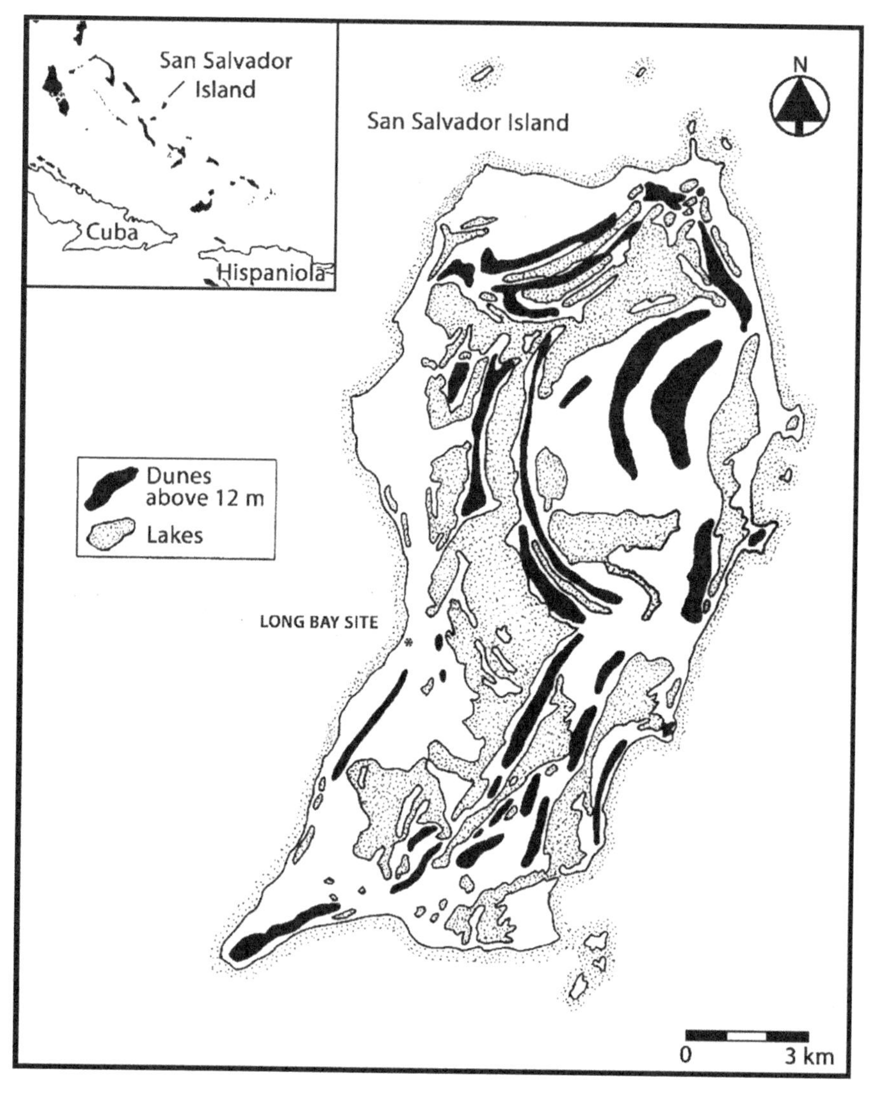

FIGURE 2.1 Map of the Bahamas and San Salvador Island
MAP DRAWN BY PERRY L. GNIVECKI

1984; Peterson 1974; Smith et al. 1985) have also produced ordnance and other artifacts of European origin.

The Long Bay Site is considered one of the earliest sites of contact between the Lucayans and the Spanish in the Bahama archipelago, although it is not known whether the European items made their way to the site through direct intercultural encounter or indirect means. The archaeologist Charles Hoffman, who excavated a portion of it between 1983–1992, sought to demonstrate that it was the Columbus landing site and theatre for direct contact between the indigenous peoples and the Spanish (Hoffman 1987a, 1987b). In the late 1990s, the construction of a house threatened to destroy the northern part of the site, so efforts to recover what the builders had not yet disturbed or removed from that area were instituted by John Winter, in 1999, and Berman and Gnivecki, in 2000. In 2008, Gnivecki mapped the whole site, and, in 2012, he and his students dug several units to the west of where Hoffman had worked. All the researchers conducted their excavations in 10 cm levels in 1×1 and 2×1 meter square units and shovel test pits. A total of 191.5 square meters has been excavated. The use of 1/16-inch mesh screen, allowed for the capture of tiny objects, such as European glass beads (2.5–3.5 mm in diameter), which otherwise may have not been recovered.

The Long Bay site inhabitants' lifeways resembled those described for other Lucayan settlements from this time period (Berman et al. 2013). They were fisher-horticulturalist-shellfish collectors who fished and collected from nearby reefs and intertidal habitats (Newsom and Wing 2004), grew root and seed crops (Berman and Pearsall 2008, 2018), and gathered and possibly managed some wild plants, including palms and fruit trees (Berman and Pearsall 2018). Food procurement, processing, and preparation implements, ceramics, woodworking tools, body adornments, and ceremonial items were manufactured from local and non-local materials. The residents engaged in down-the-line or direct trade and exchange with other Lucayans and people from islands outside the Bahama archipelago, as the site yielded a number of imported items including ceramics and stone tools. One sherd has been sourced to northern Cuba (Winter and Gilstrap 1991). Non-local indigenous (i.e., Antillean) ceramics made up 3.2 per cent and Spanish ceramics constituted 3.1 per cent of the ceramic assemblage (Bate 2011, 216).

Only a portion of the site was excavated. No discrete house structure(s) could be inferred from the postholes, which were located to the northeast and southwest of where the excavations were concentrated. Similarly, neither middens nor distinct activity areas have yet to be discerned from the artifact patterning. The Spanish objects, which cluster in the northeastern part of the site, were found at a depth of 10–40 cm below the surface intermixed with a typical Late Lucayan domestic assemblage. The assemblage included Lucayan ceramics (plain ware and red-slipped Palmetto ware sherds, Palmetto ware basketry-impressed sherds, and a handful of Palmetto ware incised or appliqué sherds) (Bate 2011), shell beads, and shell bead débitage; food debris from local sources, such as fish (e.g., parrotfish and other reef fish) (Newsom and Wing 2004) and whole shells and shellfish fragments (e.g., *Lobatus gigas*, *Codakia orbicularis*, *Nerita* sp., *Cittarium pica*), procured from in-shore and rocky intertidal areas. Non-local ceramic sherds and stone objects (e.g., greenstone fragments, including jadeite, quartz microliths and cores) were also recovered. We interpret the area where the European goods were found as generalized floor debris, the consequence of repeated discard, loss, and sweeping episodes in one or more small household areas. It would be premature to say how many households were present or if they were associated with one or more elite households, however.

The excavated European objects include six or more green glass beads and three glass bead fragments, one amber glass bead, 38 *melado* sherds, two majolica (Columbia Plain) sherds, unglazed earthenware sherds, one reconstructed half of an early-style olive jar, 10 planking nails (ship spikes), two metal hooks, four metal knife blade fragments, a bronze "D"-ring, a bronze belt buckle, a copper grommet, a *blanca* (coin), a metal button, metal fragments, and fragments of green glass (Bate 2011; Hoffman 1987a, 1987b, 241–242). The translucent green glass beads are known as *abalorios*, wire-wound beads that held little value to the Spanish (Deagan 1987, 157). Glazed and unglazed sherds of Spanish origin including *melado* and majolica wares are scattered on the surface, particularly in the southern sector of the site (Berman and Gnivecki 2000). Hoffman reconstructed part of an early-style Spanish olive jar from sherds recovered from the southeastern part of the site, adjacent to a low-lying depression that fills up with water during the rainy season (Bate 2011). During the summer of 2017, we determined that the button and green glass shards were of later origin and that Hoffman recovered more Spanish objects during the later years of his excavations (post 1987).

Chemical analyses of some of the European artifacts, conducted by the Corning Museum of Glass (Brill et al. 1987), point to their origins in the Iberian Peninsula, matching particular source areas in Portugal and Spain. Several have been keyed to specific workshops in Spain. Descriptions and the chemical analyses of the beads, "D"-ring, belt buckle, and sherds are summarized in Table 2.2.

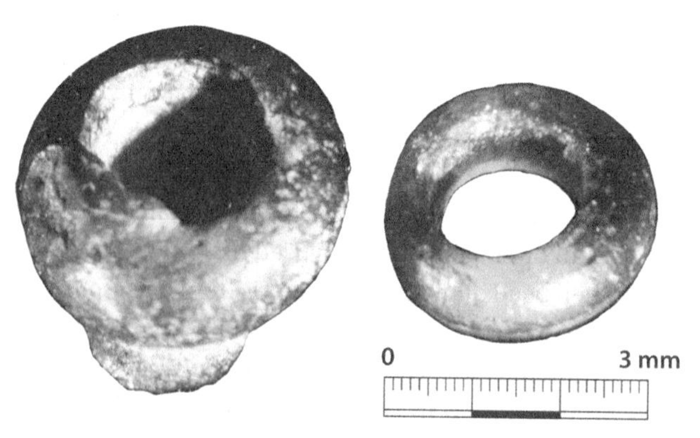

FIGURE 2.2 Green glass beads from the Long Bay Site, San Salvador Island
PHOTO COURTESY OF KATHY DOAN GERACE

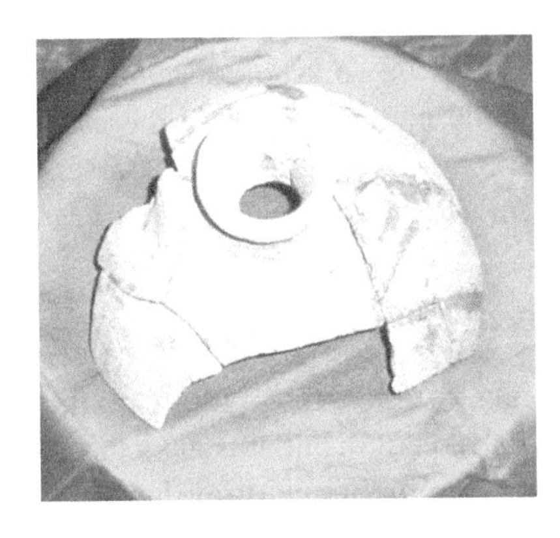

FIGURE 2.3
Reconstructed Spanish olive jar from the Long Bay Site, San Salvador Island

TABLE 2.2 European artifact descriptions

Coin (Brill 1987 et al., 255–257, 259, 280–281)
This *blanca* is a small-denomination billion coin issued between 1471–1474 during the reign of King Henry IV. It is a small denomination billion (copper alloy) coin of low value. The diagnostic images of a castle and a floret on one side and a lion on the other can be made out, although the coin is corroded. There is some evidence of lettering, as well. According to Stahl (1992, 4–5), a *blanca* such as this would have been the only small-denomination coin circulating in Spain in 1492 and 1493. Through electron microprobe analysis, the investigators determined that the coin was made of billion and contains 3.97 per cent silver. It was most likely minted in Segovia and Burgos, Spain, but the lead resembles that from the Albergaria-a-Velha galena source located near Aveiro, Portugal.

Glass Beads (Brill et al. 1987, 250–253, 259, 278)
Seven glass beads and three bead fragments were analyzed. The beads are doughnut-shaped, 2.5–3.5 mm in diameter, with circular holes for suspension measuring 1.5–2.3 mm in diameter. They were manufactured using the wire-wound technique. Hoffman (1987b, 242) states that this bead type was manufactured until 1516. The analysts described the green beads as possessing a "sparkling" color. The green color is due to the addition of a copper-containing colorant, such as scrap brass. The beads share physical and manufacturing similarities with early colonial Spanish bead types VIDle and VIDlf featured in Smith and Good (1982). Electron microprobe analysis indicates that the beads possess unusually elevated levels of lead oxide, much higher than those found in other early lead glasses. Their chemical composition is very similar to one another leading the analysts to suggest

TABLE 2.2 European artifact descriptions (*cont.*)

that they were manufactured in the same factory or workshop. The location of the manufacturing site or the source of the materials used in the bead manufacture is on the Iberian peninsula. Two of the green glass beads possess lead that is a "close match" to the galena mined at Albergaria-a-Velha, near Aveiro, Portugal.

D-ring (Brill 1987 et al. 254, 259, 280)
The bronze D-ring bears a close resemblance to fifteenth-century clothing and armor buckles pictured in Deagan (2002, 180–188), but, it is not a direct match, and Brill et al. (1987) suggest it had a decorative or maritime function. Its chemical composition was analyzed using atomic absorption and emission spectrography. The ore matches a galena ore source from Albergaria-a-Velha in Portugal. The shank is 2.7 cm in length.

Belt Buckle (Brill et al. 1987, 253, 259–260, 279)
The belt buckle is small and lacks a tongue suggesting that it might have been a mock buckle made to function as a trade gewgaw, not unlike, in concept, to other trade trinkets brought by explorers and slavers to the Americas and Africa. The buckle bears some resemblances in form and size, though, to a Spanish shoe buckle that was common in the fifteenth century and manufactured up to 1485. X-ray fluorescence determined that it was made of a lightly leaded bronze. The lead is similar to that from the Los Belgas mine in the Sierra de Gador, near Almeria, a city located in Andalusia, Spain. The length of the shank is 2.0 cm.

***Melado* and Majolica Wares** (Brill et al. 1987, 254–255, 260, 282)
The majolica was identified as Columbia Plain. Using a variety of sourcing techniques, such as x-ray diffraction and thin-section analysis, it was determined that the clay from the two sherds submitted for analysis originated from clay sources located in southern Spain. Both sherds possess a lead glaze that is indistinguishable from one another, suggesting they were manufactured from closely located workshops. The lead in the glaze is sourced to Rio Tinto, an area in southwestern Spain rich in minerals. The Colombia Plain glaze also contains tin.

Columbus' diary and other historic accounts of his first voyage serve as useful sources of information about how exchange was negotiated with indigenous peoples at this time. From them – particularly the diary – we learn what items Columbus and his men gave the Lucayans. Many of these items have also been found at sixteenth-century Spanish settlement sites elsewhere in the Caribbean suggesting that these goods were components of "gift kits" (*sensu* Brain 1975), which functioned to establish relations with indigenous peoples. The choice of the contents was influenced likely by the kinds of items that the Portuguese brought to Africa during their fifteenth-century exploratory ventures and slaving expeditions. Morison (1942, 305) notes that Columbus brought things that the Portuguese "had found to be most in demand" among the West Africans. Benjamin (2009, 83) reports that in 1479, Portuguese expeditions secured West African slaves through the exchange of cloth, brass, shells, wine, and beads. Items mentioned in the Columbus diary include strings of green and yellow glass beads, coins, broken pieces of pottery and glass cups, lace ends, red caps (bonnets), brass bells, brass jingles, and miscellaneous items, many of trifling value to the Europeans (Dunn and Kelley 1989, 65, 71, 107, 81, 85, 87, 93). Materials such as the textiles did not survive the archaeological record. In spite of numerous references to hawk bells (*cascabeles*), none has been discovered at Lucayan sites. Items that were no longer minted such as the *blanca* or manufactured, like the belt buckle, found at the Long Bay site, were also gifted or exchanged.

While it is tempting to believe that the Long Bay site was the first example of Spanish-Lucayan contact due to the congruence of the European artifacts with those mentioned in the Columbus diary, it would be foolhardy to jump at such a conclusion. The Lucayans may have gained possession of such items in other ways and the historic record presents many situations where Spanish-Lucayan interactions in the Bahama archipelago might have occurred.

Establishing when the Long Bay site was established and abandoned and pinpointing the year or years of Spanish contact, either direct or indirect, proves to be challenging, due to the wide berth of time that both the Lucayan and Spanish artifacts found at the site were manufactured and consumed. Radiocarbon assays yielded contemporary dates, giving us no radiochronometic clues. The indigenous assemblage of artifacts (ceramics, shell beads, coral artifacts, stone tools) is typical for the latter part of the Late Lucayan period, which lasted from AD 1100 into the early to mid-sixteenth century (Berman et al. 2013). A few pieces of Lucayan pottery show design affinities with late Chicoid pottery, similar to that manufactured in Hispaniola during the Late Ceramic Age and early colonial period. The Spanish objects were manufactured and circulated over several decades during the late fifteenth and early to mid-sixteenth centuries.

The use of the early-style Spanish olive jar to date the period of Spanish contact at the Long Bay site fails to provide the tight chronological definition we are seeking, but its presence at other sites in the insular Caribbean is instructive. The initial manufacture of Spanish olive jars in Spain occurred as early as 1490;

the style lasted to around 1570 when it was replaced by another form (Deagan 1987, 31). Spanish olive jars or fragments were not found in the Spanish occupation of En Bas Saline, which consisted of a small contingent of 39 men whose ship, the Santa Maria, had run aground (Deagan 2004). Columbus and his men had salvaged the ship's cargo to provision themselves during Columbus' return to Spain. Neither Goggin (1960, 11) nor Deagan (2018, personal communication) found olive jars at La Isabela (established in December 1493) although Deagan (1987, 33) reports the recovery of some sherds that have not been published casting doubt on their authenticity. Early-style olive jars were found at La Vega and Puerto Real established in 1503. Both Deagan (personal communication, 2018) and Pleguezuelo (2003, 116) believe that early-style olive jars are associated with the ascendancy of early transatlantic shipping, which would have post-dated Columbus' first voyage to the Americas.

Although only small amounts were manufactured during the sixteenth century (Deagan 1987, 160), green glass beads (*abalorios*) are believed to be a secure temporal indicator for the first half of the sixteenth century (Deagan 1987, 169; Smith 1983, 148). Like Spanish olive jars, they were not found at En Bas Saline or La Isabela. The earliest documented evidence for green glass beads (*abalorio*) is from Nueva Cádiz, Venezuela (Smith and Good 1982), which was established in 1500.

Both the historic and archaeological record suggest that the early-style Spanish olive jar and wire-wound green glass beads were present in the Caribbean in 1500. This suggests that Spanish goods were incorporated into the Long Bay site at 1500 or afterwards up to the mid-sixteenth century. Of course, all of this would be contradicted if it were found that the European items at the Long Bay site predate 1500.

The historic record suggests that the Long Bay site already may have been abandoned by the first quarter of the sixteenth century. In 1513 Ponce de León docked on San Salvador for 9–12 days to conduct ship repairs (Kelley 1991; Ober 1908), but found few people on the island. Those he observed hid from the Spanish and there are no accounts of trade and exchange between them and the indigenous inhabitants (Ober 1908, 178). In this scenario, the Lucayans would have acquired the Spanish items found at the Long Bay site sometime between 1500 and 1513.

4 Consumption and Indigenous Agency

As Lightfoot (1995), Torrence and Clarke (2000), and others have argued, prior intercultural interactions and patterns of artifact consumption provide precedents for later practices. In other words, early colonial encounters with Europeans and the exchange or acquisition of goods were probably based on existing systems of etiquette, while the consumption, distribution, and circulation of European goods were likely based on functioning economic systems. In the following section, we will examine how trade and exchange of non-local materials with foreign peoples (Keegan 1992), the experiences of raiding (Keegan 2015), symbolic associations of non-local items, and the transformation of foreign objects into native categories were integrated into Lucayan life and served as precursors for their assimilation of Spanish items.

4.1 *Trade and Exchange*

The Lucayans participated in long-distance inter-island trade and exchange, as indigenous non-local items are found throughout the whole Lucayan occupational span (Berman 2011; Keegan 1992). Chert artifacts (Berman et al. 1999), non-local pottery, and a variety of artifacts made from igneous and metamorphic materials (e.g., jadeite, diorite, basalt, green schist, quartzite, and quartz crystal) are present at nearly all Lucayan sites (Berman 2011). From Columbus' diary we determine that the Lucayans were knowledgeable of the geography and peoples of neighboring islands including Cuba and Hispaniola (Anderson-Córdova 2017; Berman 2011). Additionally, Columbus observed large canoes holding as many as 40–45 people (Dunn and Kelley 1989, 69), which are believed to have been used for long-distance travel (Keegan 1992). The Lucayans themselves alluded to their trading relationships with people of the Greater Antilles. Columbus noted that the Lucayans told him "there are many and very large ships and many traders" and "great commerce" (Dunn and Kelley 1989, 109, 113). Of course, the reference to the magnitude of the trade may have been an exaggeration on Columbus' part to impress his investors and supporters; nevertheless, the archaeological record attests to the procurement of non-local objects.

The Lucayans also disseminated local and non-local objects among themselves. Only a short time after Columbus left San Salvador, he encountered an individual who was traveling to (Fernandina) Long Island with objects obtained from San Salvador. His canoe held glass beads and two *blancas*, and items believe to be of local origin: bread, a calabash of water, red powder, and dried leaves (Dunn and Kelley 1989, 85).

The Lucayans were knowledgeable, too, of metals, which circulated throughout the Caribbean prior to the entry of the Spanish (Martinón-Torres et al. 2012; Valcárcel Rojas and Martinón-Torres 2013; Valcárcel Rojas et al. 2010). Two small unworked copper fragments were found at the North Storr's Lake site, a Late Lucayan settlement located on the eastern side of San Salvador (Shaklee et al. 2007). Unfortunately, they have not been sourced. Columbus observed gold nose rings, earrings, bracelets, anklets, and chest ornaments on Lucayan men (Dunn and Kelley 1989, 71, 73, 75, 83, 95, 109).

4.2 *Raiding*

Raiding and captive-taking were common practices in small-scale societies (Cameron 2008, 2011; Santos-Granero 2009) and it appears that the Lucayans were subject to regular attacks by people from the northwest, "who came to fight them many times" (Dunn and Kelley 1989, 71) and people from nearby islands who "tried to take them" (Dunn and Kelley 1989, 67). We do not know whether these were other Lucayans or unrelated or distantly related peoples from Florida or the Antilles. While on San Salvador, Columbus and his crew took on board six male captives (Dunn and Kelley 1989, 69). Keegan (2015) notes that the likelihood for hostilities exists in all social interactions and suggests that taking captives was a corollary expression of trade and exchange (see also Cameron 2008; Keeley 1996). The Lucayans, therefore, were no strangers to foreign intrusion and seizure. Due to their familiarity with such behavior, the appropriation of captives may have been viewed as another form of trade (in this sense) with non-local peoples.

4.3 *Symbolic Meanings*

Objects are endowed with meaning according to local cultural logics (Sahlins 1994) and European items possessed physical properties that were metaphorically connected to Lucayan cosmology. As Dietler (2005, 63) has noted, "foreign objects must be understood not only for what they represent in the society of origin, but for the culturally specific meaning and perceived utility in the context of consumption." Objects or materials that exhibited shininess, brilliance, or luminescence were (and still are) desired among the native peoples of the Americas (Quilter and Hoopes 2003; Saunders 2003) and the Taíno (Oliver 2000) and Lucayans (Berman 2011), who were closely related to and descended from the Taíno, valued such items. For numerous Amerindians, brightness signified life-giving energies (Saunders 1999, 2003). By virtue of their shininess, translucence, and or light-giving elements, most of the non-local or unusual materials or objects found at Antillean sites exemplify the "aesthetic of brilliance," a concept that epitomizes the "spiritual and creative power of light" (Saunders 2003, 15). The Taíno (Keehnen 2011) and Lucayans (Bate 2011) welcomed shiny items presented to them by the Europeans. Such objects had special resonance for the Lucayans, for like the Taíno (Ostapkowicz 2018), they fit innately into their symbolic calculus.

Olfactory, auditory, and tactile experiences play significant roles in the ways of knowing, living, and meaning-making among non-Western peoples;

for Amerindians, they constitute what Saunders (2003, 17) calls "a holistic phenomenological unity". Copper produces a unique odor, which was regarded to have been highly appealing (Berman 2011; Falchetti 2003; Valcárcel Rojas and Martínon-Torres 2013; Valcárcel Rojas et al. 2010). The analyzed metal objects found at the Long Bay site, as well as the unworked pieces of copper from North Storr's Lake, possessed high percentages of copper that may have emanated such a scent. Certain sounds also held and conveyed cosmologically-linked symbolic meanings (Hosler 1994), suggesting why the Lucayans and Taíno sought out hawk bells.

Colors, too, served as visual metaphors. The colors of many of the European objects corresponded to hues in the Lucayan spectrum and objects displaying those colors may have been accepted, even sought, on that basis, as has been observed among other Amerindians (Miller and Hamell 1986). Green, red, and yellow Spanish beads, textiles, and metal objects were most likely considered equivalent to *guanín* (Oliver 2000). They were also the colors of the parrots (considered to be a form of *guanín*) (Oliver 2000), which the Lucayans presented to the Spanish during their encounter with Columbus (Dunn and Kelley 1989). Ostapkowicz (2018, 166) argues that green glass beads, by virtue of their color, size, and physical characteristics held special significance for the Taíno and fit into a prior system of bead manufacture. According to her, the color green was associated with jadeite, which was regarded as possessing exotic status. This can be argued for the Lucayans, too, since jadeite and other greenstones do not occur naturally in the Bahama archipelago. The Lucayans obtained jadeite from distant, non-local sources, accessed only via water transport. Moreover, the size and shape of the beads lie in the range of Lucayan shell beads found throughout the Lucayan sequence (Gnivecki 2006). In sum, the materials, colors, smells, and other properties of indigenous non-local items and the European goods were associated with and served as metaphors for remoteness and distant locales (*sensu* Helms 1988), which the Spanish and their objects represented. The notion of foreignness was a routinized part of the Lucayan belief system and something that was appreciated, not feared.

4.4 *Artifact Modification*

As we have proposed, the Lucayans classified European objects in the same categories as indigenous non-local items. While most of the objects at the Long Bay site appear to have been unmodified, Brill et al. (1987, 256) found that the *blanca* was scratched and possibly hammered. They suggest that these modifications might represent attempts at perforation, so that the object could have been worn as a bead or pendant. The "D"-ring and belt buckle had the potential to be worn as pendants or other kinds of ornaments, as well. Silliman (2009) and others have noted that objects do not have to be transformed physically to become indigenous items.

Through the lens of artifact discard patterns, we see that European objects were treated the same way as indigenous local and non-local objects. At the Long Bay site, the European items were found intermingled with local and non-local (i.e., Antillean) artifacts. As another example, several *melado* sherds were found on a limestone shelf in close association with a greenstone petaloid axe and a few sherds of Lucayan pottery suggesting that the non-local artifacts, no matter their origin, were regarded in similar ways. Because they were consumed in a like manner, we suggest the items were recontextualized, i.e., transformed into Lucayan objects in use and meaning (*sensu* Silliman 2009). Samson (2010), Valcárcel Rojas, Samson, and Hoogland (2013, 29) and Hofman et al. (2014) have suggested that the occupants of El Cabo regarded European goods similarly, for they, too, were found interspersed in house sweepings with local items and food waste.

5 Discussion

The Lucayans secured European items through a variety of pathways. These, along with indigenous taste (*sensu* Stahl 2002) and past practice influenced the manner in which the objects were perceived and how they were socialized into local contexts. Some objects may never have entered Lucayan systems, such as those misplaced or discarded by the Spanish. Moreover, they may have become to be regarded suspiciously once the Lucayans recognized Spanish intentions to remove the Lucayans from their homelands. The low volume of recovered objects may be linked to a variety of factors including their rapid insertion into native exchange systems, their assignment to special curated contexts, and to recovery techniques, which, with the exception of the Long Bay site excavations, may not have been sufficiently fine-grained to retain tiny items.

Lucayan consumption of European items was motivated by indigenous economic and political practices driven by symbolic-ideological factors. The archaeological evidence demonstrates that non-local items were a regular feature of Lucayan household and most likely political economies (not examined here), and that acquisition through trade, exchange, and gift-giving with non-local peoples and inter-island down-the-line trade and exchange with indigenous peoples occurred regularly. Interacting with foreigners – even those who took them captives – was commonplace. Accustomed to non-local items and peoples, the Lucayans embraced Spanish goods because they possessed characteristics consistent with their cosmovision, making them and regarding them as their own. The remoteness of the sources, either geographical or metaphorical, rendered the objects similar to others with which they were familiar. Some of the items' sensory properties such as iridescence, color, smell, texture, and sound fit into the Lucayan symbolic reservoir.

Thomas (1991) has noted that in spite of the distinctiveness and newness of European objects in the eyes of indigenous peoples, such objects often preserve a "prior order" and are modified to resemble preexisting objects. The attempted perforation of the *blanca* is just such an example of altering a foreign object to fit a preexisting template (should it, in fact have been performed by a Lucayan). The Lucayans do not appear to have modified the other European objects found at the Long Bay site, however. Similarly, we have no evidence that European goods were embedded into the fabric of Lucayan objects, but, due to the perishable nature of much of material Lucayan culture, these may have not survived the archaeological record or may have been curated in inaccessible locations.

6 Conclusions

The Lucayans interacted with European items in an indigenous colonial space that differed geographically and politically from the Greater Antilles. The absence of close or sustained interaction between the Europeans and the Lucayans due to the geographical distance of the Bahama islands from the colonial heartland, the lack of direct colonial control, the varied, brief, intermittent nature of Lucayan-Spanish contact, and ultimately the violent conditions under which they interacted did not encourage the creation of a large body of reworked, repurposed, or hybridized items. While the European objects presented novel shapes, colors, forms, and materials, the Lucayans found the items to be analogous to materials they knew and understood symbolically and thus there may have been less desire to physically modify them. This is not to deny Lucayan agency, but is suggested as a means to explore why little to no modification is observed on the European articles found at the Long Bay site. While the biographies of European objects found in Lucayan contexts share some similarities with those from other early colonial contexts, there are differences between them. These can be attributed to historical factors. And, while it was objects that first facilitated Spanish-indigenous relations, in the end, it was the indigenous peoples, not exclusively the objects that were recontextualized, redefined, and physically reworked as commodities.

Acknowledgments

This article could not have been achieved without the support of numerous people. In particular, the authors thank Mrs. Kathy Doan Gerace and the late Dr. Donald T. Gerace for their interest in and encouragement of our research, which has allowed us a panoramic grasp of the Lucayans from their earliest peopling of the Bahama archipelago to their final days on the islands. We want to especially thank Kathy for providing the photograph of the glass beads and for helping to locate more of Charles Hoffman's field records. They revealed much more about the Long Bay site than what has appeared in publications and these findings have done much to enhance this article. We recognize and thank, too, the support provided by the staff of the Gerace Research Centre, University of the Bahamas, who have always welcomed us to San Salvador Island and gone out of their way to provide a comfortable working context. Dr. Michael Pateman, Director of the Turks & Caicos National Museum Foundation, Grand Turk, Turks & Caicos Islands has championed our research and analyses. For this, we are profoundly appreciative. We would be remiss if we did not honor the late Dr. Charles Hoffman who excavated the Long Bay site. Without his work and that of his legions of students and volunteers, the Spanish presence in the Bahamas would have continued to remain buried as a footnote in history books. Berman would like to thank Miami University's College of Arts and Science, which granted her a spring 2017 assigned research leave; this gave her the time and space to read and think through the many issues reflected in this work. Additionally, we are grateful to Eric Johnson, Miami University Numeric and Spatial Data Services Librarian, who transformed 35 year old slides into the digital images presented here. And, finally, we thank the citizens of the Commonwealth of the Bahamas, who have welcomed us for three decades. We hope that they will find our discoveries and interpretations meaningful and worthy of their generous hospitality.

References

Anderson-Córdova, Karen F. 2017. *Surviving Spanish Conquest. Indian Fight, Flight, and Cultural Transformation in Hispaniola and Puerto Rico.* Tuscaloosa: University of Alabama Press.

Bate, Emma. 2011. "Technology and Spanish Contact: Analysis of Artifacts from the Long Bay Site, San Salvador, Bahamas." PhD diss., Indiana University.

Benjamin, Thomas. 2009. *The Atlantic World. Europeans, Africans, and Their Shared History,* 1400–1900. Cambridge: Cambridge University Press.

Berman, Mary Jane. 2011. "Good as Gold: The Aesthetic Brilliance of the Lucayans." In *Islands in the Stream: Migration, Seafaring, and Interaction in the Caribbea*, edited by L. Antonio Curet and Mark W. Hauser, 104–134. Tuscaloosa: University of Alabama Press.

Berman, Mary Jane, and Perry L. Gnivecki. 1995. "The Colonization of the Bahama Archipelago: A Reappraisal." *World Archaeology* 26 (3): 421–441.

Berman, Mary Jane, and Perry L. Gnivecki. 2000. "Long Bay Site (SS9) Field Notes." Unpublished notes in possession of authors.

Berman, Mary Jane, and Deborah M. Pearsall. 2008. "At the Crossroads: Starch Grain and Phytolith Analyses in Lucayan Prehistory." *Latin American Antiquity* 19 (2): 181–203.

Berman Mary, Jane, and Deborah M. Pearsall. 2018. "Crop Dispersal and Lucayan Tool Use: Creating the Transported Landscape in the Central Bahamas. Evidence from Starch Grain, Phytolith, Macrobotanical, and Artifact Studies." Ms. On file with Mary Jane Berman.

Berman, Mary Jane, April K. Sievert, and Thomas R. Whyte. 1999. "Form and Function of Bipolar Lithic Artifacts from the Three Dog Site, San Salvador, Bahamas." *Latin American Antiquity* 10 (4): 415–432.

Berman, Mary Jane, Perry L. Gnivecki, and Michael P. Pateman. 2013. "The Bahama Archipelago." In *The Oxford Handbook of Caribbean Archaeology*, edited by William F. Keegan, Corinne L. Hofman, and Reniel Rodríguez Ramos, 264–280. New York: Oxford University Press.

Brain, Jeffrey. 1975. "Artifacts of the Adelantado." In *Conference on Historic Sites Archaeology Papers Volume 8*, edited by Stanley South, 129–134. Columbia: The South Carolina Institute of Archaeology and Anthropology, University of South Carolina.

Brill, Robert, I. Lynus Barnes, Stephen S. Tong, Emile C. Joel, and Martin J. Murtagh. 1987. "Laboratory Studies of Some European Artifacts Excavated on San Salvador, Bahamas." In *Proceedings of the 1st San Salvador Conference: Columbus and His World*, compiled by Donald T. Gerace, 247–292. Fort Lauderdale: College Center of the Finger Lakes and San Salvador: Bahamian Field Station.

Burns, Sir Alan. 1965. *History of the British West Indies, Revised 2nd Edition*. London: George Allen and Unwin.

Cameron, Catherine M. 2008. "Captives in Prehistory: Agents of Social Change." In *Invisible Citizens: Captives and Their Consequences*, edited by Catherine M. Cameron, 1–24. Salt Lake City: University of Utah Press.

Cameron, Catherine M. 2011. "Captives and Culture Change: Implications for Archaeology." *Current Anthropology* 52 (2): 169–209.

Charlton, Thomas. 1968. "Post-Conquest Aztec Commerce: Implications for Archaeological Interpretations." *Florida Anthropologist* 21 (4): 96–101.

Cipolla, Craig N. 2013. *Becoming Brothertown: Native American Ethnogenesis and Endurance in the Modern World.* Tucson: The University of Arizona Press.

Cipolla, Craig N., ed. 2017. *Foreign Objects. Rethinking Indigenous Consumption in American Archaeology.* Tucson: The University of Arizona Press.

Cobb, Charles R., ed. 2003. *Stone Tool Traditions in the Contact Era.* Tuscaloosa: University of Alabama Press.

Cusick, James G. 1991. "Culture Change and Pottery Change in a Taino Village." In Proceedings of the *13th Congress of the International Association for Caribbean Archaeology, Part 1*, edited by E.N. Ayubi and J.B. Haviser, 446–461. Aruba: Reports of the Archaeological-Anthropological Institute of the Netherlands Antilles, No. 9.

Deagan, Kathleen A. 1987. *Artifacts of the Spanish colonies of Florida and the Caribbean, 1500–1800. Vol 1: Ceramics, Glassware and Beads.* Washington D.C.: Smithsonian Institution Press.

Deagan, Kathleen A. 1988. "The Archaeology of the Spanish Contact Period in the Caribbean." *Journal of World Prehistory* 2 (2): 187–225.

Deagan, Kathleen A., ed. 1995. *Puerto Real: the Archaeology of a Sixteenth Century Spanish Town in Hispaniola.* Gainesville: University Press of Florida.

Deagan, Kathleen A. 2002. *Artifacts of the Spanish Colonies of Florida and the Caribbean 1500–1800. Volume 2: Portable Personal Possessions.* Washington D.C.: Smithsonian Institution Press.

Deagan, Kathleen A. 2004. "Reconsidering Taíno Social Dynamics After Spanish Conquest: Gender and Class in Culture Contact Studies." *American Antiquity* 69 (4): 597–626.

Deagan, Kathleen A. and José M. Cruxent. 2002a. *Columbus's Outpost Among the Taínos: Spain and America at La Isabela, 1493–1498.* Haven: Yale University Press.

Deagan, Kathleen A. and José M. Cruxent. 2002b. *Archaeology at La Isabela, America's First European Town.* New Haven: Yale University Press.

Dietler, Michael. 1998. "Consumption, Agency, and Cultural Entanglement: Theoretical Implications of a Mediterranean Colonial Encounter." In *Studies in Culture Contact: Interaction, Culture Change, and Archaeology*, edited by James G. Cusick, 288–315. Carbondale: Southern Illinois University Press.

Dietler, Michael. 2005. "The Archaeology of Colonization and the Colonization of Archaeology: Theoretical Challenges from an Ancient Mediterranean Colonial Encounter." In *The Archaeology of Colonial Encounters*, edited by Gil J. Stein, 33–68. Santa Fe: School of American Research Advanced Seminar Series.

Dunn, Oliver, and James E Kelley. 1989. *The Diario of Christopher Columbus's First Voyage to America 1492–1493.* Norman: University of Oklahoma Press.

Falchetti, Ana M. 2003. "The Seed of Life: The Symbolic Power of Gold-Copper Alloys and Metallurgical Transformations." *In Gold and Power in Ancient Costa Rica, Panama, and Columbia*, edited by Jeffrey Quilter and John W. Hoopes, 145–381. Washington DC: Dumbarton Oaks Research Library and Collection.

Gnivecki, Perry L. 1995. "Rethinking "First" Contact." In *Proceedings of the 15th Congress of the International Association for Caribbean Archaeology*, edited by Ricardo E. Alegría and Miguel Rodríguez, 209–217. San Juan: Centro de Estudios Avanzados de Puerto Rico y el Caribe.

Gnivecki, Perry L. 2006. "What Shell Beads from the Three Dog Site, San Salvador Island, Bahamas, Can Tell Us." Paper presented at the *71st Annual Meeting of the Society for American Archaeology "Current Research Bahamian Prehistory and Historical Archaeology: Papers in Memory of Charles A. Hoffman," San Juan, Puerto Rico, 26–30 April.*

Gnivecki, Perry L. 2011. "Text and Context: The Spanish Contact Period in the Bahama Archipelago." In *Proceedings of the 14th Symposium on the Natural History of the Bahamas*, edited by Craig Tepper and Ronald Shaklee, 197–211. San Salvador: Gerace Research Centre.

Goggin, John. 1960. *The Spanish Olive Jar: An Introductory Study*. Yale University Publications in Anthropology, No. 62. New Haven: Yale University Press.

Granberry, Julian. 1979. "Spanish Slave Trade in the Bahamas, 1509–1530: An Aspect of the Caribbean Pearl Industry." *Journal of the Bahamas Historical Society* 1: 14–15.

Granberry, Julian. 1980. "Spanish Slave Trade in the Bahamas, 1509–1530: An Aspect of the Caribbean Pearl Industry (Continued)." *Journal of the Bahamas Historical Society* 2: 15–17.

Granberry, Julian. 1981. "Spanish Slave Trade in the Bahamas, 1509–1530: An Aspect of the Caribbean Pearl Industry (Last Part)." *Journal of the Bahamas Historical Society* 3: 17–19.

Harisse, Henry. 1961. *The Discovery of North America (1892)*. Amsterdam: N. Israel.

Helms, Mary W. 1988. *Ulysses' Sail: An Ethnographic Odyssey of Power, Knowledge, and Geographical Distance*. Princeton: Princeton University Press.

Hofman, Corinne L., Angus A.A. Mol, Menno L.P. Hoogland, and Roberto Valcárcel Rojas. 2014. "Stage of Encounters: Migration, Mobility, and Interaction in the Precolonial and Early Colonial Caribbean." *World Archaeology* 46 (4): 590–609.

Hoffman, Charles A. 1987a. "Archaeological Investigations at the Long Bay Site, San Salvador, Bahamas." *American Archaeology* 6: 97–102.

Hoffman, Charles A. 1987b. "Archaeological Investigations at the Long Bay Site, San Salvador, Bahamas." In *Proceedings of the 1st San Salvador Conference: Columbus and His World*, compiled by Donald T. Gerace, 237–245. Fort Lauderdale: College Center of the Finger Lakes and San Salvador: Bahamian Field Station.

Hoffman, Paul E. 1990. *A New Andalusia and A Way to the Orient: A History of the American Southeast During the 16th Century*. Baton Rouge: Louisiana State University Press.

Hosler, Dorothy. 1994. *The Sounds and Colors of Power: The Sacred Metallurgical Technology of Ancient West Mexico.* Cambridge: The MIT Press.

Keegan, William F. 1992. *The People Who Discovered Columbus: The Prehistory of the Bahamas.* Gainesville: University Press of Florida.

Keegan, William F. 2015. "Mobility and Disdain: Columbus and Cannibals in the Land of Cotton." *Ethnohistory* 62 (1): 1–15.

Keehnen, Floris W.M. 2011. "Conflicting Cosmologies. The Exchange of Brilliant Objects Between the Taíno of Hispaniola and the Spanish." In *Communities in Contact: Essays in Archaeology, Ethnohistory and Ethnography of the Amerindian Circum-Caribbean*, edited by Corinne L. Hofman and Anne van Duijvenbode, 253–268. Leiden: Sidestone Press.

Keeley, Lawrence H. 1996. *War Before Civilization.* New York: Oxford University Press.

Keith, Donald H., Jim A. Duff, Steve R. James, Thomas J. Oertling, and Joe J. Simmons. 1984. "The Molasses Reef Wreck, Turks and Caicos, BWI: A Preliminary Report." *The International Journal of Nautical Archaeology and Underwater Exploration* 13 (1): 45–63.

Kelley, James E. Jr. 1991. "Juan Ponce de Leon's Discovery of Florida: Herrera's Narrative Revisted." *Revista de Historia de América* 111: 31–65.

Kopytoff, Igor. 1986. "The Cultural Biography of Things: Commoditization as Process." In *The Social Life of Things. Commodities in Cultural Perspective*, edited by Arjun Appadurai, 64–91. Cambridge: Cambridge University Press.

Kulstad-González, Pauline M. 2015. "Striking it Rich in Americas' First Boom Town: Economic Activity at Concepción de la Vega." In *Archaeology of Culture Contact and Colonialism in Spanish and Portuguese America*, edited by Pedro Paulo A. Funari and María Ximena Senatore, 313–337. New York: Springer.

Lightfoot, Kent G. 1995. "Culture Contact Studies: Redefining the Relationship Between Prehistoric and Historical Archaeology." *American Antiquity* 60 (2): 199–217.

Lightfoot, Kent G., and Antoinette Martinez. 1995. "Frontiers and Boundaries in Archaeological Perspective." *Annual Review of Anthropology* 24: 471–492.

Lightfoot, Kent G., and William Simmons. 1998. "Culture Contact in Protohistoric California: Social Contexts of Native and European Encounters." *Journal of California and Great Basin Anthropology* 20 (2): 138–170.

Martinón-Torres, Marcos, Roberto Valcárcel Rojas, Juanita Sáenz Samper, María Filomena Guerra. 2012. "Metallic Encounters in Cuba: The Technology, Exchange and Meaning of Metals." *Journal of Anthropological Archaeology* 31 (4): 439–454.

Miller, Christopher F., and George F. Hamell 1986. "A New Perspective on Indian-White Contact: Cultural Symbols and Colonial Trade." *Journal of American History* 73: 311–328.

Morison, Samuel Eliot. 1942. *Admiral of the Ocean Sea: A Life of Christopher Columbus, Volume 1.* Boston: Little, Brown, and Company.

Newsom, Lee A., and Elizabeth S. Wing. 2004. *On Land and Sea: Native American Uses of Biological Resources in the West Indies.* Tuscaloosa: University of Alabama Press.

Ober, Frederick A. 1908. *Juan Ponce de Leon.* New York: Harper and Brothers Publishers.

Oland, Maxine. 2014. "'With the Gifts and Good Treatment That He Gave them': Elite Maya Adoption of Spanish Material Culture at Progresso Lagoon, Belize." *International Journal of Historical Archaeology* 18 (4): 643–667.

Oliver, José. 2000. "Gold Symbolism Among Caribbean Chiefdoms: of Feathers, Ceibas, and Guanín Power Among Taíno Elites." In *Precolumbian Gold Technology, Style, and Iconography,* edited by Colin McEwan, 196–219. London: British Museum Press.

Ortega, Elpidio J., and Carmen Fondeur. 1978. *Estudio d la Cerámica del Periodo Indohispano de la Antigua Concepciòn de la Vega.* Santo Domingo: Funcaciòn Ortega Alvárez.

Ostapkowicz, Joanna. 2018. "New Wealth from the Old World: Glass, Jet and Mirrors in the Late 15th to Early 16th Century Indigenous Caribbean." In *Gifts, Goods, and Money. Comparing Currency and Circulation Systems in Past Societies,* edited by Dirk Brandherm, Elon Heymans and Daniela Hofmann, 153–193. Oxford: Archaeopress.

Parry, John H., and Robert G. Keith, eds. 1984. *New Iberian World: A Documentary History of the Discovery and Settlement of Latin America to the Early 17th Century, Volumes I–II.* New York: Times Books and Hector and Rose.

Peterson, Mendel. 1974. "Exploration of a 16th-Century Bahaman Shipwreck." *National Geographic Society Research Reports,* 1967, Projects 231–242. Washington, DC.

Pleguezuelo, Alfonso. 2003. "Ceramics, Business, and Economy." In *Cerámica Y Cultura: The Story of Spanish and Mexican Mayólica,* edited by Robin Farwell Gavin, Donna Pierce, and Alfonso Pleguezuelo, 102–121. Albuquerque: University of New Mexico Press.

Pohl, Frederick J. 1966. *Amerigo Vespucci, Pilot Major.* New York: Octogon Books.

Quilter, Jeffrey, and John W. Hoopes, eds. 2003. *Gold and Power in Ancient Costa Rica, Panama, and Colombia.* Washington D.C.: Dumbarton Oaks, Trustees for Harvard University.

Quinn, David B. 1979. *New American World: A Documentary History of North America to 1612, Volume I.* New York: Arno Press and Hector Bye.

Rodríguez-Alegría, Enrique. 2008. "Narratives of Conquest, Colonialism, and Cutting Edge Technology." *American Anthropologist* 110 (1): 33–43.

Rodríguez-Alegría, Enrique, Franz Scaramelli and Ana María Navas. 2015. "Technological Transformations: Adaptationist, Relativist, and Economic Models in Mexico and Venezuela." In *Archaeology of Culture Contact and Colonialism in Spanish and Portuguese America,* edited by Pedro Paulo A. Funari and María Ximena Senatore, 53–77. New York: Springer.

Rouse, Irving B. 1942. *Archaeology of the Maniabón Hills, Cuba.* Yale University Publications in Anthropology, Number 26. New Haven: Yale University Press.

Sahlins, Marshall D. 1994. "Cosmologies of Capitalism: The Trans-Pacific Sector of "'The World System'." In *Culture/Power/History. A Reader in Contemporary Social Theory*, edited by Nicholas B. Dirks, Geoff Eley, and Sherry B. Ortner, 412–455. Princeton: Princeton University Press.

Samson, Alice V.M. 2010. *Renewing the House: Trajectories of Social Life in the Yucayeque (Community) of El Cabo, Higüey, Dominican Republic, AD 800 to 1504.* Leiden: Sidestone Press.

Santos-Granero, Fernando. 2009. *Slavery, Predation, and the Amerindian Political Economy of Life.* Austin: University of Texas Press.

Sauer, Carl O. 1966. *The Early Spanish Main.* Berkeley: University of California Press.

Saunders, Nicholas J. 1999. "Biographies of Brilliance: Pearls, Transformation, and Being, c. A.D. 1492." *World Archaeology* 31 (2): 243–257.

Saunders, Nicholas J. 2003. "'Catching the Light': Technologies of Power and Enchantment in Pre-Columbian Goldworking." In *Gold and Power in Ancient Costa Rica, Panama, and Colombia*, edited by Jeffrey Quilter and John W. Hoopes, 15–49. Washington, DC: Dumbarton Oaks, Trustees for Harvard University.

Shaklee, Ronald, Gary Fry, and Thomas Delvaux. 2007. "An Archaeological Report on the Storr's Lake Site, San Salvador: 1995–2005." *Bahamas Naturalist and Journal of Science* 1: 31–39.

Silliman, Stephen W. 2005. "Culture, Contact or Colonialism? Challenges in the Archaeology of Native North America." *American Antiquity* 70 (1): 55–74.

Silliman, Stephen W. 2009. "Change and Continuity, Practice and Memory: Native American Persistence in Colonial New England." *American Antiquity* 74 (2): 211–230.

Sinelli, Peter T. 2010. "All Islands Great and Small: The Role of Small Cay Environments in Indigenous Settlement Strategies in the Turks and Caicos Islands." PhD diss., University of Florida.

Smith, Marvin. 1983. "Chronology from Glass Beads: The Spanish Period in the Southeast, c. A.D. 1513–1670." In *Proceedings of the 1982 Glass Trade Bead Conference*, edited by C. Hayes, 147–158. Rochester: Rochester Museum and Science Center Research Records, No. 16.

Smith, Marvin, and Elizabeth F. Good. 1982. *Early Sixteenth Century Glass Beads in the Spanish Colonial Trade.* Greenwood: Cottonlandia Museum Publications.

Smith, Roger C., Donald H. Keith, and Denise Lakey. 1985. "The Highbourne Cay Wreck: Further Exploration of a 16th Century Bahaman Shipwreck." *Journal of Nautical Archaeology and Underwater Exploration* 14: 63–72.

Stahl, Alan. 1992. "The Coinage of La Isabela, 1493–1498." *Numismatist* 105 (10): 1399–1402.

Stahl, Ann B. 2002. "Colonial Entanglements and the Processes of Taste: An Alternative to Logocentric Approaches." *American Anthropologist* 104 (3): 827–845.

Sullivan, Shaun D. 1981. "Prehistoric Patterns of Exploitation and Colonization in the Turks and Caicos Islands." PhD diss., University of Illinois.

Thomas, Nicholas. 1991. *Entangled Objects: Exchange, Material Culture, and Colonialism in the Pacific.* Cambridge: Harvard University Press.

Torrence, Robin, and Anne Clarke. 2000. "Negotiating Difference: Practice Makes Theory for Contemporary Archaeology in Oceania." In *The Archaeology of Difference, Negotiating Cross-Cultural Engagements in Oceania*, edited by Robin Torrence and Anne Clarke, 1–31. London: Routledge.

Turnbaugh, William A. 1993. "Assessing the Significance of European Goods in Seventeenth-Century Narragansett Society." In *Ethnohistory and Archaeology: Approaches to Postcontact Change in the Americas*, edited by J. Daniel Rogers and Samuel M. Wilson, 133–160. New York: Plenum Press.

Valcárcel Rojas, Roberto. 2016. *Archaeology of Early Colonial Interaction at Chorro de Maíta, Cuba.* Gainesville: University Press of Florida.

Valcárcel Rojas, Roberto, and Marcos Martínon-Torres. 2013. "Metals in the Indigenous Societies of the Insular Caribbean." In *The Oxford Handbook of Caribbean Archaeology*, edited by William F. Keegan, Corinne L. Hofman, and Reniel Rodríguez Ramos, 504–522. New York: Oxford University Press.

Valcárcel Rojas, Roberto, Alice V.M. Samson and Menno L.P. Hoogland. 2013. "Indo-Hispanic Dynamics: From Contact to Colonial Interaction in the Greater Antilles." *International Journal of Historical Archaeology* 17 (1): 18–39.

Valcárcel Rojas, Roberto, Marcos Martínon-Torres, Jago Cooper, and T. Rehren. 2010. "Turey Treasure in the Caribbean: Brass and Indo-Hispanic Contact at Chorro de Maíta, Cuba." In *Beyond the Blockade, New Currents in Cuban Archaeology*, edited by Susan Kepec, L. Antonio Curet, and Gabino La Rosa Corzo, 106–125. Tuscaloosa: University of Alabama Press.

Weddle, Robert S. 1985. *Spanish Sea: The Gulf of Mexico in North American Discovery 1500–1685.* College Station: Texas A&M University Press.

Winter, John, and Mark Gilstrap. 1991. "Preliminary Results of Ceramic Analysis and the Movements of Populations into the Bahamas." In *Proceedings of the 12th Congress of the International Association for Caribbean Archaeology*, edited by L.S. Robinson, 371–386. Martinique: International Association for Caribbean Archaeology.

CHAPTER 3

Treating 'Trifles': the Indigenous Adoption of European Material Goods in Early Colonial Hispaniola (1492–1550)

Floris W.M. Keehnen

1 Introduction[1]

Early colonial encounters with Europeans introduced indigenous Caribbean peoples to a wide array of foreign goods and materials. Through gift-giving and exchange, objects form vital elements for negotiating the social, cultural, and material boundaries between peoples with vastly different cultural-historical backgrounds (e.g., Cipolla 2017; Gosden 2004; Maran and Stockhammer 2012; Thomas 1991). In the Caribbean, these exotic items often possessed qualities similar to or commensurable with the preexisting values of indigenous societies, facilitating their intercultural transfer and adoption (Keehnen 2011, 2012; Oliver 2000; Saunders 1999). The blending of new and traditional material expressions ushered in a period of creativity and innovation, in which the material culture repertoires of all those involved in the colonial process increasingly transformed.

European trade goods were offered to indigenous Caribbean peoples within days after first encounter on 12 October 1492 at the island of San Salvador, The Bahamas (Dunn and Kelley 1989, 83–85; see also Berman and Gnivecki this volume). Christopher Columbus' log of his first voyage in addition to the accounts from traveling companions and other contemporaries vividly describe how such material interactions continued throughout the early colonial period. An analysis of a standard corpus of late fifteenth- and early sixteenth-century (ethno)historical sources pertaining to the Greater Antilles and Bahamas has identified a total number of 177 such (reciprocal) gift-giving, barter, and tribute events in which objects transfer between cultural groups (Keehnen and Mol 2018). The vast majority of these transactions took place within the first 5-year period of colonial interaction and these involved at least 137 different types of objects, 61 of which are of European origin.[2]

1 This chapter is largely based on the archaeological data presented and discussed in the PhD dissertation Values and Valuables by Floris Keehnen (forthcoming).

2 As part of the construction of the database used for this analysis, descriptions of transaction events have been cross-referenced where possible to account for the fact that many times the

DOI:10.1163/9789004273689_004

The nature, purpose, and desirability of early colonial trade rapidly changed over these years as Columbus' originally mercantile venture transitioned into an imperial project of conquest and colonization that incorporated all of the Greater Antilles (Deagan and Cruxent 2002a; Valcárcel Rojas et al. 2013; Wilson 1990; see also Valcárcel Rojas this volume). From a material perspective, the (minimal) variety of European-introduced objects that indigenous communities would have had direct access to is striking. In these historically documented exchanges, the most prominent European trade wares were beads and hawk bells – the basic constituents of what Jeffrey Brain (1975) defined as the standard European "gift kit". The hardness, luminosity, and lustrousness of glass and metal items were considered important material traits in precolonial Caribbean cultures and societies and would have enabled exotic things such as beads and bells to tap into indigenous systems of value easily (Keehnen 2011, 2012; Oliver 2000; Saunders 1999). Other object types mentioned include varieties of food, clothing, weapons, and personal items.

The (ethno)historical records provide valuable data on the connection and integration of the indigenous Caribbean and European material realms. However, they are less informative when it comes to the indigenous use and appropriation of European-derived objects *after* their initial reception. Did such foreign items retain the importance or connotations they were valued for when exchanged with the Spaniards? Or did their status, meanings, and functions change once absorbed within a new cultural context?

In this chapter I will present a general overview of the archaeological data currently available for the island of Hispaniola (divided between the modern nations of Haiti and the Dominican Republic) in relation to the indigenous handling and incorporation of European objects in early colonial times (1492–1550). Hispaniola was the earliest and prime locus of prolonged indigenous-European interaction in the Americas in the decades following first contact. The dynamics of progressing European colonization and indigenous responses were different here than in other parts of the Greater Antilles, where the Spanish achieved domination in much shorter time (see Valcárcel Rojas et al. 2013). Although this provides Hispaniola a unique position for research into the very beginnings of the entanglement between these two distinct cultural entities, this period of the island's history has archaeologically remained understudied. Here, I will discuss the archaeological sites of En Bas Saline (Haiti), El Variar, and Juan Dolio (Dominican Republic) to reveal and explore some of the material complexities of indigenous-European interaction in the Caribbean.

---

same event was narrated, registered, or repeated by different chroniclers. For more detailed information about the approach taken, as well as the documentary sources consulted, see Keehnen and Mol (2018). Here, the numbers are mainly meant to illustrate the rich material variety of the early colonial encounters in the Caribbean.

2 The Indigenous Archaeology of Early Colonial Hispaniola

Since the 1970s and 1980s, colonial-period archaeology in Hispaniola has largely concentrated on the island's main foci of Spanish activity, including the towns of La Isabela (Caro Alvarez 1973; Deagan and Cruxent 2002a, 2002b; Luna Calderón 1986), Concepción de la Vega (Deagan and Cruxent 2002b; Kulstad 2008, 2015; Ortega and Fondeur 1978), Puerto Real (Deagan 1995), and Santo Domingo (Olsen Bogaert et al. 1998; Ortega 1982), as well as sugar mills in Azua and Sanate (Chanlatte Baik 1978; Mañón Arredondo 1978; Tavárez María 2000) and the gold mining complex at Pueblo Viejo de Cotuí (Olsen Bogaert 2011; Olsen Bogaert et al. 2011). The extensive and pioneering work of Kathleen Deagan and colleagues (e.g., Deagan 1995; Deagan and Cruxent 2002a, 2002b), as that of numerous others, have been instrumental for our understanding of European adaptive strategies in the Caribbean and the development towards a mixed colonial society (Deagan 2003). At the same time, efforts to scrutinize indigenous experiences and transitions outside of these Spanish-based centers have been limited, in contrast to, for instance, archaeological investigations in Cuba (Domínguez 1978; Valcárcel Rojas 1997, 2016, this volume). Reasons for the current lack of data can be found in methodological challenges in site identification and the long dominant assumption of an annihilated indigenous population (Deagan 2004).

This does not mean we do not know anything about the ways indigenous communities integrated European objects into their lives. Incidental findings of Spanish-introduced materials have been reported by archaeologists excavating sites they initially regarded as precolonial. Key sites in Dominican archaeological history such as Atajadizo, Juan Dolio, and La Cucama have all yielded items of European origin, which archaeologists have discussed within their local archaeological context. Some researchers have indeed tried to come to more general interpretations of the indigenous use and valuation of European goods (Deagan 1988; García Arévalo 1978a; Vega 1979). Nevertheless, a comprehensive synthesis of the pertinent data collected over the many decades of Hispaniolan archaeology has hitherto been lacking.

A review of the available literature reveals that, based on the presence of European materials, more than thirty indigenous sites have thus far been identified as having persisted into early colonial times (Figure 3.1; Table 3.1). The core archaeological component of these sites is indigenous. However, we should be aware that the complexities of this period may compromise the way we define and single out "indigenous spaces" as opposed to "mixed" or "colonial spaces" (cf., Lightfoot et al. 1998; Silliman 2010). Indigenous territories increasingly became shared and entangled spaces as the Spanish conquest and colonization of the island progressed. In terms of European-introduced objects, it cannot always be ascertained whether these were in fact handled by indigenous peoples. Also, growing power imbalances could have jeopardized indigenous self-determination with respect to the adoption and use of foreign material culture. This and other colonial influences, as well as diverging indigenous experiences on both the individual and community levels, are not easily translated into altered material patterns.

Looking at the geographical distribution of early colonial indigenous sites presented in Figure 3.1, we see the highest concentration of sites in the Dominican Republic, in particular the country's eastern and southern halves. Currently, only two Haitian sites are known (En Bas Saline and Île à Rat), although archaeological research in the country has been minimal compared to what has been done in the Dominican Republic. The sites on the map are located both in relative proximity to early Spanish settlements, as well as in

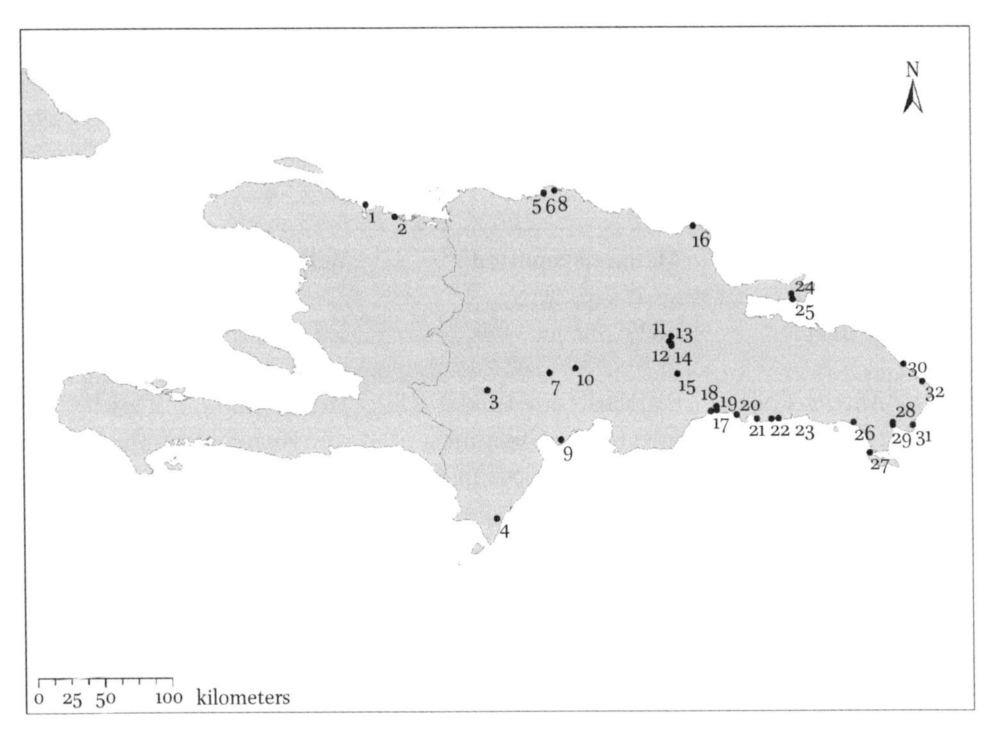

FIGURE 3.1 Map showing the locations of early colonial indigenous sites on Hispaniola: (1) Île à Rat; (2) En Bas Saline; (3) Majagual; (4) El Saladito; (5) Bajabonico; (6) El Perenal; (7) Sabana Yegua; (8) Los Balatases; (9) El Variar; (10) Guayabal; (11) Loma Piedra Imán de los Cacaos; (12) Las Lagunas; (13) Sabana del Rey (sites 12–15); (14) El Rayo; (15) Yamasá; (16) Playa Grande; (17) Antigua Calle Juan Barón; (18) Mendoza, Villa Faro; (19) Los Tres Ojos; (20) La Caleta; (21) La Cucama; (22) Guayacanes; (23) Juan Dolio; (24) Rio San Juan; (25) Anadel; (26) Boca de Chavón; (27) Punta Catuano, Isla Saona; (28) Atajadizo; (29) Boca de Yuma; (30) Punta Macao; (31) El Cabo; (32) Playa Bávaro

more distant and isolated areas in the island's northeast and southwest. About a dozen of these sites were subjected to extensive (and sometimes multiple) excavations, while other places were only surveyed or tested. Locations at which European artifacts have been found comprise indigenous (ceremonial) *plazas*, households, burials, middens, and other activity areas, as well as more secluded spots such as caves and caches.

Unfortunately, most of the European objects – the majority collected in the 1970s and 1980s – are only superficially described in the literature. Basic or even consistent artifact information (numbers, measurements, typology, etc.) is often lacking, as are context details (also unavailable due to poor stratigraphy and minimal use of fine sieving) and attempts at their social interpretation, hampering precise dating and detailed comparative analyses. In a general sense, however, many of the recovered materials typically consist of sherds of Spanish majolica, olive jar, and other types of glazed or unglazed coarse earthenware, as well as glass and metal items (Table 3.1). The most

TABLE 3.1 Overview of European materials found at early colonial indigenous sites on Hispaniola

Site	Materials reported	References
Anadel	metal pin	Krieger (1929); Vega (1979)
Antigua Calle Juan Barón	majolica; glazed and unglazed earthenware; *azulejos*; tobacco pipes; (decorated) glass fragments; nails; seals; buckles; ring bolts; coins (*maravedíes*); buttons; scabbard tips; key	Ortega (2005); Ortega and Fondeur (1978)
Atajadizo	nails; glass	Veloz Maggiolo et al. (1976)
Bajabonico	glass; earthenware	Guerrero (1999)
Boca de Chavón	horseshoe	Hatt (1932)
Boca de Yuma	Columbia Plain majolica	Goggin (1968)
El Cabo	Columbia Plain majolica; early-style olive jar; Nueva Cádiz beads; metal and glass fragments	Samson (2010)

Site	Materials reported	References
El Perenal	*unspecified*	Deagan and Cruxent (2002b)
El Rayo	Blue on White majolica; early-style olive jar; earthenware	Olsen Bogaert (2013a; 2013b)
El Saladito	glass; earthenware	López Rojas (1990)
El Variar	metal sheets and fragments	Ortega and Fondeur (1976)
En Bas Saline	musketball; copper alloy tinklers; scabbard tip fragment; Columbia Plain majolica; *melado* ware; *bizcocho* ware; glass and metal fragments	Cherubin (1991); Deagan (2004); Florida Museum of National History (2017)
Guayabal	*unspecified*	Ortega (2005)
Guayacanes	*unspecified*	De Boyrie Moya (1960)
Île à Rat	earthenware; brick	Keegan (2001)
Juan Dolio	Columbia Plain, Caparra Blue, Isabela Polychrome, Yayal Blue on White, Blue on White, Hispano-Moresque lusterware and *cuerda seca* majolica types; early-style olive jar; *melado*, Morisco green, and other types of glazed wares; transcultural ceramics; bronze buckle; metal 'threepointers' (indigenous ritual items)	Florida Museum of National History (2017); Garcia Arévalo (1978a; 1990; 1991); Garcia Arévalo and Morbán Laucer (1971); Goggin (1960; 1968); Morbán Laucer and Garcia Arévalo (1971); Ortega (2002); Ortega and Fondeur (1978); Veloz Maggiolo (1980; 1993)
La Caleta	Columbia Plain, blue on white and *cuerda seca* majolica types	Goggin (1968)
La Cucama	early-style olive jar; Isabela Polychrome and possibly other types of majolica; *melado* ware; transcultural ware; earthenware; coin (*maravedí*)	García Arévalo (1978a); Mañón Arredondo et al. (1971); Ortega (2005); Veloz Maggiolo et al. (1973)
Las Lagunas	metal book clasp	Olsen Bogaert (2015)

TABLE 3.1 Overview of European materials found at early colonial indigenous sites on Hispaniola (*cont.*)

Site	Materials reported	References
Loma Piedra Imán de los Cacaos	musketball; glass fragment; glazed and unglazed earthenwares	Olsen Bogaert (2015)
Los Balatases	metal (spear)point	Ulloa Hung and Herrera Malatesta (2015)
Los Tres Ojos	*unspecified*	Garcia Arévalo and Morbán Laucer (1971)
Majagual	metal beads	García Arévalo (1978a)
Mendoza, Villa Faro	green *lebrillo*; roof tiles; earthenware	Ortega (2005)
Playa Bávaro	early-style olive jar	Ortega (1978)
Playa Grande	Columbia Plain, Isabela Polychrome and *cuerda seca* majolica; *melado* ware; early-style olive jar; earthenware; coin (*maravedí*); glass bead; bead manufactured from sherd of glazed ware; nails; knife; metal plates; buckles; fasteners; horseshoes; glass and metal fragments; clay pipe with indigenous decoration	López Belando (2012; 2015)
Punta Catuano, Isla Saona	earthenware	Vega and Luna Calderón (2004)
Punta Macao	Columbia Plain majolica; earthenware	Atiles (2004)
Rio San Juan	iron hammer	Smithsonian National Museum of Natural History (2017); James Krakker, pers. comm. 2017
Sabana del Rey (sites 12-15)	early-style olive jar; Columbia Plain majolica; glazed ware	Olsen Bogaert (2013a; 2015)

Site	Materials reported	References
Sabana Yegua	hawk bells; metal sheets; buckles; stirrup rings; metal base of a lamp	Vega (1979)
Yamasá	majolica	García Arévalo and Morbán Laucer (1990)

common ceramic forms are household, tableware, and kitchen utility wares such as plates, (carinated) bowls, jars, and basins. Metal objects have been identified in about twenty different varieties, yet from this assortment only coins, rings, and bells also feature in (ethno)historical descriptions of indigenous-Spanish transactions (Keehnen and Mol 2018). Moreover, despite their prominence in the documentary record, brass bells and glass beads are archaeologically virtually invisible. Bells were retrieved from a cache at Sabana Yegua (Vega 1979), while blue-colored beads were found only at the villages of El Cabo and Playa Grande (Hofman et al. 2014; López Belando 2015; Samson 2010, 282–284). Finally, the European artifacts form part of a variety of material compositions, they occur in their original or a reworked state, and appear to have been treated differently across space and time (see sections below).

2.1 *Other Expressions of the Material Encounter*

Apart from the European artifacts recovered from the early colonial sites inventoried here, evidence of the incorporation of European elements within the indigenous (socio-)material world can also be seen expressed differently. A cotton belt in Vienna's Museum für Völkerkunde and the so-called beaded *cemí* in Rome's Museo Nazionale Preistorico Etnografico Luigi Pigorini are two of the finest surviving masterpieces from the Caribbean (Ostapkowicz 2013; Taylor et al. 1997; Vega 1973). Both objects were indigenous valuables of high symbolic importance that most likely originate in early sixteenth-century Hispaniola. The artworks are strikingly similar in their design and manufacture, characterized by elaborately braided cotton elements covered with meticulously executed beadwork of mainly white and red marine shells. Their uniqueness is further attested by the incorporation of European materials such as jet, brass pins, Venetian mirrors, and blue and green glass beads, which, with the exception of jet, were typical commodities of exchange in indigenous-Spanish encounters. Although many questions about the context and meaning of the two enigmatic objects are still unanswered, the integration of new and foreign materials into these aesthetically indigenous items uniquely illustrates the post-1492 transformation of the indigenous material culture repertoire.

A different way in which the presence of Europeans translated itself into the indigenous material record can be observed in a handful of caves across the Dominican Republic, where European persons, animals, and objects have been portrayed on the inner walls. Cueva de Rancho La Guardia holds pictographs of people dressed in Spanish manner (Abreu Collado 2008). In caves near the San Lorenzo Bay depictions of human figures holding spades have been found, along with those of horse or mule figures and chickens (Pagán Perdomo 1999). Horse pictographs are also seen in Cueva Mongó (Abreu Collado and Olsen Bogaert 1989) and the cave system of Pomier-Borbón, in the latter case joined by riders with lances (Guerrero and Veloz Maggiolo 1988). Representations of European ships occur in two locations: Cueva de la Arena in the San Lorenzo Bay complex (López Belando 2009) and the Cuevas del Pomier-Borbón (Schomburgk 1854). Although all of these images illustrate the indigenous engagement with 'the other', none of the places have yielded further evidence of their interaction with the Spanish.[3]

Many more aspects of the indigenous material domain were changed or affected as a result of the Spanish intrusion of the island, including foodways, ceramic production, and funerary rites. Especially when observed at indigenous sites we may assume that these alterations were not simply top-down impositions by the colonial power. At El Tamarindo, an indigenous village site close to La Isabela, residues of exotic plants and ruminant animals were found on indigenous ceramic sherds, suggesting that local communities incorporated European foods into their meals (VanderVeen 2006). No Spanish artifacts were discovered here. European faunal remains, most notably pig, were found at the sites of Antigua Calle Juan Barón, El Cabo, En Bas Saline, Loma Piedra Imán de los Cacaos, Playa Grande, and Punta Macao, although not all of the material has been confirmed to date to the late fifteenth or early sixteenth centuries (see Table 3.1). Indigenous ceramic repertoires underwent rapid changes, particularly when colonial rule manifested itself more strongly

3 It cannot be ruled out completely that these artistic expressions were made by Spaniards. Caves on Mona Island (located between the Dominican Republic and Puerto Rico) have provided evidence of European visitors interacting with indigenous wall paintings (Samson et al. 2016).

(Ernst and Hofman 2015; García Arévalo 1978b, 1990, 1991; Ortega 1980; Ortega and Fondeur 1978). While the integration of indigenous and European pottery traditions is best seen within the Spanish colonial centers of the island, so-called transcultural ceramics – or indigenous-made vessels imitating Spanish styles and/or designs – are also known from the sites of Juan Dolio and La Cucama (see Table 3.1). The Cuban site of El Yayal represents a remarkable case, where traditional zoomorphic *adornos* (molded pot handles) were shaped into horse or cow figures (Deagan 1988; Domínguez 1984). Lastly, funerary rites also show a blending of indigenous and Spanish cultural customs. From a number of indigenous sites, including Guayacanes, Juan Dolio, and La Caleta, hispanicized burial patterns are known (Boyrie Moya 1960). Generally this implies a transition from placing the body in a traditional flexed position to one in which the interred person lies stretched with the arms crossed on the chest. In addition, grave goods may change in their composition and placement.

3 The Indigenous Adoption of European Material Goods: Three Archaeological Examples

The following case studies illustrate different ways in which European material goods are manifested in early colonial indigenous sites on Hispaniola. En Bas Saline, on the northcoast of Haiti, is the largest known indigenous site of the country and one of few systematically excavated places in the Greater Antilles with a continuous indigenous occupation across the historical divide from ca. AD 1250. The town was one of the first to be in contact with the Spanish and was situated close to an important Spanish center. El Variar is situated in the southwestern mountains of the Dominican Republic, an area long ignored by the Spanish and hence home to some of the last indigenous redoubts of the island. Juan Dolio is a sizeable cemetery-habitation site on the southeast coast of the island, located in an area of increased Spanish presence after the official relocation of their main port and settlement to Santo Domingo in 1498.

3.1 *En Bas Saline*

Located some 12 kilometers east of modern Cap Haïtien, En Bas Saline is believed to have been the residence of Guacanagarí, principal *cacique* or chief of the region, who provided shelter to Columbus and his crew after a fatal shipwreck in late 1492 (Deagan 2004). The village inhabitants were left relatively undisturbed by Spanish colonists until 1503, when only 2 kilometers away

Puerto Real was founded (Deagan 1995). In the late 1970s, first archaeological work commenced, followed by large-scale excavations by Kathleen Deagan and her team in the 1980s (Cusick 1991; Deagan 1987, 1989, 1990, 2004). Although these campaigns were initially aimed at finding the short-lived Spanish fort of La Navidad, fine-grained excavation methods laid bare the post-1492 indigenous occupation of the site. The site's main features include an earthen ridge, refuse middens, and a central mound, located in the middle of a flat open area supposed to have functioned as a *plaza* or ball court. On top of the central mound two superimposed 15 meters wide house structures were identified. Three raised areas divide the *plaza* itself into two sections, interpreted as elite residence areas.

European materials at the site are scarce and generally unremarkable, comprising only 17 small fragments of glass, metal, and glazed earthenwares (Deagan 2004). Interesting, nevertheless, is that all but a piece of clear glass were found in the central mound, the residential area of the elites. The elite house further contained the highest number of indigenous ritual and ornamental items as well as the largest and most varied assemblage of pottery remains. Also the majority of European faunal remains were recovered here, including pigs (*Sus scrofa*), rats (*Rattus rattus*), mice (*Mus musculus*), and cats (*Felis domesticus*). In addition to these data, the online catalogue of the Florida Museum of National History (2017) has entries of a lead musket ball, a scabbard tip fragment, and a number of perforated copper alloy pendants. These items are not discussed in any of Deagan's publications and may have been part of earlier test excavations by William Hodges in the late 1970s.

3.2 *El Variar*

The locality of El Variar is situated near the town of Barreras in the south of the Dominican province of Azua. The only archaeological inquiry dates to 1975. Although no traces of past habitation could be discerned, a rich set of objects was found underneath a large rock of chert and limestone (Figure 3.2). The cache consisted of two small Chicoid vessels – one somewhat smaller than the other as to be placed on top of the larger one – that were traditionally used as containers of beverages or hallucinogenic powders in indigenous rituals and ceremonies (Ortega 2005, 240–244). Inside the covered pot were discovered a number of indigenous high-value objects, including an elaborate necklace of about 250 cylindrical-shaped stone and shell beads, and two anthropomorphic amulets, one made of stone, the other of shell. Interestingly, a natural resin was used to attach a small brass plate around the stone figure's neck. Also the eyes, as well as the shell figure's neck

FIGURE 3.2 Cache of artifacts found at El Variar, Azua, Dominican Republic. Museo del Hombre Dominicano, Santo Domingo, Dominican Republic
PHOTOS BY FLORIS W.M. KEEHNEN, NOT TO SCALE

showed signs of having had metal inlays or decoration. Besides, the assemblage included four brass items, two circular and two rectangular pieces, the latter perforated as to be worn as a pendant or to be inserted into the bead necklace (Ortega 2005, 240–244; see also Ortega and Fondeur 1976; Vega 1979).

3.3 *Juan Dolio*

Juan Dolio is located some 50 kilometers to the east of Santo Domingo and has an estimated indigenous occupation from AD 1000 until at least the second decade of the sixteenth century. Most archaeological work was done between the early 1950s and 1970s, with a strong focus on the approximately one hundred burials at the site's cemetery areas (Boyrie Moya 1960; Drusini et al. 1987; Veloz Maggiolo 1972). While some of the graves contained indigenous offerings, only one skeleton also held Spanish goods (García Arévalo and Morbán Laucer 1971; Morbán Laucer and García Arévalo 1971). This individual, identified as a ca. 50-year old man, reflects traditional indigenous customs, such as cranial modification and a squatted position of the interred body. The grave goods consist of large and elaborate indigenous vessels placed next to the head, and, closer to the man's waist, a Spanish bowl of a ceramic type known as Columbia Plain. Ten burials – also indigenous, based on their modified skulls – were found on a more secluded spot on the beach (Ortega 2002, 18; Veloz Maggiolo 1980, 165–166). Remarkably, all were interred in a Christian fashion, each in extended position and together placed in one straight line. Some burials contained Spanish jars and vessels, while a sub-adult was found in possession of a bronze buckle, that, evidenced by the fabric remains attached, probably formed a complete belt. The funerary pattern seems to indicate a collective burial, probably done by Spaniards, who would have allowed family members of the deceased to inter a small number of valuables. As the skeletons do not show any traces of violence, their death has been suggested the result of an epidemic outbreak.

More Spanish objects were found throughout the site, both on the surface and in lower levels (García Arévalo 1978a; García Arévalo 1991; Goggin 1968; Ortega 1982, 165; Veloz Maggiolo 1993, 171). These included hundreds of ceramic fragments, representing various types of majolica, early-style olive jar, and other glazed wares, as well as several extraordinary transcultural pieces (Figure 3.3a). Among these are two unique local manufactures of Spanish *bacines* or chamber pots, dated between 1514 and 1520, which clearly show the retention of indigenous decorative techniques and zoomorphic stylistic elements (Veloz Maggiolo 1993, 171). Another pot is typically indigenous in

FIGURE 3.3 Early colonial ceramic forms found at Juan Dolio, San Pedro de Macorís, Dominican Republic: (a) indigenous-made pots reflecting Spanish designs; (b) indigenous pot with atypical cross incision. Fundación García Arévalo, Santo Domingo, Dominican Republic
PHOTOS BY FLORIS W.M. KEEHNEN, NOT TO SCALE

form and design, but features atypical incisions showing the sign of a cross, possibly reflecting the influence of Christianity among the indigenous peoples of this region (García Arévalo 1978a, 115; García Arévalo and Morbán Laucer 1971; Morbán Laucer and García Arévalo 1971) (Figure 3.3b). An even more stunning find, perhaps, is the recovery of two metal threepointers or *cemís* (García Arévalo 1978a, 107). Threepointers were highly ritualized, animated objects within indigenous Caribbean societies, and appear in the archaeological record in various materials, sizes, and degrees of refinement (Breukel 2013; Walker 1997). These two examples, however, are completely incomparable in their way of manufacturing. The larger one is cast with a mold, the slightly smaller one hammered.

4 Material Culture Repertoires in Motion

Indigenous peoples took possession of European artifacts in a number of ways. Best documented (ethno)historically are direct exchanges with the Spanish in gift-giving and barter events. These were either politically motivated elite affairs or more informal haphazard transactions that also involved other group members. Through indigenous exchange networks European goods were distributed to communities that were not in direct contact with the Spanish (e.g., Hofman et al. 2014).[4] Other means of acquisition may have included pilfering or the collecting of items lost or otherwise left behind by Spaniards, trophies gained in combat, and shipwreck salvage (Hally and Smith 2010; see also Berman and Gnivecki this volume). Determining the distribution mechanism from a small assemblage of archaeological materials with a poor context is difficult, although the different types of recovered artifacts may provide some clues. Spanish ceramics as well as small glass and metal items including beads, hawk bells, and pins are the strongest indicators of direct indigenous-European exchanges. Less typical trade goods such as nails, horseshoes, the lamp base from Sabana Yegua, the key from Juan Barón, and the book clasp from Las Lagunas may be indications of indigenous pilfering or Spanish loss

4 Although down-the-line exchange of European objects is not easily identifiable archaeologically, it is presumed to have been common practice. Columbus directly hints at such indirect trade when navigating through the Bahamas. Here, he describes an encounter with a man in a canoe who is carrying coins (*blancas*) and glass beads previously received from the Spanish expedition to a neighboring island (Dunn and Kelley 1989, 83–85; see also Berman and Gnivecki this volume). Furthermore, European potsherds have been found on a number of Bahamian islands not known to have been visited by Spanish explorers (Keegan and Mitchell 1987).

of artifacts. Military items such as musket balls, scabbard tips, and the metal point from Los Balatases might have been taken from defeated enemies on the battlefield (see Table 3.1).

Investigating the site and find contexts of such objects is required to better grasp their place within indigenous societies. The case studies presented above have shown different ways in how their materiality is expressed in the archaeological record. In most of the sites known thus far, as is clearly demonstrated in the case of En Bas Saline, the quantity of European goods is remarkably small, despite their sometimes close proximity to Spanish towns such as Puerto Real. For Deagan, the apparently limited adoption of Spanish items into the material life of the En Bas Saline inhabitants supports the idea of the indigenous “indifference to and rejection of Spanish cultural elements and values” (2004, 621). Indeed, the lack of European artifacts in early colonial indigenous contexts cannot be easily explained, certainly not if we take into account the many and rich (ethno)historical descriptions of intercultural gift-giving and exchange. In fact, reading these sources would make it hard to believe that indigenous Caribbean peoples were not receptive of the exotic items they were offered – and actively sought after. It is possible, however, that the possession and circulation of European objects were controlled by indigenous elites. The idea of restricted access is confirmed at the site of En Bas Saline, where almost all of the European materials were found in elite contexts. Especially in the earliest years of colonial interaction, such objects would have been relatively scarce, and rather than being discarded casually, would have been incorporated into indigenous exchange networks. Likewise, Spanish chronicler Gonzalo Fernández de Oviedo narrates the redistribution of the material wealth of a deceased chief to foreign *caciques*, gifts that might well have included possessions of European origin (see Oliver 2009, 103–108).

The cache of materials found at El Variar indicates European artifacts were sometimes carefully stowed away, probably to be recovered later. The direct association of brass materials with high-value indigenous paraphernalia suggests the metal objects were given an at least similar esteem. The special valuation of European metals by indigenous Caribbean peoples has been suggested by various authors, particularly for the material's gleaming surface (also seen in glass and glazed ceramics), its peculiar smell, as well as its symbolic correlation with – and possible metaphorical substitution of – the indigenous copper alloy known as *guanín* (Keehnen 2011; Martinón-Torres et al. 2007; Oliver 2000; Saunders 1999; Valcárcel Rojas and Martinón-Torres 2013; Vega 1979). The finding of a metal base of a lamp as part of a cache found at Sabana Yegua probably best illustrates that it was, indeed, the material traits, perhaps including its oxidizing and durable properties, instead of the form or function of metal objects that attracted indigenous attention.[5] Furthermore, as mentioned earlier, brass hawk bells are described as items of special interest in the (ethno)historical sources (García Arévalo and Chanlatte Baik 2015). Columbus writes: "they desired nothing else as much as bells" and "they are on the point of going crazy for them" (Dunn and Kelley 1989, 283), a preference indigenous traders communicated by imitating the tinkling sound. Yet, notwithstanding their initial desirability, the indigenous connotation of European metals may have changed negatively in the course of the colonial process. The Spanish imposition of a tribute system within years after their arrival required each person to pay a hawk bell filled with gold every three months. In addition, a small metal disk had to be hung around the person's neck as a proof of payment (Las Casas 1986, 437). Possibly, the perforated brass items of El Variar are such tokens. The indigenous revaluation of European metals as a result of these colonial measures is not unthinkable, although such a shift cannot be discerned from the archaeological record.

Quite the opposite seems to be reflected in Juan Dolio, where the extraordinary find of the metal threepointers shows the most exquisite way in which European objects were reworked into indigenous forms. Whereas in El Variar pieces of brass were used to decorate indigenous amulets, in Juan Dolio metals are modified to create entirely new objects traditional of indigenous Caribbean society. It cannot be said, however, whether these threepointers were made by indigenous peoples having acquired European techniques, or by Spaniards perhaps aiming to gain favor from the local indigenous population. Juan Dolio shows the characteristics of having been occupied longer than most other sites, possibly indicating the existence of a Spanish outpost or plantation. Here again, the sporadic inclusion of European goods into indigenous graves suggests status differentiation in privileged access to exotic items. The deliberate deposition of European objects with the deceased likely involved a range of different meanings and motives. At least, their integral and personal connection to death and the spiritual afterlife, strongly indicates the special value indigenous peoples placed upon European goods. From the fabric remains found in association with the bronze buckle it can be assumed

5 In addition to metal objects, also glazed ceramics appear to have been valued mostly for their unique material characteristics. Rather than as complete vessels, most Spanish pottery was exchanged as (deliberately) broken pieces, as several of the (ethno)historical documents describe (Dunn and Kelley 1989, 71, 93, 109, 265; Farina and Zacher 1992, 53; MacNutt 1912, 61). Archaeologically, intact vessels have not been recovered either.

some individuals were buried in Spanish clothing, a practice that has been described for the Cuban site of El Chorro de Maíta (ca. AD 1200 to post-AD 1550) (Martinón-Torres et al. 2007; Valcárcel Rojas et al. 2010). For the Spanish, the offering of clothes and shirts was part of their effort to civilize their indigenous hosts. At the same time, wearing European dress could have been an expression of status and identity among indigenous peoples.

5 Concluding Remarks

The Spanish arrival to Hispaniola is archaeologically reflected in a limited though virtually island-wide dispersal of European goods, which appear in a variety of indigenous contexts in a range of different types and forms. European materials either appear as relatively scattered finds, as directly associated with valuable indigenous objects or as integrated parts of highly symbolic icons of indigenous culture and society. A small part of the entire assemblage consists of reworked or physically altered artifacts, some modified as to be attached to or become part of indigenous objects, as in the case of El Variar. Hybrid forms in which indigenous and European shapes, materials, and techniques are brought together, such as at Juan Dolio, are uncommon. The same holds true for weapons, tools, and other implements that are indicative of purely colonial interactions.

Most of the items recovered from sites on Hispaniola pertain to a category of goods generally associated with gift-giving and barter exchanges characteristic of the initial phase of indigenous-European interaction, a short period during which indigenous peoples freely added Spanish objects to existing material culture repertoires. Considering, in addition, the frequently coastal locations of many of these sites, such object assemblages probably point to short, occasional instances of direct interaction between local villages and bypassing Europeans, or, alternatively, down-the-line exchange through preexisting indigenous networks. More regular or sustained interaction between local communities and Spanish settlers likely took place in and around early Spanish centers such as La Isabela, Santo Domingo, and Cotuí, in the immediate vicinity of which relatively high concentrations of indigenous sites with European materials seem to appear. In these cases, colonists may have appealed to the inhabitants of nearby villages for food and other provisions, as well as labor needed for construction works or mining. The local adoption of foreign traits transformed existing repertoires and practices at locations such as Juan Dolio, where indigenous and European peoples either lived together or were in day-to-day contact. Also the cache of objects at El Variar may date to a somewhat later stage in the colonial process, reflecting a personal set of valuables taken along by an individual seeking refuge in an attempt to escape colonial power.

The symbolic-ideological qualities indigenous peoples recognized in the distinct materialities of these foreign articles facilitated their adoption and incorporation into the socio-material world of the original island inhabitants. The insertion of European artifacts into indigenous graves, caches, and (sacred) caves, as well as their connection to ceremonial *plazas* and elite households suggests a certain exclusivity in terms of their access and handling. Such a symbolic importance is further attested by the ritual portrayal of European elements on cave walls. On the other hand, the indigenous integration and recontextualization of European artifacts was part of a dynamic process of (re)negotiation in which previously ascribed meanings and attributed uses may well have been discarded or altered.

Ongoing interaction with Spaniards during the colonial period and the connection and integration of the two different material realms accelerated a process of change and entanglement that was initiated during the encounter. Indigenous peoples not only started to wear European dress, but would also have manufactured new forms of ceramics, adapted their culinary traditions, changed their burial customs, and perhaps would have been open to new beliefs. The indigenous adoption and possible rejection of European artifacts occurred on the basis of their conscious, selective, and varied treatment, and affected different domains of life and death. The transition to a colonial situation in the first decades following contact possibly limited indigenous decision-making autonomy. At the same time, the blending of indigenous and European cultural and material elements does attest to the new and flexible ways indigenous Caribbean communities were able to creatively transform their material culture repertoires over the course of the colonial process.

Acknowledgments

The research of which this chapter is the result was supported by the Netherlands Organisation for Scientific Research (PhDs in the Humanities 'Values and Valuables in the Early Colonial Caribbean', grant PGW-13-02), and is supervised by Corinne L. Hofman. I thank Hayley Mickleburgh and Andrzej Antczak for their comments on earlier versions of the manuscript.

References

Abreu Collado, Domingo. 2008. "El Arte Rupestre Y Fray Ramón Pané." *Boletín Del Museo Del Hombre Dominicano* 42: 323–330.

Abreu Collado, Domingo, and Harold Olsen Bogaert. 1989. "La Prospección: Elemento Indispensable Para La Arqueología Científica." *Boletín Del Museo Del Hombre Dominicano* 22: 65–81.

Atiles, Gabriel. 2004. "Excavaciones Arqueológicas de Punta Macao. Informe de Campo." Manuscript on file, Museo del Hombre Dominicano, Santo Domingo.

Boyrie Moya, Emile E. de. 1960. *Cinco Años de Arqueología Dominicana.* Santo Domingo: Universidad de Santo Domingo.

Brain, Jeffrey P. 1975. "Artifacts of the Adelantado." In *Conference on Historic Site Archaeology Papers 8*, edited by Stanley South, 129–138. Columbia: South Carolina Institute of Archaeology and Anthropology.

Breukel, Thomas W. 2013. "Threepointers on Trial: A Biographical Study of Amerindian Ritual Artefacts from the Pre-Columbian Caribbean." Research Master thesis, Leiden University.

Caro Alvarez, José A. 1973. "La Isabela." *Boletín Del Museo Del Hombre Dominicano* 3: 48–52.

Chanlatte Baik, Luis A. 1978. "Informe Sanate-Higuey: Arqueología Colonial." *Boletín Del Museo Del Hombre Dominicano* 10: 133–138.

Cherubin, Ginette. 1991. "Le Project 'Recherche de Navidad' Révélateur D'un Avenir Archéologique Prometteur Pour Haiti." In *Proceedings of the 14th Congress of the International Association for Caribbean Archaeology*, edited by Alissandra Cummins and Philippa King, 425–435. Barbados: The Barbados Museum and Historical Society.

Cipolla, Craig N., ed. 2017. *Foreign Objects: Rethinking Indigenous Consumption in American Archaeology.* Tucson: University of Arizona Press.

Cusick, James G. 1991. "Culture Change and Pottery Change in a Taíno Village." In *Proceedings of the 13th Congress of the International Association for Caribbean Archaeology* , edited by Edwin N. Ayubi and Jay B. Haviser, 1: 446–461. Willemstad, Curaçao: Archaeological-Anthropological Institute of The Netherlands Antilles.

Deagan, Kathleen A. 1987. "Initial Encounters: Arawak Responses to European Contact at the En Bas Saline Site, Haiti." In *Proceedings of the 1st San Salvador Conference: Columbus and His World*, edited by Donald T. Gerace, 341–359. Fort Lauderdale: College Center of the Finger Lakes and San Salvador: Bahamian Field Station.

Deagan, Kathleen A. 1988. "The Archaeology of the Spanish Contact Period in the Caribbean." *Journal of World Prehistory* 2 (2): 187–233.

Deagan, Kathleen A. 1989. "The Search for La Navidad, Columbus's 1492 Settlement." In *First Encounters: Spanish Explorations in the Caribbean and the United States,*

1492–1570, edited by Jerald T. Milanich and Susan Milbrath, 41–54. Gainesville: University of Florida Press.

Deagan, Kathleen A. 1990. "The Search for La Navidad in a Contact Period Arawak Town on Haiti's North Coast." In *Proceedings of the 11th Congress of the International Association for Caribbean Archaeology*, edited by Agamemnon G. Pantel Tekakis, Iraida Vargas Arenas, and Mario Sanoja Obediente, 453–458. San Juan: La Fundación Arqueológica, Antropológica e Histórica de Puerto Rico.

Deagan, Kathleen A., ed. 1995. *Puerto Real: The Archaeology of a Sixteenth-Century Spanish Town in Hispaniola*. Gainesville: University Press of Florida.

Deagan, Kathleen A. 2003. "Colonial Origins and Colonial Transformations in Spanish America." *Historical Archaeology* 37 (4): 3–13.

Deagan, Kathleen A. 2004. "Reconsidering Taíno Social Dynamics after Spanish Conquest: Gender and Class in Culture Contact Studies." *American Antiquity* 69 (4): 597–626.

Deagan, Kathleen A., and José M. Cruxent. 2002a. Columbus's Outpost among the Taínos: Spain and America at La Isabela, 1493–1498. New Haven: Yale University Press.

Deagan, Kathleen A., and José M. Cruxent. 2002b. *Archaeology at La Isabela: America's First European Town*. New Haven: Yale University Press.

Domínguez, Lourdes S. 1978. "La Transculturación En Cuba (S. XVI–XVII)." *Cuba Arqueológica* 1: 33–50.

Domínguez, Lourdes S. 1984. *Arqueología Colonial Cubana: Dos Estudios*. Havana: Editorial de Ciencias Sociales.

Drusini, Andrea, Ferdinando Businaro, and Fernando Luna Calderón. 1987. "Skeletal Biology of the Taíno: A Preliminary Report." *International Journal of Anthropology* 2 (3): 247–253.

Dunn, Oliver, and James E. Kelley, eds. 1989. *The Diario of Christopher Columbus's First Voyage to America 1492–1493*. Abstracted by Fray Bartolomé de Las Casas. Translated by Oliver Dunn and James E. Kelley. Norman: University of Oklahoma Press.

Ernst, Marlieke, and Corinne L. Hofman. 2015. "Shifting Values: A Study of Early European Trade Wares in the Amerindian Site of El Cabo, Eastern Dominican Republic." In *GlobalPottery 1: Historical Archaeology and Archaeometry for Societies in Contact*, edited by Jaume Buxeda i Garrigós, Marisol Madrid i Fernández, and Javier Garcia Iñañez, 195–204. Oxford: Archaeopress.

Florida Museum of National History. 2017. "Artifact Gallery." Accessed June 1. http://www.flmnh.ufl.edu/histarch/artifactGallery.asp.

García Arévalo, Manuel A. 1978a. "La Arqueología Indo-Hispana En Santo Domingo." In *Unidad Y Variedad: Ensayos Antropológicos En Homenaje a José M. Cruxent*, edited by Erika Wagner and Alberta Zucchi, 77–127. Caracas: Instituto Venezolano de Investigaciones Científicas.

García Arévalo, Manuel A. 1978b. "Influencias de La Dieta Indo-Hispanica En La Cerámica Taína." In *Proceedings of the 7th International Congress for the Study of the Pre-Columbian Cultures of the Lesser Antilles*, edited by Ripley P. Bullen, 263–277. Montreal: Centre de Recherches Caraïbes, Université de Montréal.

García Arévalo, Manuel A. 1990. "Transculturation in Contact Period and Contemporary Hispaniola." In *Archaeological and Historical Perspectives on the Spanish Borderlands East*, edited by David Hurst Thomas, 269–280. Columbian Consequences, Vol. 2. Washington, DC: Smithsonian Institution Press.

García Arévalo, Manuel A. 1991. "Influencias Hispánicas En La Alfarería Taína." In *Proceedings of the 13th Congress of the International Association for Caribbean Archaeology*, edited by Edwin N. Ayubi and Jay B. Haviser, 363–383. Willemstad, Curaçao: Archaeological-Anthropological Institute of the Netherlands Antilles.

García Arévalo, Manuel A., and Luis A. Chanlatte Baik. 2015. "Los Cascabeles O Sonajeros Taínos." In *Proceedings of the 25th International Congress for Caribbean Archaeology*, edited by Laura del Olmo, 609–613. San Juan: Instituto de Cultura Puertorriqueña, el Centro de Estudios Avanzados de Puerto Rico y el Caribe y la Universidad de Puerto Rico, Recinto de Río Piedras.

García Arévalo, Manuel A., and Fernando Morbán Laucer. 1971. "Localizan Fosa Que Pertenecía A Los Indígenas." *El Caribe*, January 19.

García Arévalo, Manuel A., and Fernando Morbán Laucer. 1990. "La Plaza O Batey Aborigen de Yamasá." *Boletín Del Museo Del Hombre Dominicano* 23: 79–96.

Goggin, John M. 1960. *The Spanish Olive Jar: An Introductory Study*. New Haven: Department of Anthropology, Yale University.

Goggin, John M. 1968. *Spanish Majolica in the New World: Types of the Sixteenth to Eighteenth Centuries.* New Haven: Department of Anthropology, Yale University.

Gosden, Chris. 2004. *Archaeology and Colonialism: Cultural Contact from 5000 BC to the Present.* Cambridge: Cambridge University Press.

Guerrero, José G. 1999. "Una Lectura Arqueo-Histórica Del Contacto Temprano Indo-Europeo: El Caso de La Isabela, Primera Villa de América." *Boletín Del Museo Del Hombre Dominicano* 27: 97–109.

Guerrero, José G., and Marcio Veloz Maggiolo. 1988. *Los Inicios de La Colonización En América: La Arqueología Como Historia.* San Pedro de Macorís: Universidad Central del Este.

Hally, David J., and Marvin T. Smith. 2010. "Sixteenth-Century Mechanisms of Exchange." *Journal of Global Initiatives: Policy, Pedagogy, Perspective* 5 (1): 53–65.

Hatt, Gudmund. 1932. "Notes on the Archaeology of Santo Domingo." *Geografisk Tidsskrift* 35 (1/2): 9–17.

Hofman, Corinne L., Angus A.A. Mol, Menno L.P. Hoogland, and Roberto Valcárcel Rojas. 2014. "Stage of Encounters: Migration, Mobility and Interaction in the Pre-Colonial and Early Colonial Caribbean." *World Archaeology* 46 (4): 590–609.

Keegan, William F. 2001. "Archaeological Investigations on Ile À Rat, Haiti." In *Proceedings of the 18th Congress of the International Association for Caribbean Archaeology, Volume 2*, edited by The International Association for Caribbean Archaeology, Region Guadeloupe, 233–239. St. George, Grenada.

Keegan, William F., and Steven W. Mitchell. 1987. "The Archaeology of Christopher Columbus' Voyage Through the Bahamas, 1492." *American Archaeology* 6 (2): 102–108.

Keehnen, Floris W.M. 2011. "Conflicting Cosmologies: The Exchange of Brilliant Objects between the Taíno of Hispaniola and the Spanish." In *Communities in Contact: Essays in Archaeology, Ethnohistory & Ethnography of the Amerindian Circum-Caribbean*, edited by Corinne L. Hofman and Anne Van Duijvenbode, 253–268. Leiden: Sidestone Press.

Keehnen, Floris W.M. 2012. "Trinkets (f)or Treasure? The Role of European Material Culture in Intercultural Contacts in Hispaniola during Early Colonial Times." Research Master thesis, Leiden University.

Keehnen, Floris W.M., and Angus A.A. Mol. 2018. "The Roots of the Columbian Exchange: An Entanglement and Network Approach to Early Caribbean Encounter Transactions." Unpublished manuscript, Faculty of Archaeology, Leiden University.

Krieger, Herbert W. 1929. *Archaeological and Historical Investigations in Samaná, Dominican Republic*. Washington, D.C.: United State Government Printing Office.

Kulstad, Pauline M. 2008. "Concepción de La Vega 1495–1564: A Preliminary Look at Lifeways in the Americas' First Boom Town." Master thesis, University of Florida.

Kulstad, Pauline M. 2015. "Striking It Rich in the Americas' First Boom Town: Economic Activity at Concepción de La Vega (Hispaniola) 1495–1564." In *Archaeology of Culture Contact and Colonialism in Spanish and Portuguese America*, edited by Pedro P.A. Funari and Maria X. Senatore, 313–337. Cham: Springer.

Las Casas, Bartolomé de. 1986. *Historia de Las Indias*. Edited by André Saint-Lu. Vol. 1. Caracas: Biblioteca Ayacucho.

Lightfoot, Kent G., Antoinette Martinez, and Ann M. Schiff. 1998. "Daily Practice and Material Culture in Pluralistic Social Settings: An Archaeological Study of Culture Change and Persistence from Fort Ross, California." *American Antiquity* 63 (2): 199–222.

López Belando, Adolfo José. 2009. "El Arte Rupestre En El Parque Nacional Los Haitises." Rupestreweb. http://www.rupestreweb.info/haitises2.html.

López Belando, Adolfo José., ed. 2012. "El Sitio Arqueológico de Playa Grande, Rio San Juan, María Trinidad Sánchez: Informe de Las Excavaciones Arqueológicas Campaña 2011–2012." Manuscript on file, Museo del Hombre Dominicano, Santo Domingo.

López Belando, Adolfo José. 2015. "Excavaciones Arqueológicas En El Poblado Taíno de Playa Grande, República Dominicana." In *Proceedings of the 25th Congress of the International Association for Caribbean Archaeology*, edited by Laura Del Olmo, 254–279. San Juan: Instituto de Cultura Puertorriqueña, el Centro de Estudios

Avanzados de Puerto Rico y el Caribe y la Universidad de Puerto Rico, Recinto de Río Piedras.

López Rojas, Elba. 1990. "Informe de Un Viaje de Prospección Al Parque Nacional Jaragua – 1987." *Boletín Del Museo Del Hombre Dominicano* 23: 41–53.

Luna Calderón, Fernando. 1986. "El Cementerio de La Isabela: Primera Villa Europea Del Nuevo Mundo: Estudio de Antropología Física." *Primera Jornada de Antropología* 1 (1): 10–17.

Mañón Arredondo, Manuel J. 1978. "Importancia de Los Ingenios Indo-Hispánicos de Las Antillas." *Boletín Del Museo Del Hombre Dominicano* 10: 139–164.

Mañón Arredondo, Manuel J., Fernando Morbán Laucer, and Aida Cartagena Portalatín. 1971. "Nuevas Investigaciones de Áreas Indígenas Al Noreste de Guayacanes Y Juan Dolio." *Revista Dominicana de La Arqueología Y Antropología* 1 (1): 81–113.

Maran, Joseph, and Philipp W. Stockhammer, eds. 2012. *Materiality and Social Practice: Transformative Capacities of Intercultural Encounters*. Oxford: Oxbow Books.

Martinón-Torres, Marcos, Roberto Valcárcel Rojas, Jago Cooper, and Thilo Rehren. 2007. "Metals, Microanalysis and Meaning: A Study of Metal Objects Excavated from the Indigenous Cemetery of El Chorro de Maíta, Cuba." *Journal of Archaeological Science* 34: 194–204.

Morbán Laucer, Fernando, and Manuel A. García Arévalo. 1971. "Describen Hallazgos Objetos Indígenas." *Listín Diario*, January 21.

Oliver, José R. 2000. "Gold Symbolism among Caribbean Chiefdoms: Of Feathers, Çibas and Guanín Power among Taíno Elites." In *Precolumbian Gold: Technology, Style and Iconography*, edited by Colin McEwan, 196–219. London: British Museum Press.

Oliver, José R. 2009. *Caciques and Cemí Idols: The Web Spun by Taíno Rulers Between Hispaniola and Puerto Rico*. Tuscaloosa: University of Alabama Press.

Olsen Bogaert, Harold, ed. 2011. "Sitio Arqueológico No. 11: Investigación Estructuras Coloniales." Manuscript on file, Museo del Hombre Dominicano, Santo Domingo.

Olsen Bogaert, Harold. 2013a. "Prospección Y Registro de Sitios Arqueológicos En Pueblo Viejo de Cotuí 2003–2010." *Boletín Del Museo Del Hombre Dominicano* 45: 121–150.

Olsen Bogaert, Harold., ed. 2013b. "Investigación Sitio Arqueológico No. 17: El Rayo." Manuscript on file, Museo del Hombre Dominicano, Santo Domingo.

Olsen Bogaert, Harold. 2015. "Inventario de La Selección de Bienes de Las Investigaciones Arqueológicas En El Proyecto Pueblo Viejo de Cotuí, Y Su Entorno." Manuscript on file, Museo del Hombre Dominicano, Santo Domingo.

Olsen Bogaert, Harold, Francisco Coste, and Jiménez L. Abelardo 2011. "Estructuras Coloniales En La Mina de Oro de Pueblo Viejo, Cotuí, Provincia Sánchez Ramírez." *Boletín Del Museo Del Hombre Dominicano* 44: 39–73.

Olsen Bogaert, Harold, Eugenio Pérez Montás, and Esteban Prieto Vicioso, eds. 1998. *Arqueología Y Antropología Física En La Catedral de Santo Domingo*. Santo Domingo: Centro de Altos Estudios Humanisticos y del Idioma Español.

Ortega, Elpidio J. 1978. "Informe Sobre Investigaciones Arqueológicas Realizadas En La Región Este Del País, Zona Costera Desde Macao a Punta Espada." *Boletín Del Museo Del Hombre Dominicano* 11: 77–105.

Ortega, Elpidio J. 1980. *Introducción a La Loza Común O Alfarería En El Período Colonial de Santo Domingo*. Santo Domingo: Fundación Ortega Álvarez.

Ortega, Elpidio J. 1982. *Arqueología Colonial En Santo Domingo*. Santo Domingo: Taller.

Ortega, Elpidio J. 2002. *Artefactos En Concha: Arqueología En Coral Costa Caribe, Juan Dolio, R.D.* Santo Domingo: Fundación Ortega Álvarez.

Ortega, Elpidio J. 2005. *Compendio General Arqueológico de Santo Domingo, Vol. 1.* Santo Domingo: Academia de Ciencias de la República Dominicana.

Ortega, Elpidio J., and Carmen G. Fondeur. 1976. *Primer Informe Sobre Piezas Metálicas Indígenas En Barrera.* Santo Domingo: Centro Dominicano de Investigaciones Antropológicas.

Ortega, Elpidio J., and Carmen G. Fondeur. 1978. *Estudio de La Cerámica Del Período Indo-Hispana de La Antigua Concepción de La Vega*. Santo Domingo: Taller.

Ostapkowicz, Joanna. 2013. "'Made … With Admirable Artistry': The Context, Manufacture and History of a Taíno Belt." *The Antiquaries Journal* 93: 287–317.

Pagán Perdomo, Dato. 1999. "El Estudio Del Arte Rupestre En La Isla de Santo Domingo." *Boletín Del Museo Del Hombre Dominicano* 27: 19–43.

Samson, Alice V.M. 2010. *Renewing the House: Trajectories of Social Life in the Yucayeque (Community) of El Cabo, Higüey, Dominican Republic, AD 800 to 1504.* Leiden: Sidestone Press.

Samson, Alice V.M., Jago Cooper, and Josué Caamaño-Dones. 2016. "European Visitors in Native Spaces: Using Paleography to Investigate Early Religious Dynamics in the New World." *Latin American Antiquity* 27 (4): 443–461.

Saunders, Nicholas J. 1999. "Biographies of Brilliance: Pearls, Transformations of Matter and Being, C. AD 1492." *World Archaeology* 31 (2): 243–257.

Schomburgk, Robert H. 1854. "Ethnological Researches in Santo Domingo." *Journal of the Ethnological Society of London* 3: 115–122.

Silliman, Stephen W. 2010. "Indigenous Traces in Colonial Spaces: Archaeologies of Ambiguity, Origin, and Practice." *Journal of Social Archaeology* 10 (1): 28–58.

Smithsonian National Museum of Natural History. 2017. "Department of Anthropology Collections." Accessed June 1. http://collections.nmnh.si.edu/search/anth/.

Tavárez María, Glenis. 2000. "El Ingenio Diego Caballero: Aspectos Históricos Y Culturales de Una Factoría Azucarera Del Siglo XVI." *Boletín Del Museo Del Hombre Dominicano* 28: 65–75.

Taylor, Dicey, Marco Biscione, and Peter G. Roe. 1997. "Epilogue: The Beaded Zemi in the Pigorini Museum." In *Taíno: Pre-Columbian Art and Culture from the Caribbean*, edited by Fatima Bercht, Estrellita Brodsky, John A. Farmer, and Dicey Taylor, 158–169. New York: The Monacelli Press.

Thomas, Nicholas. 1991. *Entangled Objects: Exchange, Material Culture and Colonialism in the Pacific*. Cambridge: Harvard University Press.

Ulloa Hung, Jorge, and Eduardo Herrera Malatesta. 2015. "Investigaciónes Arqueológicas En El Norte de La Española, Entre Viejos Esquemas Y Nuevos Datos." *Boletín Del Museo Del Hombre Dominicano* 46: 75–107.

Valcárcel Rojas, Roberto. 1997. "Introducción a La Arqueología Del Contacto Indo-Hispánico En La Provincia de Holguín, Cuba." *El Caribe Arqueológico* 2: 64–77.

Valcárcel Rojas, Roberto. 2016. *Archaeology of Early Colonial Interaction at El Chorro de Maíta, Cuba*. Gainesville: University of Florida Press.

Valcárcel Rojas, Roberto, and Marcos Martinón-Torres. 2013. "Metals in the Indigenous Societies of the Insular Caribbean." In *The Oxford Handbook of Caribbean Archaeology*, edited by William F. Keegan, Corinne L. Hofman, and Reniel Rodríguez Ramos, 504–522. Oxford: Oxford University Press.

Valcárcel Rojas, Roberto, Marcos Martinón-Torres, Jago Cooper, and Thilo Rehren. 2010. "Turey Treasure in the Caribbean: Brass and Indo-Hispanic Contact at El Chorro de Maíta, Cuba." In *Beyond the Blockade: New Currents in Cuban Archaeology*, edited by Susan Kepecs, L. Antonio Curet, and Gabino La Rosa Corzo, 106–125. Tuscaloosa: University of Alabama Press.

Valcárcel Rojas, Roberto, Alice V.M. Samson, and Menno L.P. Hoogland. 2013. "Indo-Hispanic Dynamics: From Contact to Colonial Interaction in the Greater Antilles." *International Journal of Historical Archaeology* 17 (1): 18–39.

VanderVeen, James M. 2006. "Subsistence Patterns as Markers of Cultural Exchange: European and Taíno Interactions in the Dominican Republic." PhD diss., Indiana University.

Vega, Bernardo. 1973. "Un Cinturón Tejido Y Una Careta de Madera de Santo Domingo, Del Período de Transculturación Taíno-Español." *Boletín Del Museo Del Hombre Dominicano* 3: 199–226.

Vega, Bernardo. 1979. *Los Metales Y Los Aborígenes de La Hispaniola*. Santo Domingo: Museo del Hombre Dominicano.

Vega, Bernardo, and Fernando Luna Calderón. 2004. "Descubrimiento de La Primera Plaza Indígena En La Isla Saona." *Boletín Del Museo Del Hombre Dominicano* 36: 31–34.

Veloz Maggiolo, Marcio. 1972. *Arqueología Prehistórica de Santo Domingo*. Singapore: McGraw-Hill Far Eastern Publishers.

Veloz Maggiolo, Marcio. 1980. *Vida Y Cultura En La Prehistoria de Santo Domingo*. San Pedro de Macorís: Universidad Central del Este.

Veloz Maggiolo, Marcio. 1993. *La Isla de Santo Domingo Antes de Colón*. Santo Domingo: Banco Central de la República Dominicana.

Veloz Maggiolo, Marcio, Elpidio J. Ortega, Renato O. Rimoli, and Fernando Luna Calderón. 1973. "Estudio Comparativo Y Preliminar de Dos Cementerios Neo-Indios: La

Cucama Y La Unión, República Dominicana." *Boletín Del Museo Del Hombre Dominicano* 3: 11–47.

Veloz Maggiolo, Marcio, Iraida Vargas, Mario Sanoja, and Fernando Luna Calderón. 1976. *Arqueología de Yuma (República Dfdominicana)*. Santo Domingo: Taller.

Walker, Jeffrey B. 1997. "Taíno Stone Collars, Elbow Stones, and Three-Pointers." In *Taíno: Pre-Columbian Art and Culture from the Caribbean*, edited by Fatima Bercht, Estrellita Brodsky, John A. Farmer, and Dicey Taylor, 80–91. New York: The Monacelli Press.

Wilson, Samuel M. 1990. *Hispaniola: Caribbean Chiefdoms in the Age of Columbus*. Tuscaloosa: University of Alabama Press.

CHAPTER 4

Contact and Colonial Impact in Jamaica: Comparative Material Culture and Diet at Sevilla la Nueva and the Taíno Village of Maima

Shea Henry and Robyn Woodward

1 Introduction

For the many indigenous cultures encountered in the Americas by the Renaissance voyages of discovery, and particularly those on the islands of the Caribbean, the arrival of Europeans on their shores led to rapid demographic and cultural decline. Introduction of European diseases, violent confrontations, enslavement, Crown-sanctioned forced labor, and the destruction of traditional cultural patterns resulted from this devastating contact and colonialism. But to simplify these initial encounters into narratives of conquest and devastation is to ignore the profound social change to both the indigenous peoples of the Caribbean and settler European groups that this encounter provoked (Deagan 2004, 597; Patterson 2010, 133). Fifty years of historical and archaeological research has explored both indigenous and European responses to issues of cultural survival and continuity, resistance and power negotiations, accommodation, acculturation, transculturation, and ethnogenesis. This research has demonstrated that depending on the time, geographic setting, and context of these intercultural encounters, there will be significant variations in responses by both the indigenous peoples and European settlers (Deagan 2004, 598). This chapter looks at both the transformation of Iberian material culture, social practices, and diet in households of the elite and non-elite residents of Sevilla la Nueva, the first Spanish capital on the island of Jamaica, as well as the concurrent social adjustment and resistance to the Iberian colonizing efforts that occurred in the adjacent indigenous Taíno village of Maima.

The Spanish colony of Sevilla la Nueva has been explored and analyzed archaeologically on and off through the past century. Excavations reveal a large-scale attempt at building the colony into an extensive and productive trading port, capable of supporting further colonization throughout the Caribbean. Excavations conducted at Sevilla la Nueva show the extent of the construction and expectations of this colony, including the building of a town, governor's fort, and abbey. The role and presence of the indigenous so-called 'Taíno' at

 | DOI:10.1163/9789004273689_005

this colony is evident through the presence of a particular style of colonoware known as "New Seville ware" (Woodward 2006a, 2006b). In addition to the material culture excavated from Sevilla la Nueva, historical records indicate that the indigenous peoples were present working under the forced labor system of the *encomienda*. In 2014 and 2015 (Burley et al. 2017a, 2017b), excavations at the nearby Taíno village of Maima have added to the overall picture of a site impacted by a rich precontact history, protracted contact period, and an ultimately impactful and devastating colonial period.

This chapter will explore the three time periods represented at Maima and Sevilla la Nueva, and the changing and unchanging material culture represented in each. First, the precontact material culture and diet of Maima will be reviewed, creating a baseline from which the later contact and colonial periods effected. Then, the brief contact era, represented by the marooning of Spanish sailors in 1503 within just a few kilometers of Maima. We then explore the colonial era with the founding of the first Spanish capital of Jamaica, Sevilla la Nueva, followed by an exploration and overview of the material culture, specifically colonoware, found at the colonial capital. Finally, we place circumstances seen archaeologically at Maima and Sevilla la Nueva into the broader Caribbean contact and colonial experience. Through the archaeological studies at Sevilla la Nueva and Maima we get a rare glimpse at the effects of both initial contact and colonialism on this indigenous Jamaican village.

2 History of Indigenous and Proto-Historic Jamaica

As compared to the eastern part of the Caribbean, where human occupation began around 4000 BC, Jamaica was settled comparatively late, less than 1000 years before the arrival of the Spanish. The first settlers on the island were characterized by their Ostionoid ceramics, who expanded across the Jamaica Channel from Hispaniola to the south coast of Jamaica by AD 650 (Rouse 1992, 110). The so-called 'Western Taíno' culture gradually replaced the initial Ostionoid culture series in Jamaica and parts of Cuba after AD 880 and was predominant on those islands until the arrival of the Spanish (Rouse 1992, 96). The Late Ceramic Age indigenous peoples of Jamaica had a distinctly different material culture from that of their predecessors, which is characterized by their own ceramic tradition, White Marl style, that fits within the regional Meillacoid series (Meillacan Ostionoid sub-series) (AD 950–1500). Despite their isolation, and different ceramic styles, these so-called 'Western Taíno' of Jamaica shared some linguistic and cultural traits with the Classic Taíno peoples of Hispaniola and Puerto Rico (Atkinson 2003, 1); thus it might be assumed that as on Hispaniola, ceramic production, basket and mat making, and weaving and spinning of cotton was done by the women in the community (Deagan 2004, 601). Defining the indigenous peoples of the Caribbean under one name however has come under scrutiny of late (Curet 2014; Keegan and Hofman 2017) with scholars leaning more towards an individual look at differing cultures around the Caribbean rather than one defining 'Taíno' identity.

Despite not defining the indigenous Caribbean peoples under one title, they do share certain cultural traits. They have similar stone tool traditions and agricultural practices, such as the mounding of fields into *conucos*, for the cultivation of cassava. In addition to cassava, they introduced and cultivated sweet potatoes, beans, peppers, squash and peanuts (Rouse 1992). Once the *conucos* were built, the only labor required was periodic planting and harvesting which appeared to have been done by all members of the community (Lee 1980, 2; Rouse 1992, 170). Guava, mamey, pineapple and tobacco were all additional Taíno cultigens, and cotton was cultivated and traded between the islands (Deagan 2004). Of these domesticates, only cotton and tobacco were exploited by the Europeans on a widespread commercial basis over the five centuries (Rouse 1992, 12). To supplement their protein intake the indigenous inhabitants harvested a diverse array of both inshore and deep-water species of fish and shellfish and trapped *hutias*, a member of the rodent family, which was the only terrestrial mammal indigenous to the island (Faerron 1985, 2; Wing 2001).

Social organization among the Western Taíno of Jamaica centered around polities of allied villages with perhaps eight to ten principal chiefs, or *caciques*, although Wesler suggests that structures may have included simple and complex chiefdoms at the time of contact (Wesler 2013, 253; Wilson 2007, 110). Little is written about the Taíno mythology that was practiced on the island or if both men and women participated in community rituals as was the case on Hispaniola. The Taíno population of Jamaica in 1494 was estimated to be at least 60,000 although some anthropologists believe the island had a population base of at least 100,000 (Wilson 2007).

Spanish engagement with the Taíno of Jamaica began on May 5, 1494 when Columbus sailed into the bay of Santa Gloria (later St. Ann's Bay) while on his second voyage to the Americas. Columbus gave the Taíno assurances of good faith and peace and passed out trade trinkets as a gesture of goodwill. Over the succeeding days, with but one instance, they traded small trinkets for gifts of food as they sailed west along the north coast of the island (Padrón 2003, 3). Over the course of the next decade there are no reports of other Spaniards visiting the island, save for Alonso de Hojeda's provisioning run along the south coast of the island in 1502, most probably because Columbus had noted in his report to the Crown in 1494 that there was no gold on the island (Morison 1942, 643; Wright 1921, 71).

In June 1503 on the final and fateful leg of his fourth voyage of exploration Columbus was forced to seek shelter in Santa Gloria as the two remaining ships of his squadron were too waterlogged to make it back to Santo Domingo on Hispaniola. For more than a year Columbus and his men endured, at times, uneasy relations with the local Taíno, with whom they traded for food (Morison 1963; Padrón 2003, 8). Columbus himself was silent about this time in his journals, however, his eldest son Diego Colón who accompanied him on the voyage kept an historic account of the year's events. During this time, he noted in his journal that the island was thickly populated and the Taíno village of Maima was about a league distance from his beachhead and named Aguacadiba as the village with whom they traded for hawk bells, small glass beads and lace tips for cassava, maize, *hutia* and fish (Morison 1942, 643; 1963, 356). Despite being forced into dependent transactions for their very survival, Columbus made only sparse ethnographic observations about the communities that surrounded their beached ships (Morison 1963, 367; Wesler 2013, 253).

In 1508, Diego Colón was appointed Governor of the Indies. In an effort to forestall any further erosion of his family's claims in the New World, he ordered a former military officer Juan de Esquivel, to take 60 settlers and establish a settlement, in the bay of Santa Gloria, which was known to have both a sheltered harbor and a large, peaceful indigenous populace (Padrón 2003, 51–52). Esquivel was charged with establishing agricultural and ranching properties with an aim at producing supplies for local markets and colonizing efforts in Central America (Wright 1921, 71). As a native of Sevilla, Esquivel named his settlement, Sevilla la Nueva (Padrón 2003, 52).

Initially Esquivel enjoyed the confidence of the Crown and he reportedly showed enthusiasm for the conversion of the indigenous peoples and had distributed land and indigenous laborers to some of his men through the feudal institution of *encomienda* (Rouse 1992, 19). The *encomenderos*, the Spanish who received these entitlements, could extract tribute from their indigenous workers in the form of food, precious metals, or direct labor services in exchange for protection and instruction in the Catholic religion and civilization (Woodward 2006a, 63; Yaeger 1989, 843). In his reports to the king, Colón indicates that Esquivel had promoted agricultural endeavors and introduced cattle, sheep and sugar cane to the island (Padrón 2003, 54). However, in another report made by Pedro de Mazuelo, the settlement treasurer remarked that given their brutal treatment, the settlers would likely run out of indigenous labor in two years (Padrón 2003, 149). As a result of this cruel behavior, the local indigenous groups rose up against the Spanish and the *encomienda* system. In response, Esquivel rounded up the local *caciques*, killing them in a show of power and control, after which there were no further rebellions.

Resulting from the inhumane treatment of the indigenous peoples and the reports the Crown was receiving about the labor shortfalls, in 1513 the King ordered a *residencia* (routine review of an official's tenure) of Esquivel's governorship. This report faulted him for his management of the indigenous population that had occurred two years into his administration, and claimed he had initiated a system of favoritism with respect to land and labor grants (Padrón 2003, 53–54). He was dismissed and in 1514 they conferred the position of Governor of Jamaica on Francisco de Garay a successful gold miner, ship owner, slave trader, entrepreneur, and administrator on Hispaniola between 1594–1513 (Padrón 2003, 53). In late 1514, prior to returning to the Caribbean, Garay met with King Ferdinand and entered into a five-year *asiento* partnership agreement with his Royal patron with regards to the economic development of the island (Floyd 1973, 137; Weddle 1985, 97).

On his arrival Garay undertook an accurate census of the island's indigenous population to determine the number of indigenous laborers available for distribution to the colonists. Although this report has not survived, it is understood that in his capacity as *repartidor*, he redistributed a number of indigenous laborers to numerous officials and new settlers, as well as, assigning men to the new royal *estancias* (farms), in which he was a partner (Padrón 2003, 150; Wright 1921, 73). Jamaica was very prosperous under Garay's administration, the Spanish population continued to expand, and he established two more towns on the island, Oristán on the south coast, and Melilla, 12 to 14 leagues east of Sevilla la Nueva, neither of which has been found. He built the first sugar mill on the island and was in the process of building a second when he left the island in 1523 to pursue his claim on a portion of Mexico. His departure drained the island of manpower and the ships needed to transport its products to regional markets.

In the absence of Garay's capital and able administration the fate of Sevilla la Nueva was in the hands of the island's treasurer, Pedro de Mazuelo, who proceeded to manipulate the affairs of the island to benefit his plantation on the south side of the island. During the 25 years that Sevilla la Nueva was occupied, archival sources document the employment of indigenous labor in the fields, tending animals on the *estancias* as well as building the church (Padrón 2003; Woodward 2006a; Wright 1921; Wynter 1984).

3 Excavations of Sevilla la Nueva and Maima

Even though Sevilla la Nueva was only established for 25 years, archaeological investigations over the past 70 years of this site have demonstrated that the early sixteenth-century Spanish colonial remains are both diverse and well preserved, albeit deeply buried under thick layers of alluvium (Cotter 1948, 1970, n.d.; Hammond 1970; Lakey et al. 1983; López y Sebastián 1982, 1986a, 1986b, 1987; Osborne 1973; Woodward 1988, 2006a, 2006b). Both Cotter and Osborne assumed they were dealing with single event deposits, so they employed basic horizontal controls during their excavations, and Cotter kept adequate notes over the course of the eighteen years he worked on the site. The Spanish Archaeological Mission under the direction of Sr. López y Sebastián worked on the site for eight years, excavating some 327 2 m^2 or 4 m^2 units but other than three short preliminary reports that identified the site, they did not publish their research on the artifacts or file copies of their notes/maps with the Jamaica National Heritage Trust. The archaeological program initiated by Woodward and a team from Simon Fraser University and the Jamaican National Trust included topographical mapping, electromagnetic conductivity survey and testing, auger coring and areal excavations. The research design and sampling techniques for all the post-2001 excavation units at Sevilla la Nueva and Maima included strict horizontal controls and arbitrary vertical controls fine screening of all the deposits leading to a more robust data base from which it has been possible to draw more nuanced interpretations.

Far from being the "black hole" of precontact archaeology of the Caribbean and circum-Caribbean, Jamaica actually has a rich database of surveys and excavations from all regions of the island, although until the last few decades much of this has been conducted and published locally rather than disseminated to a wider audience. Carbon-14 dates obtained from a number of indigenous sites in Jamaica suggest that they were occupied well into the Spanish period (Wesler 2013, 255, 259). Data from other documented contact period indigenous sites suggest that European objects (fragments of glass, ceramic metal) were only minimally present. Sampling designs of many of the early projects on Jamaica did not go beyond rough sorting or ceramic and faunal remains, making recovery of all the European plant and animals remains unlikely. The absence of European materials on these sites however has helped to reinforce the assumption that the Taíno either abandoned their villages shortly after contact or did not survive long enough to generate detectable archaeological deposits (Deagan 2004, 603).

A number of indigenous sites were noted on the hills surrounding St. Ann's Bay, including a large site, a "quarter league" southwest of the Spanish Governor's fortress that has been identified as Maima (Burley et al. 2017a; Burley et al. 2017b). Some work was done on the eastern edge of this site by Spanish archaeologists in the early 1980s that was not reported on and the material covered in the project were not fully studied (López y Sebastián 1986a). As indigenous involvement at Sevilla la Nueva was clearly demonstrated by the presence of later period White Marl pottery and a type of colonoware (Woodward 2006a 2006b), the potential to correlate archaeological features from both the Spanish and adjacent indigenous site with the extraordinarily rich reserve of archival documentation prompted a re-examination of materials to gain a better understanding of how the Spanish adapted to their new surroundings and how the indigenous peoples responded to both Spanish usurpation of their traditional lands and the imposition of the *encomienda*.

Over the past 40 years the organization of archaeological materials into functional categories for the purposes of quantitative analysis has enabled archaeologists to both organize and compare materials from analytical units of like functions such as households, workshops, churches etc. to address questions about labor, exchange, diet, and gender (South 1977, Deagan 2004, 611). For the purpose of this chapter only household units with their emphasis on living quarters and workspaces were considered for comparative analysis. From the Spanish town site, then the elite residence of the governor's fort, excavated by Cotter in the 1950s and the remains of a non-elite residence (Spanish House Area 6) that was identified in 2009, adjacent to the Spanish meat processing feature (Industrial Area) were included in this study (Figure 4.1).

During excavations carried out in 2014 and 2015 at Maima, now a part of the Seville Heritage Park, a team of archaeologists from Simon Fraser University identified the indigenous village as the Maima noted in the Columbus chronicles (Burley et al. 2017a). Five house units were excavated in 2015 with three (houses 7, 8, and 10) producing significant artifacts and material culture representative of the late precontact period, and two (houses 7 and 10) including artifacts from the contact era. House 10 in particular includes indications of being an elite household as well as including a large portion of contact era artifacts. Elite artifacts from the precontact era include a number of ceramic faces, and the highest percentage of decorated ceramic fragments. While these indications may point to an elite household, further research and data would be needed to explore the nature and extent to which House 10 held social or economic status within the community.

Precontact data shows that the indigenous villagers at Maima produced ceramics similar to those found at other late precontact Jamaica sites across

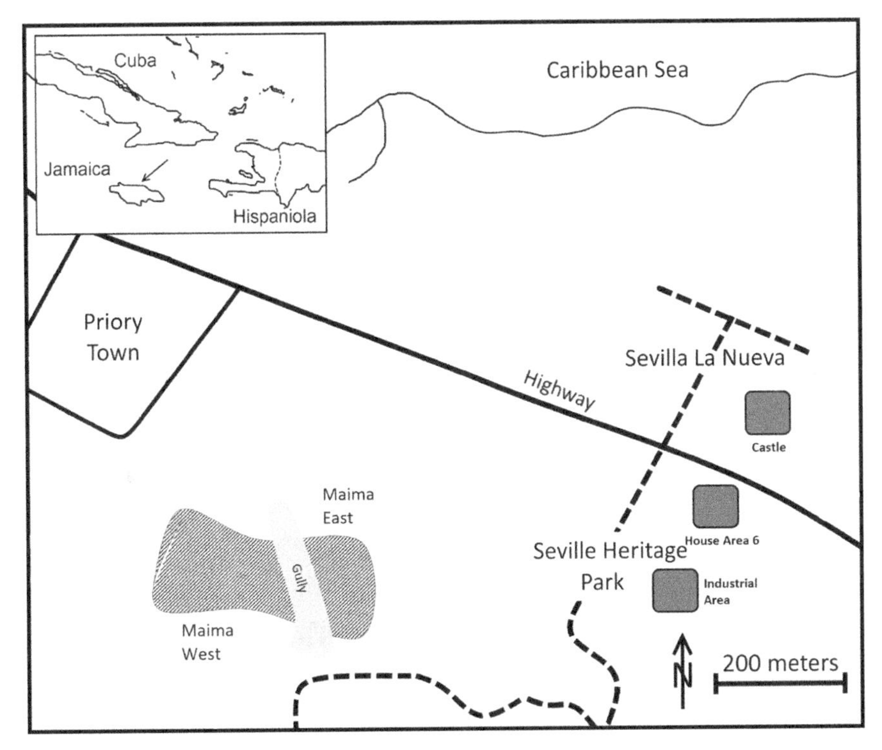

FIGURE 4.1 Map of Sevilla la Nueva and Maima, Jamaica, with areas excavated, modified from Burley et al. 2017a

Jamaica and the surrounding region (Figure 4.2). The characteristic filleted rims, vessel forms, geometric designs, and surface decoration are all present at Maima (Burley et al. 2017b, 38–39). Faunal remains were found in small amounts, but those recovered from midden contexts throughout the site indicate that fish and shellfish were the primary source of protein, coming mostly from the shallow reef within a kilometer of the village (Henry 2017). Hutia, a medium sized rodent, were also eaten but represent only 15% of the total faunal bone assemblage. Age profiles of the hutia at Maima indicate that they were likely kept in some form of domestication as primarily young immature hutia are represented, consistent with herd management behavior (Wing 2001, 2008). Historical accounts from Diego Colón note the Taíno bringing the Spanish hutia upon their arrival, suggesting that they were easily accessible yet prized enough to offer the newcomers (Morison 1963). The Spanish sailors likened the rodent to rabbit but claimed that it did not satisfy their strong desire for meat (Morison 1963, 356).

The artifacts excavated at Maima also show a general continuity between the precontact, contact, and colonial eras. Fish and shellfish data indicate that

FIGURE 4.2 Examples of traditional Taíno pottery recovered from excavations at Maima, Jamaica

the villagers collected most of their meat products from the nearby reef, with 89% of the total fauna coming from a shallow marine environment (Henry 2017). As Maima is approximately 1 km from a reef located in what is now St Ann's Bay it is likely that it was there that the Taíno fished. The percentages of fish and shellfish taken from reef environments does not change over time nor does it change within the levels associated with contact era artifacts. The diet remains the same throughout the rest of the site until the abandonment of the village. Other artifact types, including ceramics and lithics also maintain continuity through the entire occupancy of the site.

The contact era artifacts are the most representative of the European material culture at Maima. After Columbus and his crew landed in what is now St. Ann's Bay, historic records from Diego Colón and Diego Mendez state that the marooned Spanish traded items from their ship in exchange for food, information, canoes, and other supplies (Morison 1963, 356). At Maima in particular, European items were traded primarily for food. European artifacts found at Maima fit with these historic accounts with the uncovering of Spanish nails, three glass fragments, and two pieces of unidentified metal (Burley et al. 2017b). It is possible that these artifacts were deposited at Maima during the initial contact in 1503, or during the later colonization in 1509, however these are all objects that the Spanish would have had in that initial contact.

Artifacts that can be placed solidly into the colonial era at Maima are the few fragments of European domestic animals, imported by the Spanish for Sevilla la Nueva. These faunal remains include two sheep/goat metatarsals and a single cow tooth. The marooned Spanish that arrived at the initial contact did not carry these domesticates with them and therefore must represent a later time period. There are no cut or butcher marks on either of these bones and it is possible that they do not represent food remains, rather were taken as oddities during the colonial era.

4 Colonoware and Transculturated Material Culture

The site of Sevilla la Nueva contains the multiple habitation and work areas that were constructed for the colony. The excavations conducted over the past 70 years include the sugar mill, artisan's quarters, governor's fort, abbey, households, and wells. These excavations produced thousands of artifacts representing material from industrial, utilitarian, elite, non-elite, household circumstances among many others. This chapter draws data specifically from three areas of the site, the governor's fort, Spanish house area 6 and the industrial area. At each of these locations indigenous or Meillacoid ceramics, colonoware, and Spanish ceramics, in differing percentages were found. Each area had the potential to have indigenous laborers and Spanish colonists present.

TABLE 4.1 Spanish and Taíno pottery recovered from three excavation areas at Sevilla la Nueva, Jamaica

Description	Governor's Fort	%	Spanish Area 6 House	%	Industrial Area	%
Spanish Majolicas	191	6.83	55	8.99	50	1.58
Misc. Glazed Spanish Wares	76	2.72	145	23.7	123	3.9
Unglazed Spanish Earthenwares	101	3.61	15	2.41	49	1.5
Olive Jar	591	21.18	31	5.1	69	2.28
Sugar Moulds					2426	76.8
Roof Tiles			79	12.9	44	1.39
Taíno Ceramics	1781	63.77	287	46.9	369	11.69
Manioc Griddle	16	0.57			21	0.66
New Seville Ware	37	1.32			6	0.2
Total	2793	100	612	100	3157	100

Beyond ceramic artifacts, the majority of the material culture represented at the site is Spanish.

In addition to the Spanish material culture, Meillacoid ceramics have been identified in nearly every location around the site. Through an analysis of the rim and decorated sherds in the Sevilla la Nueva collection, there is no evidence that ceramic styles underwent any kind of change during the colonial era (Woodward 2006b). The filleted rims and decoration styles noted on the ceramic fragments found at Sevilla la Nueva match entirely with those excavated at Maima in the more recent excavations. These unaltered and unchanged Meillacoid ceramics at Maima indicate first that there was contact and connection between Maima and Sevilla la Nueva as these ceramics likely passed between the forced laborers going back and forth to their home, and second the continued use of traditional styles rather than completely integrating Spanish material culture.

The colonoware, known from this site as New Seville ware (Woodward 2006a, 2006b), is represented by the use of traditional indigenous ceramic materials and methods, manufactured into Spanish styles. New Seville ware is characterized by the sandy, pale brown to yellowish-brown paste. They are constructed through hand-formed coils, as opposed to the wheel thrown European ceramics. Thirty-six sherds and seven vessels can be identified as New Seville ware, representing domestic ceramics including bowls and cups (Figure 4.3). The appropriation of local materials and ceramic manufacturing

FIGURE 4.3 New Seville ware recovered from Governors Fort at Sevilla la Nueva, Jamaica

techniques is directly representative of the *encomienda* forced labor system in place at Sevilla la Nueva. Local potters were being used as craft producers as part of their service to the colony. The two examples pictured in figure 4.3, the jug and pedestal cup, were both found at the governor's fort, and were likely used as tableware for the governor (Woodward 2006b, 171). Two vessels found at the governor's fort are incised with traditional Meillacoid decorations similar to those found at Maima (Figure 4.3). These decorations act as a stamp put on these wares from the indigenous potters forced to make them.

The majority of colonoware at Sevilla la Nueva was found at the governor's fort, however, the Meillacoid ceramics are much more represented throughout the site. We can trace the presence of the indigenous peoples across the site by where and how much of Meillacoid traditional ceramics and colonoware is found. The most are found at the castle, where the indigenous peoples would be laboring to build the ornate and complex governors fort, which acted as his residence. A large number of Meillacoid ceramics were also found in the industrial area, also likely from the indigenous laboring at the sugar mill and in the artisan's quarters where the bricks and decorations were made for the castle and abbey. The Meillacoid ceramics found at the house site are possibly representative of the indigenous women that were taken as wives by the Spanish (Woodward 1988). According to Spanish records, single male colonists were encouraged to marry indigenous women in order to grow the population of the colony and ensuring its progress and success (Padrón 2003, 58). However, no colonoware was found in domestic areas but only in the governor's fort, and industrial area. In that way it seems to suggest that the colonoware was created for the Spanish and not for use by the indigenous peoples. In particular, it was used most in the elite area of the colony by the governor specifically.

At Maima, only a few small fragments of Spanish roof tile and no pieces of colonoware matching that of Sevilla la Nueva were found. The roof tile was located in house unit 10 along with two glass fragments, three European domestic animal faunal fragments, and nine nails. While we know from historic records, and these fragmentary pieces of European goods found at Maima, that it was occupied concurrently with Sevilla la Nueva, the Maima villagers did not significantly incorporate either European material culture or colonoware, into their daily lives. Since the majority of contact and colonial era artifacts found at Maima were recovered from house unit 10, which is possibly representative of an elite household, it is possible that the unique Spanish artifacts were being held by the elite, a pattern noted in other sites in the Caribbean (Ernst and Hofman this volume; Hofman et al. 2014; Keehnen this volume; Samson 2010; Valcárcel Rojas this volume; Valcárcel Rojas et al. 2013).

5 Discussion and Summary

The idea that the indigenous peoples of the Caribbean were devastated immediately and without reaction, the "fatal impact" model, was a commonly held misconception for decades within archaeological and particularly historical research (Silliman 2005). Despite this, more recent researchers have been finding that the indigenous Caribbean islanders utilized a number of behaviors and adaptations to counter this destructive and impactful colonialism. The study of contact and colonial era, from the perspective of the indigenous islanders, is a dynamic and changing field, with various patterns and models being observed. As this volume displays, the response to contact and colonialism is dynamic and never the same in one place. This is particularly true in their study of concurrent and nearby indigenous settlements in Cuba in which one area resisted the inclusion of Spanish material culture, while another did not (Valcárcel Rojas et al. 2013). Other scholars and research done in the Caribbean note the incorporation of Spanish artifacts into the daily lives and cosmologies of the indigenous peoples (Deagan 1988, 1996; Keehnen 2010). Deagan (2004) in her study of the indigenous village of En Bas Saline in present-day Haiti, introduces a colonial model in which the indigenous peoples were taken from their villages for part of the year, made to work for the Spanish, then were able to return to their villages for the remainder of the year. She notes that despite sharing their time between their home communities and the Spanish settlements, there was little European material culture found at the indigenous village. This pattern is observable at other indigenous Caribbean sites (Deagan 2011; Hauser and Armstrong 2012).

The continuity observed in the archaeology at Sevilla la Nueva and Maima between the precontact, contact, and colonial eras provides evidence less of transculturation but of forced labor and resistance between these colliding and vastly different cultures. The Maima village, by all archaeological accounts, represents a settlement similar to those surrounding it from the time. They made pottery similar in style and design to other White Marl Jamaican sites. They grew crops we expect to see at indigenous villages, raised and/or hunted hutia, and gathered fish and shellfish from the local reef. This way of life and subsistence strategy maintained itself through the contact and colonial eras.

Contact at Maima, though short lived at most sites and hard to locate archaeologically (Deagan 2004; Valcárcel Rojas et al. 2013), was sustained for a full year, while the marooned Spanish sailors relied on the indigenous peoples for subsistence. Despite a sustained contact, a single year is as unlikely to be represented in the archaeological record as a single moment of contact. However, the small number of Spanish nails and glass fragments found at Maima point to this initial, though difficult year for both the Spanish and the indigenous peoples. If not for the founding of Sevilla la Nueva, life may have gone back to normal at Maima, but only five years later the Spanish returned to the area to found their island capital, with the reasoning that there were close sources of indigenous labor (Woodward 2006a).

Colonialism had the most devastating and untimely destructive impact on Maima. With the enforcing of the *encomienda,* villagers were removed from their home and made to live and work at Sevilla la Nueva. Ceramic evidence from the settlement show that the indigenous peoples were not only working at the colony, but also living there based on the household items found including cassava griddles, *metates*, and Meillacoid-style pottery. As Sevilla la Nueva was an agricultural and resupply stop and did not involve mining or resource extraction, the *encomienda* likely involved agricultural and household labor. It is unclear from the historical records whether men were taken away from the island and sent to mine precious metals on other islands or whether they were made to work in the fields. It is also unclear whether women were forced into labor through the *encomienda* system or through marriage to the Spanish settlers, which was encouraged by Crown and colonial leadership to increase the settlement population (Padrón 2003, 150; Woodward 2006a).

The women that were made wives and/or domestic laborers to the Spanish were likely the manufacturers of the colonoware observed at Sevilla la Nueva, leaving their mark on the Spanish material culture, similar to that seen at Puerto Real (Deagan 2004). Whether they were able to return to their village of Maima, or whether other indigenous peoples working under the *encomienda* returned, they did not return with Spanish items. They were also likely not using the colonoware made at Sevilla la Nueva as most of it was found at the elite Spanish residence by the governor and not by the indigenous peoples making it.

This rejection of Spanish material culture at Maima is adding to a growing body of literature suggesting that the indigenous peoples resisted Spanish influence in their lives outside of the Spanish settlements. While indigenous influences on Spanish material culture can be seen through the creation of colonoware at Sevilla la Nueva, a passive resistance to the colonialism that was quickly devastating their population was occurring at Maima. Ultimately, and likely quickly, Maima was abandoned, leaving behind few traces of the devastating European contact that forever altered their lives and culture.

This type of passive resistance, the non-incorporating of European culture into the indigenous goods, whether modified or unmodified can be seen at contact and colonial sites throughout the region (Deagan 2004, 2011; Hauser and Armstrong 2012). It is unclear whether the Maima villagers were intentionally not incorporating European material culture and design into their village life or whether they did not have the opportunity, having been taken away from their homes and made to live at Sevilla la Nueva. However, putting intentionality aside as Deagan (2011) suggests, the end result is the same, the material culture of Maima remained unchanged until the abandonment of the village. The abandonment itself being a further act of resistance, escaping not only the *encomienda* but the Spanish cultural influence and devastation.

Evidence at Maima of the precontact, contact, and colonial eras make it a unique and important site for the study of the impact of colonialism in Jamaica and the Caribbean. The villagers of Maima were impacted by a long protracted contact as well as a quickly devastating colonial period that lead to the ultimate abandonment of their village. In that short time however, the villagers showed a resistance to the culture and influence of those colonizing them. By being subjected to a strained transculturation at the Spanish settlement through a forced labor system, but not bringing that culture back to the indigenous village shows a purposeful rejection of Spanish material culture and style. This response to contact and colonialism adds to the ongoing research being done on the differing and dynamic reactions by the indigenous peoples of the Caribbean to the devastating impact of colonization.

References

Atkinson, Lesley-Gail. 2003. "Jamaican Redware Revisited." Presented at *Archaeological Society of Jamaica Symposium: Zemis, Yabbas and Pewter: The Diversity of Jamaican Archaeology, Kingston, Jamaica, 3 April.*

Burley, David V., Robyn P. Woodward, Shea Henry, and Ivor C. Connolley. 2017a. "Jamaican Taino Settlement Configuration at the Time of Christopher Columbus." *Latin American Antiquity* 28 (3): 337–352.

Burley, David V., Robyn P. Woodward, Shea Henry, and Ivor C. Connolley. 2017b. *Report on the Survey and Excavations at Maima, A Late Prehistoric/Proto-Historic Taino Village in St. Ann's Bay, Jamaica.* Report Submitted to Jamaican National Heritage Trust, Kingston.

Curet, Antonio L. 2014. "The Taino: Phenomena, Concepts, and Terms." *Ethnohistory* 61 (3): 467–495.

Cotter, Charles S. 1948. "The Discovery of Spanish Carvings at Seville." *Jamaica Historical Review* 1 (3): 227–233.

Cotter, Charles S. 1970. "Sevilla Nueva: The Story of an Excavation." *Jamaica Journal* 4: 15–22.

Cotter, Charles S. N.d. "Field notes of excavations at Sevilla la Nueva."

Deagan, Kathleen A. 1988. "The Archaeology of the Spanish Contact Period in the Caribbean." *Journal of World Prehistory* 2 (2): 187–233.

Deagan, Kathleen A. 1996. "Colonial Transformation: Euro-American Cultural Genesis in the Early Spanish-American Colonies." *Journal of Anthropological Research* 52 (2): 135–160.

Deagan, Kathleen A. 2004. "Reconsidering Taíno Social Dynamics after Spanish Conquest: Gender and Class in Culture Contact Studies." *American Antiquity* 69 (4): 597–626.

Deagan, Kathleen A. 2011. "Native American Resistance to Spanish Presence in Hispaniola and La Florida, ca. 1492–1650." In *Enduring Conquests: Rethinking the Archaeology of Resistance to Spanish Colonialism in the Americas,* edited by Matthew Liebmann and Melissa S. Murphy, 41–56. Santa Fe: School for Advanced Research Press.

Faerron, Judith C. 1985. "The Taínos of Hispaniola." *Archaeology Jamaica* 85 (1): 1–4.

Floyd, Troy S. 1973. *Columbus Dynasty in the Caribbean,* 1492–1526. Albuquerque: University of New Mexico Press.

Hammond, Phillip C. 1970. "Stratigraphic and Electronic Survey: 1970 Season. Seville (Sevilla la Nueva), St. Ann's Bay, Jamaica." W.I. Kingston, Jamaica, on file at the Jamaica National Heritage Trust Commission: 93.

Hauser, Mark W. and Douglas V. Armstrong. 2012. "The Archaeology of Not Being Goverened: A Counterpoint to a History of Settlement of Two Colonies in the Eastern Caribbean." *The Journal of Social Archaeology* 12 (3): 310–333.

Henry, Shea. 2017. "Late Pre-Contact Era Taino Subsistence Economy and Diet: Zooarchaeological Perspectives from Maima." PhD diss., Simon Fraser University.

Hofman, Corinne L., Angus A.A. Mol, Menno L.P. Hoogland, and Roberto Valcárcel Rojas. 2014. "Stage of Encounters: Migration, Mobility and Interaction in the Pre-Colonial and Early Colonial Caribbean." *World Archaeology* 46 (4): 590–609.

Keegan, William F. and Corinne L. Hofman. 2017. *The Caribbean Before Columbus.* Oxford: Oxford University Press.

Keehnen, Floris W.M. 2010. "Conflicting Cosmologies: The Exchange of Brilliant Objects between the Taíno of Hispaniola and the Spanish." in *Communities in Contact: Essays in Archaeology, Ethnohistory, and Ethnography of the Amerindian Circum-Caribbean,* edited by Corinne L. Hofman and Anne van Duijvenbode, 253–268. Leiden: Sidestone Press.

Lakey, Denise, Bruce F. Thompson, Thomas J. Oertling, and Robyn P. Woodward. 1983. *The 1981 Survey of Sevilla la Nueva.* College Station: Department of Anthropology, Texas A&M University.

Lee, James W. 1980. "Arawak Burens." *Jamaica – Archaeology* 80: 1–11.

López y Sebastián, Lorenzo E. 1982. "Sevilla la Nueva (Jamaica): Un Proyecto de Arqueología Colonial." *Revista Espanola de Antropología Americana* XII: 292–300.

López y Sebastián, Lorenzo E. 1986a. "Cultural Heritage of Jamaica. Sevilla la Nueva – Archaeology." Jamaica National Heritage Trust. UNDP/UNESCO NATIONAL

PROJECT JAM/86/001. Report on file at the Jamaic National Heritage Trust, Kingston, Jamaica.

López y Sebastián, Lorenzo E. 1986b. "El Proyecto Sevilla la Nueva, Jamaica. Primera Fase." *Revista Espanola de Antropología Americana* XVI: 295–302.

López y Sebastián, Lorenzo E. 1987. "Asentamientos Europes en America: El Caso de Sevilla la Nueva (Jamaica)." Report on file at the Jamaica National Heritage Trust, Kingston, Jamaica.

Morison, Samuel E. 1942. *Admiral of the Ocean Sea: A Life of Christopher Columbus.* Boston: Little, Brown, and Company.

Morison, Samuel E. 1963. *Journals and Other Documents on the Life and Voyages of Christopher Columbus.* New York: Heritage Press.

Osborne SJ, F.F. 1973. "The Spanish Church at Seville." *Archaeology Jamaica* 73: 3–11.

Padrón, Francisco M. 2003. *Spanish Jamaica.* Kingston: Ian Randle Publishers.

Patterson, Thomas C. 2010 "Archaeology Enters the 21st Century" in *Handbook of Postcolonial Archaeology,* edited by Jane Lydon and Uzma Z. Rizvi, 133–140. Walnut Creek: Left Coast Press.

Rouse, Irving. 1992. *The Taínos: Rise and Decline of the People who Greeted Columbus.* New Haven: Yale University Press.

Silliman, Stephen W. 2005. "Culture Contact or Colonialism? Challenges in the Archaeology of Native North America." *American Antiquity* 70 (1): 55–75.

South, Stanley A. 1977. *Method and Theory in Historical Archaeology.* New York: Academic Press.

Valcárcel Rojas, Roberto, Alice V.M. Samson, and Menno L.P. Hoogland. 2013. "Indo-Hispanic Dynamics: From Contact to Colonial Interaction in the Greater Antilles." *International Journal of Historical Archaeology* 17 (1): 18–39.

Weddle, Robert S. 1985. *Spanish Sea. The Gulf of Mexico in North American Discovery 1500–1685.* College Station: Texas A&M University Press.

Wesler, Kit W. 2013. "Jamaica." In *The Oxford Handbook of Caribbean Archaeology*, edited by William F. Keegan, Corinne L. Hofman, and Reniel Rodriguez Ramos, 250–263. Oxford: Oxford University Press.

Wilson, Samuel M. 2007. *The Archaeology of the Caribbean.* Cambridge: Cambridge University Press.

Wing, Elizabeth S. 2001. "Native American Use of Animals in the Caribbean." In *Biogeography of the West Indies: Patterns and Perspectives*, edited by Charles A. Woods and Florence E. Sergile, 481–518. Boca Raton: CRC Press.

Wing, Elizabeth S. 2008. "Pets and Camp Followers in the West Indies." In *Case Studies in Environmental Archaeology*, edited by Elizabeth Reitz, C. Margaret Scarry, and Sylvia J. Scudde, 405–426. New York: Springer.

Woodward, Robyn P. 1988. "The Charles Cotter Collection: A Study of the Ceramic and Faunal Remains from Sevilla la Nueva." Master thesis, Texas A&M University.

Woodward, Robyn P. 2006a. "Medieval Legacies: The Industrial Archaeology of an Early Sixteenth-Century Sugar Mill at Sevilla La Nueva, Jamaica." PhD diss., Simon Fraser University.

Woodward, Robyn P. 2006b. "Taíno Ceramics from Post-Contact Jamaica." *The Earliest Inhabitants: The Dynamics of the Jamaican Taíno Culture*, edited by Lesley-Gail Atkinson, 161–176. Kingston: University of the West Indies Press.

Wright, Irene A. 1921. "The Early History of Jamaica (1511–1536)." *English Historical Review* 36 (141): 70–95.

Wynter, Sylvia. 1984. *New Seville: Major Dates 1509–1536; New Seville: Major Facts, Major Questions.* Kingston: National Historic Trust.

Yaeger, Timothy J. 1989. "Encomienda or Slavery? The Spanish Crown's Choice of Labor Organization in Sixteenth-Century Spanish America." *Journal of Economic History* 50: 842–859.

CHAPTER 5

European Material Culture in Indigenous Sites in Northeastern Cuba

Roberto Valcárcel Rojas

The handling of European objects by the indigenous peoples of the Antilles has been generally perceived as the naive and enthusiastic reception of an exotic and complex materiality by primitive and technologically-backward people. The picturesque image of gold exchanged for ceramic fragments, hawk bells, or pins, is recurrent. Also recalled is when the *cacique* Guacanagarix wore the gloves Christopher Columbus gave him, or the moment when the bravest of the indigenous chiefs in Hispaniola, Caonabo, was deceived and let Alonso de Ojeda put shining brass shackles on him. In recent years, in archaeology and anthropology, a vision has been constructed that intends to explain some of these situations by contextualizing them in systems of value to local societies. Many of the European objects were received in this way because they found room in the indigenous symbolic setting, or shared similarities with their concepts of the sacred and that imbued them with supernatural power (Keehnen 2012; Oliver 2000; Valcárcel Rojas and Martinón-Torres 2013). On the other hand, their transfer often worked as a complement to the establishment of ties of friendship and alliance, both of which indigenous peoples and Europeans were interested in promoting (Szászdi León-Borja 2015).

The barter or *rescate* of indigenous goods in exchange for European objects, as well as the act of gift-giving on both sides, is associated with certain objects, forms of interaction, and periods. It is in this sense that the interest of the Antillean communities in European metals such as brass can be explained. Brass resembled *guanín*, an indigenous alloy of gold and copper considered highly valuable since it came from very distant places – Colombia, to be precise. It was associated with mythic narratives, and with the concepts of brilliance and the numinous that marked the sacred in the perspective of diverse indigenous societies from the Antilles and South America (Oliver 2000; Saunders 1999; Valcárcel Rojas 2016a, 206–222; Valcárcel Rojas and Martinón Torres 2013). The acquisition of these objects at the beginning of the encounters and conquest was framed by processes where the indigenous peoples preserved their capacity for negotiation and choice, and in circumstances in which their traditions, symbolic conceptions, and value systems were fully functioning.

Nonetheless, the transfer of objects through *rescate* and gifting was not the only way for European material culture to enter the Antillean indigenous universe. This chapter reflects on how this process was readjusted when the colonization strategy, strengthened by the experiences in Hispaniola, focused on the rest of the Greater Antilles. The expansion to these islands – which are the object of this analysis – was undertaken with a perspective that reduces the time for negotiation and contact with the indigenous inhabitants, moving quickly to take control of the population, and to the imposition of forced labor (Valcárcel Rojas 2016a, 24). In this environment, the European objects, the circumstances of their transfer, and the attitudes of the indigenous themselves at the moment of receiving them and using them, were all different. This results in diminished indigenous access to the goods used by the Europeans for gifts and *rescate* – part of the so-called gift kit – and the expansion of the use of other kinds of artifacts.

We assess this idea using historical and archaeological data from the Greater Antilles, primarily Hispaniola, Puerto Rico, and Cuba, as these are the most accessible. Some results of an ongoing study on archaeological material from the northeast of Cuba, developed by the author as part of the ERC-synergy project NEXUS1492, are also presented.

1 From Gifts and Barter to the *Encomienda*

The first processes of transfer or acquisition of European material culture by the indigenous peoples of the Caribbean took place in an environment of relationships where European domination had not yet been imposed. This was the so-called *situation of contact* (Valcárcel Rojas 2016a, 21). Specifically, the societies in contact carry with them their own experiences of socialization and interaction with strangers, as well as criteria of value originating in their respective socio-cosmic universes, which necessarily determine how the interactions occur, their content, and their feasibility (Mol 2007).

Both indigenous peoples and Europeans recognized the importance of these processes of interaction with others, and handled their own protocols – previously developed in their respective cultural environments – that included the giving of goods as gifts or as a part of exchanges. This course of action should have facilitated relationships, by indicating an understanding of the importance of the counterpart, and demonstrating one's own status and the capacity to own goods. Although the true intention of the gestures was often not understood by either side, generally, the act of giving goods was reciprocated. Afterwards, the traded objects flowed through both groups' interaction networks in an independent and parallel manner. In the case of the indigenous communities, the European objects tended to be controlled by the elites and quickly reached their exchange networks, often preceding the arrival of the Europeans themselves (Wilson 1990, 4).

At the same time, giving and receiving originated with the purpose of obtaining goods from the other party. Thus, the value of the objects that were given and obtained was more important than the prestige and networking that the exchange may have offered the traders. This is the action of *rescate* itself, and it is based on the great differences of value that each side attributed to the goods in movement; each side perceived that they obtained something much more valuable than what they gave. *Rescate* generally takes place between societies with different levels of technological development. Often the more technologically advanced party will subordinate the other through coercive and violent acts that may include plundering and stealing (Lacueva 2012, 547).

In the Columbian perspective, following the schemes used by the Portuguese in Africa, the commercial activity related to the *rescate* of gold, spices, and goods of European interest, would have to have a sufficiently broad and productive scale to become a key part of the economic activity in recently discovered lands. The colonizers' primary goal was not populating these lands. Instead, they wanted to implement a commercial-military system oriented at resource extraction, including slavery and the sale of indigenous peoples. For the purpose of using them in the *rescates* or trade on Hispaniola, goods that had proven useful in former transactions undertaken by the Europeans in other places were imported, such as glass beads, hawk bells, and garments, among others. In practice, objects of many kinds were used, mainly those of ceramic and metal that often were damaged or were fragments of larger pieces. Columbus attempted to regulate and achieve the control of the *rescates* with the goal of channeling to himself and the monarchs the greater profit, and used gift-giving selectively in order to build ties and denote his position.

When the indigenous peoples of Hispaniola rejected ties to the Europeans, and it became clear that their incorporation into the commercial system Columbus intended to establish was not viable, as it was beyond their productive capacities to procure the gold expected of them, a general tribute was imposed in the wake of military subjugation of the island (Cassá 1992, 179–186). Henceforth, the Europeans moved to the appropriation of indigenous labor, a move finally formalized as the *encomienda* system between 1503 and 1505 (Valcárcel Rojas 2016a, 23). Under these new circumstances, *rescate* as a commercial option, and object exchange as a means to foster or sustain alliances and friendships, declined and disappeared. According to some scholars, the end of these kinds of interactions occurred around 1497 when tribute collection collapsed and the crisis of indigenous structures and their leadership and population became evident (Keehnen 2012, 128).

The *encomienda* implied an assignment-subordination of a group of indigenous peoples to a Spanish colonist (*encomendero*), so they would work for his benefit several months per year. The indigenous peoples were supposed to receive religious and civilizing education in return. According to Bartolomé de Las Casas (1875c, 78), in 1503 the Queen ordered that the *indios* from Hispaniola be paid for their work. Payment was done with objects such as beads, *espejuelos* (fragments of mirror), and combs, although rarely did the Spaniards really comply with the payment. The indigenous peoples called this payment the *cacona*.

In 1512, the Laws of Burgos – "for good treatment of the *indios*" – attempted to regulate this process, specifying forms of monetary compensation and giving of goods to the *indios*. The law specified that the indigenous peoples should be given land to start living close to the Spaniards, and that, in order to facilitate their subsistence, the *encomendero* should provide the community under his control with maize to sow, a dozen hens, and a rooster. The *encomendero* would also give a hammock to each *indio* and a gold *peso* per year to acquire clothes; the *cacique* was required to receive better treatment in regard to his payment and clothing (Muro 1956). Las Casas (1875c, 435) comments that with such a poor payment it was impossible to buy clothes. Diverse historical references regarding Cuba, Hispaniola, and Puerto Rico, show that during the term of labor, especially in mining and agriculture, the *encomenderos* supplied the *indigenes* with food and the necessary tools for work, although some cases are mentioned where the indigenous peoples had to use their own wooden and lithic tools and supply their own food (Sued Badillo 2001, 325).

2 Things for *Indios*

In the Antilles, the use of objects for *rescate* or trade with the *indios* as well as for gifts, was a practice derived from European commercial strategies, particularly, the ones practiced by the Portuguese in Africa. In fact, Columbus requested merchandise for the *rescate* when he presented his project of traveling to the Indies to the King of Portugal, as well as to the monarchs of Castile and Aragon. The required objects were: hawk bells, brass chamber pots, brass sheets, bead strings, mirrors, scissors, knives, needles, pins, linen shirts, colored coarse cloths, and colored bonnets (Las Casas 1875a, 219, 237).

In the diary of Columbus' first voyage (1492), he registers *rescates* and gifts in the Bahamas, Cuba and on Bohío Island, later called Hispaniola. The interest of the indigenous peoples in everything coming from the Europeans stands out. On his second voyage, the main goals were the development of commercial activity, and the establishment of spaces where these activities could be carried out. It is for this reason that this voyage was meticulously planned, including the material resources to be used. On this occasion, the *rescates* were important not only to obtain gold – of which they obtained significant amounts – but also food (Álvarez Chanca 1977, 92; De Cúneo 1977, 30).

In the case of the Spanish, particularly their circum-Caribbean voyages, the *rescate* system functioned within the plan of discovery and to a lesser degree during the action of conquest. Nonetheless, the *rescate* system tended to diminish and disappear when settlement was considered, and especially, once control over local societies had been achieved. In fact, there is an entire group of voyages, undertaken between 1499 and 1502, that are denominated "Lesser voyages" or "Andalusian voyages of discovery and *rescate*" aimed at obtaining geographic information and discovering new lands. These were complemented by the action of bartering with the *indios* with the goal of producing economic dividends, but also facilitating communication and friendly relations. The *rescate*, especially of gold, pearls, *guanín*, and slaves, was regulated in regard to the areas where it could be practiced, the taxes to be paid for the benefits obtained, and the circumstances in which it was to be carried out (Alonso 2005; Gutiérres 2009). Logically, apart from the *rescate*, the *indigenes* could also have obtained diverse objects abandoned by Europeans or salvage them from shipwrecks.

The set of objects to be used in *rescates* and gifts, or the gift kit, according to Brain (1975, 130), continued to develop throughout the sixteenth century and it would be used in multiple areas of the New World into the seventeenth century. This set would mainly consist of glass beads and hawk bells, with variations that included the incorporation of artifacts in regard to regional or cultural preferences. As a system, it would become more nuanced according to the characteristics of the indigenous societies and their level of autonomy; in some cases, it would play a fundamental role in the transformation of these societies. The entry of new European powers into the discovery, conquest, and colonization processes would also give the system diverse forms,[1] influencing the formation of large indigenous exchange networks, the transformation

1 The diverse expressions of trade and gifts, depending on the geographical area, historical moment, type of context and other factors, generate diverse archaeological patterns (see Dalton-Carriger 2016; McEwan and Mitchem 1984; Worth 2016).

of the links between communities and the balance of leadership and power between these societies (Gassón 1996; Smith 1984).

In documentary terms, Hispaniola is the place in the Greater Antilles where the issues of *rescate* and gifting can be best studied. Information about Puerto Rico, Cuba, and Jamaica is relatively scarce, compared to Hispaniola. The data for Cuba and Jamaica principally come from Columbus' voyages (Álvarez Chanca 1977; Colón 1961; De Cúneo 1977; Las Casas 1875b, 62). Until 1508, when the conquest of these islands began, colonial actions remained focused on Hispaniola, and contact with the peoples of these outlying territories was sporadic, which meant that the *rescates* occurred on a limited scale.

The action of conquest in these outlying territories followed a different scheme from the one on Hispaniola, partly due to the experiences of dominion found there. By combining negotiations with military actions, a situation of Spanish predominance was attained more swiftly, while the *rescate* and gift giving acquired a secondary character. The main goal in these cases was not obtaining goods by means of *rescate*, but controlling the indigenous peoples and using them for gold extraction, or to produce resources and services that were necessary for the colonial project. The control of the workforce was reached through diverse methods, generally through the mediation of *caciques*. The *encomienda* was imposed in a relatively short time: in 1509 in Puerto Rico, in 1513 in Cuba, and in 1515 in Jamaica (Cassá 1992, 224–237; Fernández 1966; Marrero 1993, 168).

With the beginning of the *encomiendas,* the transfer of material culture to the indigenous peoples seems to focus on giving clothes as a payment for services. In 1506, on Hispaniola, shirts and other garments were given to the *caciques* Yaguax and Caicedo as payment for their work on constructions in the city of Santo Domingo (Mira Caballos 2000, 107). In Cuba, Gonzalo de Guzmán rewarded his *indios* in Bocas de Bani with 50 shirts and tools for the service they gave in the recovery of goods from a shipwreck (Real Academia de la Historia 1888:4, 233). The payment of *cacona* in Cuba was still mentioned in 1544 (Sarmiento 1973, 99).

A detailed record of payments to the indigenous peoples under the *encomienda* in Puerto Rico, between 1510 and 1519, is well known (Tanodi 1971). The *cacona* given to different *caciques* and their people, consisted mainly of garment elements: bonnets, pointed hoods, doublets, smocks, petticoats, long underwear, *servillas* (a thin-soled shoe), shirts, *zaragüelles* (knee length, open bottomed shorts), breeches, head cloths, shoes, espadrilles, and belts. Combs, mirrors and glass bead strings were also given in lesser quantities. This final group of objects was almost exclusively given to women, and in most cases, to female *caciques, caciques*' wives, and their captains. Their use as payment is recorded during the first years, and then gradually decreases. There are records of shipments of hawk bells to the island during this period, but they are not referred to as part of the *cacona*.

The existence of payments in the context of the *encomienda* does not exclude gift giving and perhaps bartering situations. Nonetheless, this seems to be rare, and the materiality used points to interests that are closer to the perspectives of appearance and life that were imposed by Europeans, and that the indigenous peoples were progressively incorporating. For example, in Cuba, Diego Velázquez bequeathed clothes to his *indios* in his will, and, in 1533, garments were given in Hispaniola to the rebel *cacique* Enriquillo and his group of renegade *indios* during the peace negotiations the Spaniards held with him (Fernández de Oviedo 1851, 146; Torres de Mendoza 1880, 518).

3 The Archaeological Record

A large part of the indigenous sites with European material identified in the Antilles are found in Cuba and Hispaniola. In the case of The Bahamas, the scarce material found (ceramic fragments, a coin, nails, glass beads, knife blades, etc.), mainly at the Long Bay and Three Dog sites, has been considered as typical of an interaction where the *indigenes* seem to have had some autonomy (see Berman and Gnivecki, this volume). They are explained as: (1) typical of initial *rescate* activity; (2) goods obtained in the context of indigenous exchange networks, which quickly incorporated exotic objects from the Spaniards; and, (3) those recovered by the *indigenes* from a Spanish shipwreck, or from a place where the Spaniards abandoned them (Blick 2014; Gnivecki 2011).

In Puerto Rico, according to Anderson-Córdova (2005, 350–351), and Deagan (1988, 205), few sites with this peculiarity have been confirmed. The sites on Mona Island are notable for the variety of artifacts that have been found (from glass beads to coins), and because they suggest diverse forms of indigenous manipulation across a long period of time (from AD 1493 to 1590) (Cooper et al. 2016; Samson and Cooper 2015). In Jamaica, there is mention of three indigenous sites with European materials, all of them close to the Spanish settlement of Sevilla la Nueva (Deagan 1988, 205). Recent studies in one of these sites, Maima, suggest forms of resistance associated with a limited use of Spanish material culture (see Henry and Woodward, this volume).

Data from Hispaniola regarding anywhere from 18 to 32 sites (see Keehnen, this volume; Valcárcel Rojas 2016b, 221) are scarce and in only a few cases come from detailed studies (Ernst and Hofman 2015; Hofman et al. 2014; Keehnen

2012, this volume; Valcárcel Rojas et al. 2013; Vega 1987; Samson 2010). These sites can be summarized as follows:

- There was a limited presence of European objects, with fragmented ceramics and diverse metal fragments or small metal objects (sheets, rings, pins, buckles) occurring most frequently. Fragments of glass objects are reported but in most cases their chronology is unknown. Some authors mention frequent reports of glass beads (García Arévalo 1978), however, they are very rare in known sites (see Keehnen this volume). There are few references to weapons or tools (García Arévalo 1978; Ortega 2005). Keehnen (2012, 158) relates the limited presence of European objects to the quick incorporation of these artifacts into indigenous networks of interaction and exchange, producing a wide spatial dispersion. The idea of a quick incorporation and wide distribution of these items throughout the island, in virtue of the intensity of exchange processes with the indigenous population, has also been used to explain their scarcity in the Spanish settlement of La Isabela, where a great quantity of these objects was brought during Columbus' second voyage (Deagan 2002).
- There were European ceramic vessels and fragments, together with indigenous vessels, as funerary offerings in indigenous burial grounds (García Arévalo 1978); depositing of European objects (brass sheets, rings, buckles, hawk bells) together with valuable indigenous objects in protected places of potential ceremonial meaning (Ortega and Fondeur 1976; Vega 1987, 30–31); incorporation of these artifacts and materials (mirrors, jet beads, brass pins and metal sheets) into indigenous objects of great symbolic importance such as idols and cotton belts (Ostapkowicz 2013, 303; Vega 1987, 31); manufacture of ritual objects from Spanish materials (lead three-pointers) (García Arévalo 1978). These cases indicate the importance of the artifacts and European materials by virtue of their exoticism and potential link to indigenous concepts of the sacred, as well as a use focused on the elites and the ceremonial and ritual worlds.
- There were rare imitations of European ceramic vessel forms using indigenous technology and local materials.

In the case of Cuba, about 30 indigenous sites with European artifacts are known[2] (Valcárcel Rojas 2016b, 221). The four that are not in the east of the island are characterized by small quantities of artifacts, which are mostly

2 There is information on European artifacts from other indigenous sites but because of the chronology of the materials and their stratigraphic positioning their use by the *indigenes* is doubtful.

ceramics (Knight 2010; Pendergast 2003; Tomé and Rives 1987). In the eastern region, the sites are mainly concentrated in the northern area, in the modern Holguín Province where 20 sites are reported. In the remaining areas of eastern Cuba, the common pattern is the presence of ceramic. Occasionally, horseshoes and fragments of diverse metal artifacts are found in small quantities (Martínez Arango 1997; Morales Patiño and Pérez de Acevedo 1945; Romero 1981).

In the following, I assess the available information from the sites in present-day Holguín Province (Rouse 1942; Valcárcel Rojas 1997) and the direct analysis from seven of these sites of European artifacts and artifacts associated with the interaction between indigenous and Europeans. This is one of the Cuban regions where the indigenous societies with Meillacoid pottery settled early on (Aguas Gordas site: MO-399, 1000 ± 105 BP; 2 sigma cal AD 801–1258) and reached greater demographic and cultural power. Although Spanish settlements were not founded there until the eighteenth century, there are signs of the existence of several *encomiendas* and the presence of indigenous descendants until the nineteenth century (Valcárcel Rojas 2016a, 67–71).

At these 20 sites, there are 43 artifacts generated by the indigenous use of European material or the imitation of European shapes or elements, 2201 European objects, and 10 non-Antillean indigenous objects. These last were potentially imported during the colonization process of the island. There are also objects that could not be quantified, so for many types of objects the quantity was greater than what was analyzed. For instance, many artifacts could often not be identified because of their deteriorated condition.

These materials come from surface finds and controlled excavations. Several are associated with middens and in one case (El Chorro de Maíta) with funerary context. Eleven of these sites have clear evidence of a settlement that was initiated prior to European arrival, although it is possible that this was also the case in the remaining sites. These are mainly habitational sites, some with occupations of several centuries and signs of settlements of large proportions and socio-political importance, in particular Potrero de El Mango and El Chorro de Maíta (Persons 2013; Valcárcel Rojas 2002).

At eight sites, the indigenous manipulation of European material is evidenced (beads manufactured from majolica fragments, modified ceramics, an indigenous axe made of iron, pendants made of metal sheets, etc.). Also, these sites have objects that were crafted to imitate European shapes or elements (imitation of vessel shapes and candlesticks) (Figure 5.1). Most of these materials are found at only two sites: El Yayal and El Chorro de Maíta. The most common are the imitated European vessel forms and the modified European ceramic fragments for diverse purposes.

FIGURE 5.1 Indigenous vessel that copies a European form. El Yayal site, Cuba
PHOTO BY ROBERTO VALCÁRCEL ROJAS

Ceramics are the most reported objects, being found at 16 of the 20 sites. Of these the early-style olive jar, Columbia Plain majolica, and *melado*-type lead glazed coarse earthenware stand out (Figure 5.2). Whole vessels have been identified only in a few cases, although the great volume of fragments found at some sites suggests that the indigenous peoples had access to whole vessels with some frequency. Most common are glazed ceramics.

Glass objects are infrequent. Glass beads are found at only four sites; usually one to two pieces. At the Alcalá site alone, nine pieces of Chevron and Nueva Cádiz beads, and a necklace of 103 unidentified blue glass beads were found. In El Chorro de Maíta 105 red coral beads and one jet bead were found. The latter were associated with human burials and with circumstances that suggest their use as part of the evangelization process of the indigenous population (Valcárcel Rojas 2016a, 258).

Aglets and pins are rare. They were only found at two sites, and in the case of El Chorro de Maíta, they seem to be linked to the use of clothes by the indigenous peoples (Valcárcel Rojas 2016a, 222). Equally scarce is the evidence of

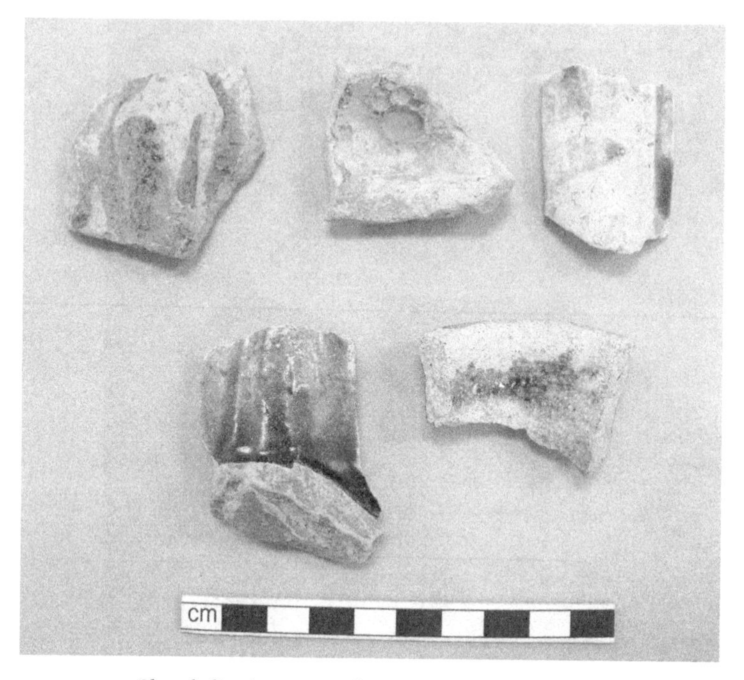

FIGURE 5.2 Glazed olive jar ceramic fragments. El Chorro de Maíta, Cuba
PHOTO BY ROBERTO VALCÁRCEL ROJAS

other objects of personal use such as buckles, belt hooks, and rings. There are 52 sheets of brass or non-ferrous metal distributed across seven sites, which indicates popularity in the use of these objects and material (Figure 5.3).

Rumbler bells (hawk bells) appear at four sites, with El Yayal having eight of the 13 recorded pieces. This site also provided six of the nine identified coins, all from the sixteenth century. In regards to architectural and furniture elements, as well as locks, El Yayal together with Alcalá and El Chorro de Maíta (the latter in lesser proportion), have most of the pieces, although they are also present at other sites. In total, there are seven locations where these objects are reported. Architectural and furniture elements are the most common metallic pieces found (85 objects), especially nails (74 of them).

Tools were found at eight sites, for a total of 38 pieces. The most common among the pieces are: knives, scissor blades, and iron chocks. Among the identified artifacts, there are hoes, axes, chisels, and pliers (Figure 5.4). There are eight sites with 38 armament-related artifacts – mainly fragments of edged weapon blades. As for horse riding gear, there are 52 pieces distributed

FIGURE 5.3 Sheets of non-ferrous metal, possibly brass. El Yayal site, Cuba
PHOTO BY ROBERTO VALCÁRCEL ROJAS

FIGURE 5.4 European tool. El Yayal site, Cuba
PHOTO BY ROBERTO VALCÁRCEL ROJAS

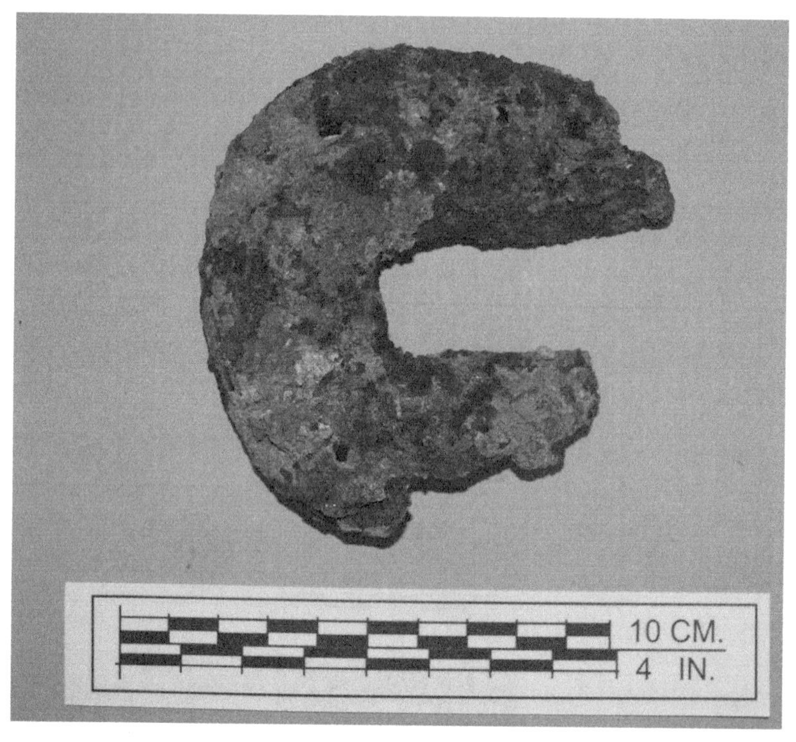

FIGURE 5.5 Horseshoe. Alcalá site, Cuba
PHOTO BY ROBERTO VALCÁRCEL ROJAS

across 6 sites, where the predominant objects are horseshoes or their fragments (Figure 5.5).

Non-Antillean indigenous evidence was found only at two sites. The evidence found in El Chorro de Maíta, which consist of *guanín*, Mexican Red Painted and Aztec IV ceramics, and non-Antillean indigenous pottery (the majority of all the recorded objects) seems to be linked to acts of importation promoted by the Europeans. Its use in several burials, with diverse indicators of high status, is associated with the processes of interaction with local elites promoted by the Spaniards. The presence of human remains of non-Antillean or non-local indigenous individuals, considered slaves, could also be linked with the presence of these objects (Valcárcel Rojas 2016a, 300).

The materials from these 20 sites are numerous and show great diversity. This fact suggests different schemes of use and acquisition of this materiality, potentially related to the distinct functions and chronologies of the sites; it also points to diverse modes of interaction with the Europeans, particular to the circumstances of conquest and colonization in Cuba. In any case, this is a panorama that is different from what is known – still very tentatively – from the other islands, especially Hispaniola.

Despite differences in the types and quantities of objects found at the studied sites in the northeastern part of Cuba, there are some general tendencies that are useful to assess processes of interaction. Except for brass sheets and non-ferrous metal, the components that are usually considered as part of the gift kit (glass beads, hawk bells, rings, and personal accessories), are limited in terms of quantity and reporting at sites. Although it cannot be dismissed that some of these and other objects relate to *rescate* or indigenous acquisition in circumstances of autonomy, it is likely that at some sites their presence is due to links with certain elite indigenous individuals receiving payment for services or the promotion of religious attitudes; e.g., glass bead necklaces at Alcalá and coral bead necklaces at El Chorro de Maíta. In any case, the secondary character of the action of *rescate* is shown as well as the probability that the presence of these artifacts indicates other functions.

Artifacts not typical of the gift kit, at least in the Antillean case, such as weapons, tools, and architectural and horse riding elements, are quite common and are found even at sites with small collections. This points to a type of interaction that was frequent and where these objects were important. It also indicates that these were objects that were used in places where Europeans and indigenous peoples interacted in a regular basis, or in indigenous villages that were under European control, rather than being used more causally for exchange or gift-giving. In their diversity, they point to a profile associated with the process of colonization that could be correlated to spaces of labor and settlement. In the rural universe, farms and mines developed material culture that could be transferred to the indigenous peoples and their villages; even in some cases, indigenous settlements were modified and refashioned as mining camps and annexes to Spanish farms. There are historical records, as the one mentioned above, regarding the payment of Gonzalo Guzmán to his *indios* that evidence the giving of tools to the natives.

Considering the historical and archaeological information on the *encomiendas* in this region, and the distribution patterns of material culture mentioned above, it is possible that many of the objects found at the sites in northeast Cuba were part of the *cacona*. In this sense, the evidence for clothing used among the indigenous individuals buried at the cemetery of El Chorro de Maíta is important (Valcárcel Rojas 2016a, 222).

In the cases where an assessment can be made of the way Spanish materiality was used, an attitude of sacralization or valuation of exoticism does not arise. These European-indigenous artifacts regularly appear with indigenous objects and frequently were treated as waste. In El Chorro de Maíta, where the population was left under an *encomienda* regime and a cemetery of colonial character was established, the way in which materiality and religious, funerary, and European civilizing codes were treated points to an adjustment to the norms of the colonial setting. European material culture is apparently integrated as part of the dress code along with Christian religious paraphernalia (coral beads and jet). A portion of the objects incorporated into the burials, those of non-Antillean indigenous character (such as *guanín*), are related to local symbolic conceptions and are particular to the elite, who incorporated Christian religiosity in a way that diversified the funerary ritual.[3] This is an acquisition that belongs to a process that does not exclude old codes, but from which there emerges a new individual and a new ethno-cultural entity: the *indio* (Valcárcel Rojas 2016a, 323).

The sites of El Yayal, Alcalá, and El Chorro de Maíta, concentrate the largest and most diverse collections. They seem to follow a pattern where the access to European material is related to the labor functions of these populations and spaces, and to the possibility that these belong to *encomienda* contexts. In fact, the study of diverse lines of evidence has determined this to be true at El Chorro de Maíta (Valcárcel Rojas 2016a).

This is not necessarily the nature of the remaining sites, although the fact that they have a less important material record does not exclude the possibility that their inhabitants may have been under this type of system. As the *encomienda* moved the labor force out of its villages, it may have been the case that some communities, which lived under this system, incorporated limited Spanish materiality into their original life spaces. Likewise, it is possible that any of these sites may have incorporated Spanish material in circumstances prior to the establishment of the *encomienda*. In the case of El Yayal and Alcalá, there are historical references that suggest that they could have kept functioning as spaces with indigenous presence or that of their descendants, beyond the end of the *encomiendas* in 1553. This may have influenced the complexity and peculiarity of their archaeological record (Valcárcel Rojas 1997). In these circumstances, European material culture might be connected to the presence of Spaniards themselves at such sites.

The behaviors identified at the sites in Holguín Province indicate the importance of this region in the conquest and colonization processes in Cuba, as

3 The incorporation of European objects to indigenous burials, signaling the social position and reinforcing local traditions, is widely documented in other parts of the continent (Hally 2008; McEwan and Mitchem 1984). The process is modified or totally transformed, to the same extent that the control of the Europeans over the indigenous populations increases (Graham 1995, 1998; McEwan 2001); this does not exclude situations of persistence, resistance, and syncretism.

well as the intensity of the link between Europeans and indigenous peoples. The discovery of this evidence is undoubtedly associated with the prevalence of archeological research in this area, and such evidence probably is not exclusive to this region. This outlook indicates that forms of interaction and use of European materiality that included independent ways of accessing it, and its incorporation into local cultural codes, were possible. However, these forms did not have the same relevance that can be observed in Hispaniola, and were soon substituted by others that were imposed or generated in the context of circumstances of Hispanic domination, which could have been channeled by the indigenous peoples to facilitate their existence and adaptation to the colonial context.

4 Conclusions

Available historical data, particularly from Puerto Rico, as well as archaeological information from Hispaniola and that derived from the Cuban case, support the idea that on Puerto Rico and Cuba transfer schemes of material culture to the *indigenes* were different from those on Hispaniola. A similar situation probably occurred on Jamaica. This is due to a colonization process that did not have the goal of obtaining goods through *rescates*, but rather, by extracting gold and other resources through the concentrated labor of the local population. In these circumstances, the link with the indigenous peoples originated from a position of domination, and the material culture that passed to the indigenous peoples had other characteristics and was transferred through different channels. In contrast with many parts of North and South America, in the Antilles the *rescate* did not develop into the consolidation of systems of interaction and the creation of new and large indigenous exchange networks, but rather, it collapsed or readjusted with the imposition of forced labor.

The act of *rescate* and gifting ceases or diminishes when the *indigenes* perform forced work and their community is disarticulated. This indicates that the gift kit loses relevance, or that its objects were given from a different perspective: mainly as payment for work or service provided. On the other hand, the indigenous perspective of the Spaniards and their material culture was marked mainly by a context of dominance. This created potentially different attitudes among the dominated groups, which sought more to take advantage of the technology and other resources of the Spanish (e.g., use of tools and weapons [reported by indigenous rebels], and consumption of pork [*Sus scrofa*]), or to adapt to their codes of ornamentation and appearance (e.g., use of clothing), instead of the symbolic acquisition of their goods. However, both schemes may have coexisted at certain times, or a circumstance may have arisen that did not totally abandon indigenous cultural conceptions, but evidences their transformation. A rejection of the Hispanic artifacts must also have been a common approach.

The archaeological pattern observed in northeast Cuba suggests that on other islands where colonization occurred with similar characteristics, there could be similar archaeological contexts. This could have also occurred in spaces of *encomienda* on Hispaniola, or at archaeological sites associated with periods where this system became predominant.

Considering these details, it becomes evident that, in the case of the Greater Antilles, the discovery of European materials at indigenous sites, including the case of Hispaniola, cannot always be interpreted as an expression of *rescate* or trade activities. Neither can the presence of brass sheets, beads, or hawk bells always be associated with the gift kit. The analysis of these archaeological contexts and artifacts requires a temporal and cultural assessment that is historically contextualized. This will not only prevent mistakes, but will allow us to appreciate the true complexity of the impact made by the Europeans in the region, and the richness of indigenous responses and attitudes.

Acknowledgments

The research leading to the results presented in this chapter is part of the ERC-Synergy NEXUS1492 project and was funded by the European Research Council under the European Union's Seventh Framework Programme (FP7/2007–2013)/ERC grant agreement no. 319209. The team of the Departamento Centro Oriental de Arqueología de Holguín provided their collaboration in the study of the Cuban material. I want to thank Jana Pesoutova and Jaime Pagán Jiménez for their support and help in accessing documentary sources from Puerto Rico, as well as Corinne Hofman and Floris Keehnen for their invitation to the symposium for which this text has been created. Valeria Corona and Konrad Antczak are thanked for translating the original Spanish text into English. In regards to the translation the help of Vernon James Knight is also appreciated.

References

Alonso, Icíar A. 2005. "Explorar, conocer: los intérpretes y otros mediadores en los viajes andaluces de descubrimiento y rescate." In *Estudios sobre América: Siglo*

XVI–XX, edited by A. Gutiérrez Escudero and M.L. Laviana Cuetos, 515–528. Sevilla: Asociación Española de Americanistas.

Álvarez Chanca, Diego. 1977. "Carta de Diego Álvarez Chanca al ayuntamiento de Sevilla." In *El segundo viaje de descubrimiento*, edited by Fernando Portuondo, 57–97. La Habana: Editorial de Ciencias Sociales.

Anderson-Córdova, Karen F. 2005. "The Aftermath of Conquest. The Indians of Puerto Rico during the Early Sixteenth Century." In *Ancient Borinquen. Archaeology and Ethnohistory of Native Puerto Rico*, edited by Peter E. Siegel, 335–352. Tuscaloosa: The University of Alabama Press.

Blick, Jeffrey P. 2014. "El caso de San Salvador como sitio del desembarco de Colón en 1492. Principios de Arqueología histórica aplicados a las evidencias actuales." *Cuba Arqueológica* 7 (2): 29–49.

Brain, Jeffrey P. 1975. "Artifacts of the Adelantado." In *Conference on Historic Site Archaeology Papers* 8, edited by Stanley South, 129–138. Columbia: The South Carolina Institute of Archaeology and Anthropology, University of South Carolina.

Cassá, Roberto. 1992. *Los indios de Las Antillas*. Madrid: Editorial Mapfre.

Colón, Cristóbal. 1961. *Diario de Navegación*. La Habana: Comisión Nacional Cubana de la Unesco.

Cooper, Jago, Alice V.M. Samson, Miguel A. Nieves, Michael J. Lace, Josué Caamaño-Dones, Caroline Cartwright, Patricia N. Kambesis and Laura del Olmo Frese. 2016. "The Mona Chronicle: The Archaeology of Early Religious Encounter in the New World." *Antiquity* 90 (352): 1045–1071.

Dalton-Carriger, Jessica N. 2016. "New Perspectives on the Seventeenth-Century Protohistoric Period in East Tennessee: Redefining the Period through Glass Trade Bead and Ceramic Analyses." PhD diss., University of Tennessee.

Deagan, Kathleen A. 1988. "The Archaeology of the Spanish Contact Period in the Caribbean." *Journal of World Prehistory* 2 (2): 187–233.

Deagan, Kathleen A. 2002. "La Isabela y su papel en el paradigma inter-atlántico: la colonia española de la isla Española (1493–1550) desde la perspectiva arqueológica." Paper presented at *The XV Coloquio de Historia Canario-Americana, Las Palmas de Gran Canaria*.

De Cúneo, Miguel. 1977. "Carta de Miguel de Cúneo." In *El segundo viaje de descubrimiento*, edited by Fernando Portuondo, 19–56. La Habana: Editorial de Ciencias Sociales.

Ernst, Marlieke and Corinne L. Hofman. 2015. "Shifting values: a study of Early European trade wares in the Amerindian site of El Cabo, eastern Dominican Republic." In *Global Pottery 1. Historical Archaeology and Archaeometry for Societies in Contact edited by* Jaume Buxeda i Garrigós, Marisol Madrid and Javier G. Iñañez, 195–204. Oxford: British Archaeological Reports, International Series 2761.

Fernández de Oviedo y Valdés, Gonzalo. 1851. *Historia General y Natural de Las Indias, Islas y Tierra Firme de la Mar Océano* Vol.1. Madrid: Imprenta de la Real Academia de la Historia.

Fernández, Eugenio. 1966. "La encomienda y la esclavitud de los indios de Puerto Rico. 1508–1550." *Anuario de Estudios Americanos* 23: 377–443.

García Arévalo, Manuel A. 1978. "La Arqueología indohispana en Santo Domingo." In *Unidad y variedades. Ensayos en homenaje*, edited by José M. Cruxent, 7–127. Caracas: Centro de Estudios Avanzados.

Gassón, Rafael A. 1996. "La evolución del intercambio a larga distancia en el nororiente de Suramérica. Bienes de intercambio y poder político en una perspectiva diacrónica." In *Chieftains, Power and Trade: Regional Interaction in the Intermediate Area of the Americas*, edited by Carl H. Langebaek and Felipe C. Arroyo, 133–154. Bogotá: Universidad de los Andes.

Gnivecki, Perry L. 2011. "Text and context: the Spanish contact period in the Bahama archipelago." Paper presented at the 14th Symposium on the Natural History of the Bahamas, San Salvador.

Graham, Elizabeth. 1995. "A Spirited Debate: Why Did the Maya Convert to Catholicism?" *Rotunda* 28 (2): 18–23.

Graham, Elizabeth. 1998. "Mission Archaeology." *Annual Review of Anthropology* 27: 25–62.

Gutiérres, Antonio E. 2009. "Las capitulaciones de Descubrimiento y Rescate. La Nueva Andalucía." *Araucaria. Revista Iberoamericana de Filosofía, Política y Humanidades*21: 257–276.

Hally, David J. 2008. *King: The Social Archaeology of a Late Mississippian Town in Northwestern Georgia.* Tuscaloosa: University of Alabama Press.

Hofman, Corinne L., Angus A.A. Mol, Menno L.P. Hoogland, and Roberto Valcárcel Rojas. 2014. "Stage of Encounters: Migration, Mobility, and Interaction in the Precolonial and Early Colonial Caribbean." *World Archaeology* 46 (4): 590–609.

Keehnen, Floris W.M. 2012. "Trinkets (f)or Treasure?. The role of European material culture in intercultural contacts in Hispaniola during early colonial times." Unpublished Research Master thesis, Leiden University.

Knight, Vernon J. 2010. "La Loma del Convento: Its centrality to current issues in cubanArchaeology." In *Beyond the Blockade. New Currents in Cuban Archaeology*, edited by Susan Kepecs, L. Antonio Curet and Gabino La Rosa Corzo, 26–46. Tuscaloosa: The University of Alabama Press.

Lacueva, Jaime J. 2012. "De Sevilla al Nuevo Mundo (1491–1521). La Real Hacienda y el negocio de los metales preciosos."*Boletín de la Real Academia Sevillana de Buenas Letras: Minervae Baeticae* 40: 543–578.

Las Casas, Bartolomé de. 1875a. *Historia de Las Indias* Vol. 1. Madrid: Imprenta de Miguel Ginesta.

Las Casas, Bartolomé de. 1875b. *Historia de Las Indias* Vol. 2. Madrid: Imprenta de Miguel Ginesta.

Las Casas, Bartolomé de. 1875c. *Historia de Las Indias* Vol. 3. Madrid: Imprenta de Miguel Ginesta.

Martínez Arango, Felipe. 1997. *Los aborígenes de la cuenca Santiago de Cuba*. Miami: Ediciones Universal.

Marrero, Levi. 1993. *Cuba: Economía y Sociedad. Antecedentes. Siglo XVI (la presencia europea)*. Vol. 1. Santo Domingo: Editorial Playor, S.A.

McEwan, Bonnie G. 2001. "The Spiritual Conquest of La Florida." *American Anthropologist* 103 (3): 633–644.

McEwan, Bonnie G. and Jeffrey M. Mitchem. 1984. "Indian and European acculturation in the Eastern United states as a result of trade." *North American Archaeologist* 6 (4): 271–285.

Mira Caballos, Esteban. 2000. *Las Antillas Mayores 1492–1550. Ensayos y documentos.* Madrid: Iberoamericana.

Mol, Angus A.A. 2007. "Universos socio-cósmicos en colisión: descripciones etnohistóricas de situaciones de intercambio en Las Antillas Mayores durante el período de protocontacto." *El Caribe Arqueológico* 10: 13–30.

Morales Patiño, Oswaldo and Roberto Pérez de Acevedo. 1945. "El Período de Transculturación Indo-hispánica.Contribuciones del Grupo Guama." *La Habana* 4: 5–34.

Muro Orejón, Antonio. 1956. "Ordenanzas reales sobre los indios. (Las Leyes de 1512–13), transcripción, estudio y notas." *Anuario de Estudios Americanos* 13: 417–471.

Oliver, José R. 2000. "Gold Symbolism Among Caribbean Chiefdoms: Of Feathers, *Cibas*, and *Guanín* Power among Taíno Elites." In *Precolumbian Gold. Technology, Style and Iconography* edited by Colin McEwan, 196–219. London: British Museum Press.

Ortega, Elpidio J. 2005. *Compendio General Arqueológico de Santo Domingo* 1. Santo Domingo: Academia de Ciencias de República Dominicana.

Ortega, Elpidio J. and Carmen G. Fondeur. 1976. *Primer informe sobre piezas metálicas indígenas en Barrera*. Santo Domingo: Centro Dominicano de Investigaciones Antropológicas.

Ostapkowicz, Joanna. 2013. "'Made…With Admirable Artistry': The Context, Manufacture and History of a Taíno Belt." *The Antiquaries Journal* 93: 287–317.

Pendergast, David, Jorge Calvera Rosés, Juan E. Jardines Macías, Elizabeth Graham, and Odalys Brito. 2003. "Construcciones de madera en el mar. Los Buchillones, Cuba." *El Caribe Arqueológico* 7: 24–32.

Persons, Ashley. 2013. "Pottery, people, and place: examining the emergence of political authority in Late Ceramic Age Cuba." Unpublished PhD diss., The University of Alabama.

Real Academia de la Historia. 1888. *Colección de Documentos Inéditos relativos al Descubrimiento, Conquista y Organización de las Antiguas posesiones españolas de*

Ultramar. Colección de Documentos Inéditos. Segunda serie, tomo 4, volumen 2 de la isla de Cuba. 25 vols. Madrid: Sucesores de Rivadeneyra.

Romero, Leandro. 1981. "Sobre las evidencias de contacto y transculturación en el ámbito cubano." *Santiago* 44: 71–108.

Rouse, Irving. 1942. *Archaeology of the Maniabón Hills, Cuba*. Yale University Publications in Anthropology 26. New Haven: Yale University Press.

Samson, Alice V.M. 2010. *Renewing the House. Trajectories of social life in the yucayeque (community) of El Cabo, Higüey, Dominican Republic, AD 800 to 1504*. Leiden: Sidestone Press.

Samson, Alice V.M. and Jago Cooper. 2015. "La historia de dos islas en un mar compartido: Investigaciones pasadas y futuras en el pasaje de La Mona." *Boletín del Museo del Hombre Dominicano* 46: 23–48.

Sarmiento, Domingo F. 1973. "Carta del Obispo al Emperador dando cuenta de la visita hecha á las villas é iglesias, y del estado en que se hallan. Año de 1544." In *Documentos para la Historia de Cuba* vol. 1, edited by Hortensia Pichardo, 96–101. La Habana: Editorial de Ciencias Sociales.

Saunders, Nicholas J. 1999. "Biographies of Brilliance: Pearls, Transformations of Matter and being, c. AD 1492." *World Archaeology* 31 (2): 243–257.

Smith, Marvin Thomas. 1984. "Depopulation and culture change in the early historic period Interior Southeast." Unpublished PhD diss., University of Florida.

Sued Badillo, Jalil. 2001. *El Dorado Borincano. La economía de la conquista, 1510 -1550*. San Juan: Ediciones Puerto.

Szászdi León-Borja, István. 2015. "Los pactos de hermandad entre los indios taínos y los conquistadores españoles." *Clío* 165: 13–31.

Tanodi, Aurelio. 1971. *Documentos de la Real Hacienda de Puerto Rico* (1510–1519). Río Piedras: Centro de Investigaciones Históricas, Universidad de Puerto Rico.

Tomé, José and Alexis Rives. 1987. *Carta Informativa no.83* (*2da. Época*). La Habana: Departamento de Arqueología. Academia de Ciencias de Cuba.

Torres de Mendoza, Luis. 1880. *Colección de Documentos Inéditos relativos al Descubrimiento, Conquista y Organización de las Antiguas posesiones españolas de América y Oceanía sacados de los archivos del Reino y muy especialmente del de Indias*. Serie 1, Vol. 1. Madrid: Imprenta de Manuel G. Hernández.

Valcárcel Rojas, Roberto. 1997. "Introducción a la arqueología del contacto indohispánico en la Provincia de Holguín, Cuba." *El Caribe Arqueológico* 2: 64–77.

Valcárcel Rojas, Roberto. 2002. *Banes Precolombino. La ocupación agricultora*. Holguín: Ediciones Holguín.

Valcárcel Rojas, Roberto. 2016a. *Archaeology of Early Colonial Interaction at El Chorro de Maíta, Cuba*. Gainesville: University Press of Florida.

Valcárcel Rojas, Roberto. 2016b. "El mundo colonial y los indios en las Antillas Mayores. Repensando su estudio arqueológico." *Boletín del Museo del Hombre Dominicano* 47: 359–376.

Valcárcel Rojas, Roberto and Marcos Martinón-Torres. 2013. "Metals in the Indigenous Societies of the Insular Caribbean." In *The Oxford Handbook of Caribbean Archaeology*, edited by William F. Keegan, Corinne L. Hofman and Reniel Rodríguez Ramos, 504–522. New York: Oxford University Press.

Valcárcel Rojas, Roberto, Alice V.M. Samson and Menno L.P. Hoogland. 2013. "Indo-Hispanic Dynamics: From Contact to Colonial Interaction in the Greater Antilles." *International Journal of Historical Archaeology* 17: 18–39.

Vega, Bernardo. 1987. "Los metales y los aborígenes de La Española." In *Santos, shamanes y zemíes*, edited by Bernardo Vega, 31–57. Santo Domingo: Fundación Cultural Dominicana.

Wilson, Samuel M. 1990. *Hispaniola. Caribbean chiefdoms in the Age of Columbus.* Tuscaloosa: The University of Alabama Press.

Worth, John E. 2016. "Interpreting Spanish Artifact Assemblages in the Mid-Sixteenth-Century Southeast: The View from the 1559–1561 Tristán de Luna Settlement on Pensacola Bay." Paper presented at the symposium *Documenting Early European/ Native American Contacts and their Repercussions in the Southeast: A Symposium honoring Marvin T. Smith, 73rd Annual Meeting of the Southeastern Archaeological Conference, Athens, GA, October 27*. Available in https://www.researchgate.net/ publication/309550137

CHAPTER 6

Breaking and Making Identities: Transformations of Ceramic Repertoires in Early Colonial Hispaniola

Marlieke Ernst and Corinne L. Hofman

1 Introduction

The first interactions between Spaniards and the peoples of the New World on the island of Hispaniola (presently Haiti and the Dominican Republic) set the stage for the course of colonization in the rest of the Americas (Hofman et al. 2018). Outcomes of the first encounters included miscommunication, misunderstanding, conflict, enslavement, and a range of other intercultural interactions. Intermarriages between Spanish men and Amerindian women, slavery, the taking of concubines, as well as exchange of goods and food items occurred on a regular basis (Deagan 1988, 2004; Sauer 1966; Valcárcel Rojas et al. 2013). These exchanges resulted in a process of transculturation;[1] a creative, ongoing, process of appropriation, revision, and survival both in social and material dimensions (Ortiz [1940] 1995). Transculturation did not only occur between the indigenous peoples of Hispaniola and the Spanish. In 1503, the Spanish obtained legal justification to move indigenous peoples across the islands. Indigenous slavery was thereby officially sanctioned by the Crown (Anderson-Córdova 1990, 2017; Hofman et al. 2018; Rivera-Pagán 2003). One of the destinations of these indigenous enslaved laborers was Hispaniola (Anderson-Cordova 1990; Rivera-Pagán 2003; Sued Badillo 2001). By 1505, enslaved Amerindians were supplemented by enslaved Africans (Rivera-Pagán 2003; Olsen Bogaert et al. 2011a). Through time, increasingly more Africans

1 Colonialism was typically associated with domination, ignoring competition, and especially alliance. However, a more nuanced understanding of colonial interaction is achieved when considering postcolonial concepts of agency and identity (Valcárcel Rojas 2012). It is here that models of transculturation (Ortiz [1940] 1995), ethnogenesis (Deagan 1996, 1998; Voss 2008), creolization (Hannerz 1987), and hybridity and hybridization (Bhabha 1994; van Dommelen 1997) have become prominent in archaeology and studies of colonialism. These models provide interpretive frameworks for the sociocultural factors that led to the emergence of new identities and cultural expressions. Transculturation was retained as a concept here as it was developed within a Caribbean setting as it covers the colonial process as a whole, and includes the transformative effects that all parties experienced as a consequence of the mutual influences on each other.

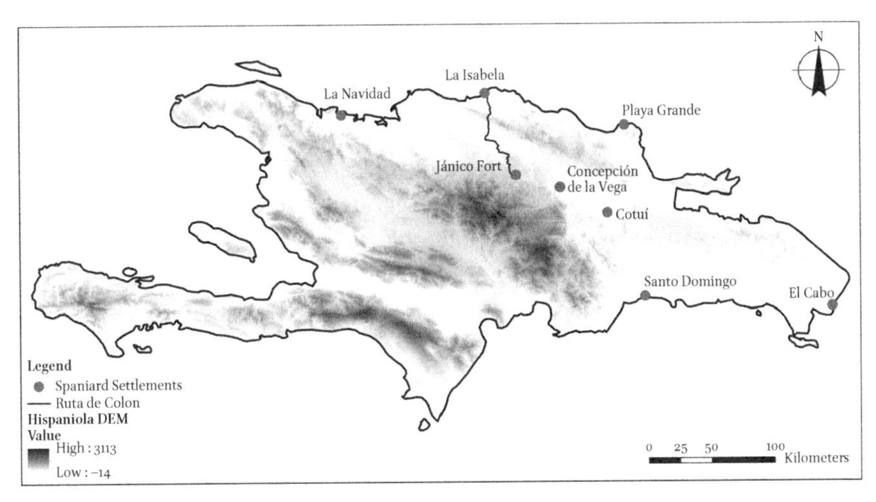

FIGURE 6.1 Map of Hispaniola
MAP BY EDUARDO HERRERA MALATESTA. DEM BASED ON AN ASTER GDEM IMAGE: HTTPS://GDEX.CR.USGS.GOV/GDEX/. ASTER GDEM IS A PRODUCT OF NASA AND METI

were imported resulting in new blends of enslaved Amerindian and African communities within the framework of Spanish colonialism.

In this chapter, we examine the material implications of these blended communities through (1) the incorporation of European earthenwares in the indigenous settlements of El Cabo and Playa Grande, in southeastern and northern Dominican Republic, respectively, and (2) the reflection of Amerindian, Spanish, and African interactions in the ceramic manufacturing traditions in the early Spanish colonial sites of Cotuí and Concepción de la Vega, in central Dominican Republic (Figure 6.1).

2 Methodology

This chapter works with the assumption that people and objects maintain dialectic relationships. Objects are seen as active agents in creating social relationships (Giddens 1979; Gosden 2006; Gosselain 2000; Hoskins 2006).

A low-tech ceramic analysis[2] was carried out on four assemblages from the Dominican Republic. This analysis addressed stylistic, morphological, and technological features of the ceramics. To study the incorporation of

2 A "low-tech" (or "no-tech") analysis can be considered of simple techniques and tools (10x hand lens, a simple microscope, and pliers). This way you can rapidly and easily study large collections of ceramic materials in the field. As these techniques are non-destructive, it is

European earthenwares at the indigenous sites of El Cabo and Playa Grande, a typological identification of the European sherds was conducted based on vessel shapes and decoration modes. The low-tech fabric analysis was aimed at documenting the texture of the clay bodies and inclusions in order to establish whether sherds belonged to one or more vessels or if they were traded as individual sherds. The latter possibility is suggested by early historical accounts stating that sherds were used as trade objects (Eliot 2001). Establishing the provenance of the European ceramics was not the focus of this study, therefore mineralogical characterization of these sherds was not carried out. Microscopic analysis (10X hand-held and 20X Leica DM300 lens) was performed to identify intentional modifications of the European sherds. Abrasion and perforation are ways to intentionally modify sherds to transform them into other objects. Modifications of European objects by indigenous peoples have been documented elsewhere; e.g., abraded sherds used as miniature pot lids, spindle whorls, game pieces or valuables, and needles of a navigator's compass converted into pendants (Knight 2010; Roe and Ortiz 2011; Torres and Carlson 2011).

Spatial analyses were performed to identify distribution patterns, contexts, and relationships between the European and Amerindian ceramics in El Cabo and Playa Grande. Ceramic spatial distributions were important to study potential differential treatment of the European and Amerindian wares before and after disposal. In El Cabo, it was possible to study post-depositional processes to determine if the European and Amerindian ceramics shared similar life cycles before and after deposition. In particular, it was possible to address variation in and consequences of trampling rates between the two wares. For this purpose, 56 Chicoid[3] sherds were selected for comparison with 45 European ceramics. Under similar conditions, we would expect sherds of both wares to break in the same manner if they were of the same quality (Nielsen 1991; Schiffer 1983). Sherd hardness is an important variable in terms of trampling damage; soft materials are more readily damaged than hard materials. Sherd hardness was measured on the Moh's scale (Rice 1987). Due to different excavation strategies and limited access to the materials of Playa Grande it was not possible to perform the exact same analysis as was done for the sherds of El Cabo.

The non-European ceramics of the Spanish colonial sites of Concepción de la Vega and Cotuí were classified by vessel shape and wall profile, lip shape and rim profile, wall thickness, orifice diameter and percentage of rim present,

also a good alternative to study materials in museums or with other restrictions when it comes to technological or compositional analysis (Rice 1987).

3 Chicoid is a ceramic series characterized by incision and punctuation as well as zoomorphic and anthropomorphic modeled appliqués as described by Rouse 1992.

decorations, vessel interior and exterior Munsell colors, firing atmosphere, surface finish, and the presence of slips, or appendages (Hofman 1993).

Manufacturing techniques of the locally-made ceramics were studied according to the *chaîne opératoire* approach. This approach enables the study of the manufacture process of material culture, and reveals the technical choices selected by the potters (Bar-Yosef and van Peer 2009; Farbstein 2011; Lemmonier 1986, 1992, 2002; Leroi-Gourhan 1964; O'Shepard 1963; Roux 2016). The *chaîne opératoire* reflects on the potters' various technical traditions, networks, and strategies (Leroi-Gourhan 1964; Roux 2016). Manufacturing processes are not static and may change in response to a variety of internal and external factors, including invasion and conquest and innovation and inspiration through new cultural forms. Invaders or in-marrying partners may bring new objects or novel ideas that result in changes to established manufacturing techniques, which in turn may spread through socioeconomic networks or human dispersal (Gosselain 2000; Rice 1984). It is probable that interactions between Amerindians, Spaniards, and Africans resulted in changes in the *chaîne opératoire*. Some steps in the *chaîne opératoire* of ceramics are more prone to change while others are more conservative and difficult for the potter to change. Ceramic ethnographic studies show that the conceptualization and forming methods of vessels are highly conservative, while it is easier to vary the raw materials, vessel shape, surface finishing, decoration, and firing techniques (Balfet 1984; Gosselain 2000; Hernández Sánchez 2011; van der Leeuw 1993; Orton et al. 2005; Rice 1984). The techniques used are determined by the potter's conceptualization of pottery technologies. These technologies involve motor skills and specialized gestures deeply rooted within the learned behaviors of the potters (Löbert 1984; Rice 1987; Skibo and Schiffer 2008). This pottery technology is what potters express as essential characteristics of their wares, especially in terms of how they think pots should be made and what they look like.

Low-tech analysis of the *chaîne opératoire* of the non-European sherds of La Vega and Cotuí focused on forming and firing techniques (Figure 6.2). Forming techniques were studied by looking for traces of coiling or wheel throwing on the surface and broken edges of the sherds. The study of firing techniques focused on the firing atmosphere and sherd hardness (Rice 1987). The presence of stone tools for the surface finishing during ceramic manufacture and the recovery of a potter's wheel in Concepción de la Vega indicate that at least some ceramics were made on site. Objectives of the analysis were to identify choices selected in the manufacture of non-European ceramics produced at Cotuí and Concepción de la Vega and to compare them with precolonial manufacturing techniques in Hispaniola to detect continuities and changes.

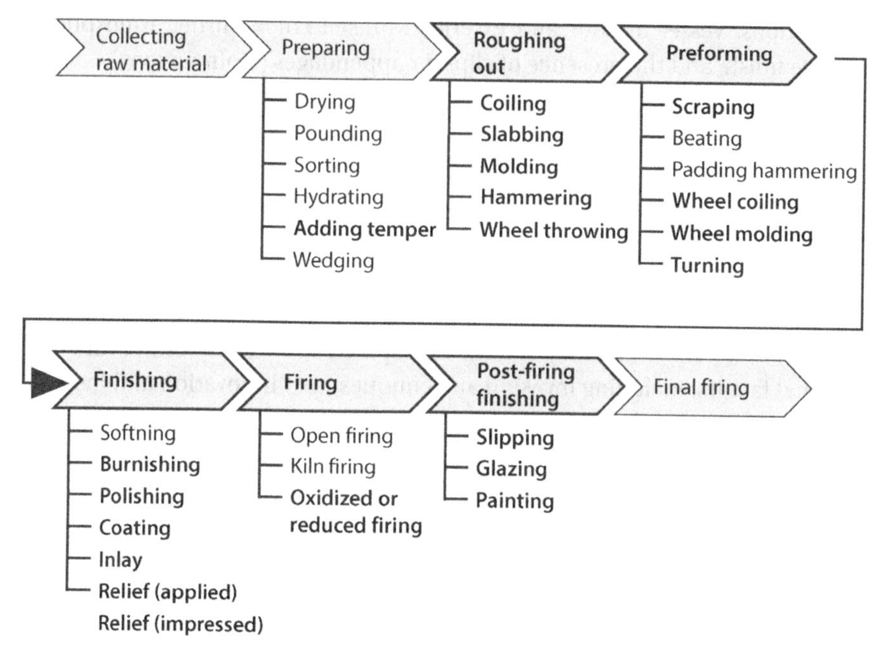

FIGURE 6.2 The *chaîne opératoire* of ceramics. In bold are the steps for which this project has data
ORIGINAL DESIGN BY MARLIEKE ERNST AND FIGURE BY MENNO L.P. HOOGLAND

3 European Earthenwares at Indigenous Sites

3.1 *El Cabo*

The archaeological site of El Cabo is situated in the Higüey region on the southeastern coast of the Dominican Republic in the Altagracia Province. Excavation of the site took place between 2005 and 2008 by a team from Leiden University, in collaboration with the Museo del Hombre Dominicano, under the direction of Menno Hoogland and Corinne Hofman (Hofman et al. 2008; Samson 2010). Radiocarbon samples provided a range of dates between the seventh and early sixteenth centuries AD.

The material assemblage associated with the later component of the site consists primarily of Chicoid ceramics and associated materials (Hofman et al. 2008; Samson 2010). In specific areas of the site materials were mixed with early European colonial materials, including 100 earthenware pieces, five glass beads, a few fragments of glass, animal bones, and fragments of unidentifiable iron objects (Ernst and Hofman 2015; for distribution map of the European material see Valcárcel Rojas et al. 2013) (Figure 6.3). Forty-five European sherds

FIGURE 6.3 Olive jar sherds from El Cabo, Dominican Republic
PHOTO BY MENNO L.P. HOOGLAND

were selected for analysis: 4 rim sherds, 1 complete handle, 3 handle fragments, and 37 vessel wall fragments. The handle was identified as a vertical handle (French 2005, 25). Five sherds were recognized as pieces of plates and the other 40 sherds were portions of an independent restricted vessel with a composite contour. Most sherds were green tin-glazed, and five were decorated with a white/gray glaze. The green tin-glazed sherds were from an olive jar that could be identified as an early style jar (AD 1500–1570) based on the handle type. The rims were identified as a Type A1 Rim according to Goggin's (1960) system. The white-glazed sherds are most likely Columbia Plain sherds from a plate. The low-tech fabric analysis confirmed that all 45 sherds came from only two vessels, one olive jar, and one Columbia Plain plate. Whether the ceramics ended up at El Cabo as whole pots or as broken pieces is not clear. Microscopic analysis of the surface and the plane of the cracks of the sherds showed that none were abraded or otherwise intentionally modified, suggesting that individual sherds were not re-used post-breakage.

All sherds were recovered from a small area in the main excavation unit. This area is characterized by considerable sweeping accumulations, with possible incidences of primary context finds. The deposits in the unit were shallow, suggesting that they do not represent the main or final dumping areas of the site (Samson 2010). The distribution of the colonial materials can be linked directly to one of the house trajectories in the habitation area. The material was clustered at the back of the final structure in the house trajectory, dated to the early sixteenth century (end date AD 1502). This house trajectory is associated with some of the most elaborate finds in the excavations. The colonial materials were found together with Chicoid ceramics, a large stone three-pointer (or trigonolith), and a small shell mask (*guaíza*). The exceptional combination of these artifacts was interpreted as belonging to the residence of an elite member(s) of the El Cabo community (Samson 2010). It appears that the colonial materials exhibit a clear distribution pattern within the disposal area (in contrast to the indigenous ceramics). The olive jar was placed at the center of the deposit and flanked by two pieces of Columbia Plain. The European beads and the ornamental glass fragment were located east of the olive jar. They appear to be small, one time (not reentered) deposits, referred to as "time capsules" (Samson 2010). The distribution pattern could indicate a ritual disposal (see also Fontijn 2002). However, it can also mean that the colonial materials were not of use anymore, lost their value, and were thrown away at different times (Samson 2010; see also Keehnen this volume on En Bas Saline).

Based on European artifact frequency, distribution, and limited early Spanish presence in the area, El Cabo has been interpreted as a contact site, whereby European materials are poorly represented indicating short or indirect interactions (Ernst and Hofman 2015; Hofman et al. 2014; Samson 2010; Valcárcel Rojas et al. 2013). The finding of European materials reflects a short period during which Spanish objects were incorporated into the local material culture and indigenous practices (Samson 2010; Valcárcel Rojas et al. 2013). The close association of European and indigenous items suggests an acceptance and integration of these foreign objects by the local community during the early colonial period (Hofman et al. 2014; Valcárcel Rojas et al. 2013).

3.2 *Playa Grande*

The Playa Grande site is situated in the town of Rio San Juan, on the north coast of the Dominican Republic. When the Spanish arrived, the community belonged to the region of Cuhabo in the province of Hyabo and was tribute to the *cacicazgo* of Magua, under the leadership of the *cacique* Guarionex (López Belando 2013; Mártir Angleria 1964). Archaeological research at the site has been undertaken since 1978 and extensive excavations took place between November 2011 and April 2012, led by Adolfo López, delegated by the Museo del Hombre Dominicano and the Instituto de Investigaciones Antropológicas de la Universidad Autónoma de Santo Domingo (López Belando 2012, 2013; Olsen Bogaert 2004). The site was continuously occupied between the eighth and seventeenth centuries AD (López Belando 2012, 2013). Currently it is not possible to link the most recent radiocarbon date of ca. AD 1680 to the indigenous component of the site. The material record shows that the site was still in use after the early colonial period. Later stages are, until now, not likely to correspond to indigenous occupations on the island. Four spatial areas were identified in the indigenous site: a habitation area, a burial area, an agricultural zone, and a sweeping area (López Belando 2013).

Material culture associated with Playa Grande mainly consists of Ostionoid, Meillacoid,[4] and Chicoid ceramics. Many tools were recovered, manufactured from different materials, including jadeite and flint (Knippenberg 2012). The percentage of Ostionoid, Meillacoid, and Chicoid ceramics found in the different strata increases and decreases significantly as the settlement develops. The initial occupation is associated with a majority of Meillacoid, some Ostionoid, and very little Chicoid ceramics. By the final occupation, the percentages of Meillacoid and Chicoid ceramics were almost identical (37% each), but still some Ostionoid ceramics were present (López Belando 2012). In addition, Spanish materials were recovered from the site including a 1505 *maravedí* coin (minted in Seville), horseshoes, bronze buckles, iron nails, an iron knife, fragments of glass, a blueish glass bead, and fragments of European glazed ceramics (López Belando 2013).

In total, 122 European sherds were recovered, however, only 77 are considered to be from the initial encounter period. The remaining sherds were found near the surface, including probable eighteenth- and nineteenth-century ceramics believed to have washed ashore from shipwrecks off the Playa Grande coast (López Belando 2012). Initial identification of the European sherds was based on the type of glaze, as sherd size and the almost lack of rims did not allow for a further identification. Six glaze types are present in the assemblage. Most sherds were green tin-glazed, followed by glazes of green with white, white, white with blue and purple, white with blue and brown, and brown. The green-glazed ceramics were from olive jar vessels, the base shape did not allow for a more specific identification within the olive jar typology. One of the green-glazed sherds differed from the others by paste and thickness, allowing a possible identification as Green Bacin/Green Lebrillo (Deagan 1987). The one rim present was glazed green on the outside and white on the inside and top, possibly from a jar. The white-glazed ceramic belongs to the Columbia Plain majolica group. The plates and bowls glazed white with blue and purple are typical of Isabela Polychrome majolica. The white with blue and brown belong to the *cuerda seca* ceramics and the brown-glazed to the *melado* style (Deagan 1987).

4 Ostionoid and Meillacoid ceramic series as described by Rouse 1992.

One of the Isabela Polychrome sherds shows signs of reworking. The edges were sanded producing a circular form with a diameter of 3 cm. From other areas in the Caribbean we know that colonial sherds were sometimes modified into spindle whorls (Torres and Carlson 2014), potlids (Roe and Montañez 2014), game pieces (Roe and Montañez 2014), or buttons (Deagan 1999). The size of this sherd suggests that it is was most likely not abraded for functional reasons.

At Playa Grande the European ceramics seem to occur throughout the settlement, perhaps reflecting an easy access to the materials by the indigenous inhabitants through trade and exchange, both through lines of indigenous interactions as well as intercultural contact between indigenous peoples and Spaniards.

The region of Cuhabo experienced far more Spanish influence than El Cabo did. This region paid tribute to Concepción de la Vega, and Spaniards traveled this area a lot more. The region is situated in the north of the island and was subjected to many initial Spanish explorations as well as established tribute systems (Las Casas [1527] 1974; López Belando 2012; Mártir Angleria 1964; Veloz Maggiolo and Ortega 1980). It is therefore likely that the European ceramics present in the site are a result of direct interactions between Amerindians and Spaniards.

4 Indigenous Pottery in Early Spanish Sites

4.1 *Concepción de la Vega*

The colonial town of Concepción de la Vega is one of the most important archaeological sites for the early colonial period of the island. Situated in the Cibao Valley, it played an important role in suppressing Amerindian resistance. Concepción de la Vega consisted of a military fort, a monastery, and a residential town in which many economic activities took place. The economy of the town included gold minting, sugar production, and cattle herding. The first fort at La Vega was founded in 1494 in the territory of *cacique* Guarionex, possibly near the town where the *cacique* lived. Due to increasing resistance to the Spanish by Guarionex's people, La Vega was abandoned by 1498. A new fort was built in 1512 and was in use until 1562 when it was destroyed by an earthquake (Cohen 1997; Deagan 1999; Kulstad 2008; Shephard 1997). The current research focuses on the second fort that was excavated between 1976 and 1996, under the direction of José González and a team from the University of Florida and the Florida Museum of Natural History (Cohen 1997; Deagan 1999).

The assemblage consists of a mix of Spanish, Amerindian, and transcultural artifacts. The great abundance of material culture indicates wealth within the community, represented by the high percentage of Spanish imported ceramics (both utilitarian and table wares) and the many elaborate glass and metal artifacts (Deagan 2002; Kulstad 2008). The discovery of a potter's wheel (Card 2007; Ortega and Fondeur 1978) suggests that some wheel-thrown ceramics were made locally, possibly to somewhat maintain an Iberian lifestyle.

The non-European ceramics of La Vega reveal some continuation in the production and use of precolonial Chicoid and Meillacoid pottery during the contact period, including applications of *adornos* (modeled clay applications under the vessel's rim often depicting zoomorphic or anthropomorphic figures). Other continuities in the production of Amerindian ceramics included methods of vessel forming, selection of clays and tempers, and vessel shapes and wall thicknesses.

However, some major changes from the precolonial record were recorded as well; vessel conceptualization, vessel shape, pre-firing decorations, surface treatment, and use show new introductions. There are clear traces of the introduction of the potter's wheel and some new raw materials used as temper.

At La Vega, new vessel shapes emerged resembling Spanish (plate, jars, and jugs) and African (*olla*) shapes. The most striking forms are those that do not resemble only precolonial Hispaniolan, European, or African shapes and decorations, and have been labeled as transcultural ceramics (Figure 6.4). Transcultural ceramics show possible influences from Europe, elsewhere in the

FIGURE 6.4 New forms and decorations showing Amerindian-African and European intercultural dynamics from Concepción de la Vega, Dominican Republic
PHOTOS BY MARLIEKE ERNST AND FIGURE BY MENNO L.P. HOOGLAND

Caribbean, Central and South-America, and/or Africa. Vessels are often red- or white-slipped, or some combination, with elaborate designs. Occasionally, the slipped decoration is combined with fine-line incisions or applications, and more seldom with small pieces of quartz impressed in the clay. Coiling remains the most ubiquitous forming technique, however, the use of the potter's wheel is evident within the non-European ceramics.

At La Vega, substantial amounts of Chicoid *adornos* were uncovered. However, most of the modeled appliqués resemble a geometric modeling similar to a more stylized form of a Chicoid frog *adorno*. This stylized form might correspond to the decrease of Amerindian decorative forms in the ceramics. Pastes, wall thicknesses, and vessel shapes, are similar to the precolonial ceramics. However, they are less decorated than before the conquest, showing a decrease in the number of incised vessels. An exception to this is the presence of ceramics decorated with a multi-banded pattern of very fine lines, created by dragging a comb-like tool over the surface (Figure 6.5). These multi-banded, intersecting, wavy patterns are common decorations in West and Central Africa. Variations of these designs commonly adorn ceramics, pipe bowls, circular divination boards, and carved wooden doors (DeCorse and Hauser 2003; Ernst 2016; Ogundiran 2007; Schreg 2010; Weaver 2015). The comb-dragged, multi-banded patterns at La Vega occur on the vessel's wall, the rim, and occasionally on the inside of the vessel's base. The rims of this type not only bear incisions, but are also decorated with a pattern created by finger pinching, which is considered to be both an Amerindian as well as an African decorative technique (Weaver 2015). From these rims, it is evident that vessels were very large, possibly associated with sugar production that was introduced to the Caribbean at La Vega (Moya-Pons 1978). Traces of manufacturing techniques show that

FIGURE 6.5 Pottery sherd from La Vega with a multi-banded pattern of fine lines, created by dragging a comb-like tool over the surface
DRAWING BY MARLIEKE ERNST AND FIGURE BY MENNO L.P. HOOGLAND

these vessels were made on the potter's wheel. Perhaps this type of ceramic shows the adaption of the potter's wheel by African potters. However, Spanish Moorish influence should not be excluded.

4.2 *Cotuí, an Early Colonial Mining Camp*

Site 11 in the Cotuí region is also known as the colonial mining camp of the first gold mine exploited by the Europeans in the New World. Gold was obtained by the colonizers at Cotuí by at least 1505, when Fray Nicolás de Ovando sent an expedition to manage the mines in the area. The gold of Cotuí was minted at Concepción de la Vega. Excavations at the site were conducted between July and December 2010 under the direction of Harold Olsen Bogaert and the Museo del Hombre Dominicano. Site 11 is comprised of two buildings: the church and the mining camp (consisting of three plots of stone buildings). In addition, eight postholes were identified. The area contained many traces of firing, and possibly represents a perishable booth or place to prepare food for the miners, or a place for the processing or removing of metals during the sixteenth century (Olsen Bogaert et al. 2011a, 2011b).

Artifacts from the site included both Spanish and Amerindian materials. Most of the artifacts were ceramics. Various stone tools were recovered, including some probably used in the making of pottery, indicating on-site pottery production. Many luxury metal and glass objects reflect the importance of the mine to the Spaniards (Olsen Bogaert et al. 2011a, 2011b). In total, 59.05 kg of ceramics were analyzed, out of which 21.66 kg were identified as non-European (Ernst 2016). The area with the highest concentration of non-European ceramics corresponds with the eight postholes, although non-European sherds were also found across much of the site with the exception of the colonial structures (Olsen Bogaert et al. 2011a, 2011b). Most of the non-European ceramics were recovered from the 12–25 and 25–46 cm deep strata (Ernst 2016). Deeper layers produced elevated percentages of European ceramics dating to the fifteenth and sixteenth centuries. This stratigraphic arrangement suggests a process whereby the initial widespread use of imported European earthenwares was gradually replaced by locally produced pottery (Olsen Bogaert et al. 2011a).

The Cotuí non-European ceramic assemblage shows that vessel conceptualization, surface treatment, pre-firing decorations, and use changed during the transculturation process after European colonization. In some cases, changes were documented in raw material collection, vessel forming, and firing. Vessel shape and decorations were especially subject to change. None of the elaborate vessel shapes and decorations previously present in Meillacoid and Chicoid ceramics were present in the Cotuí assemblage. The only observed continuities were the raw materials, paste preparation, coiling forming techniques, surface treatment, and the use of applications on the vessel surface (Ernst 2016; Ting et al. 2018).

Vessel shapes unknown to precolonial Hispaniola were introduced resembling both African and Spanish vessels (the African *olla* and the Spanish *loza común bacín*). Nonetheless, these vessels were still made with the traditional coiling technique characteristic of both traditional Amerindian and African ceramic manufacturing techniques. The addition of appliqués directly under the vessel's rim also coincides with Amerindian ceramic manufacture. Nine traditional *adornos* were found, which is remarkably low when compared to their occurrence in precolonial assemblages in Hispaniola. The majority of the appliqués, however, resemble the same geometric modeling as shown in La Vega. Pastes, wall thicknesses, and vessel shapes are similar to the precolonial ceramics. Vessel decorations are plainer than precolonial versions. About 30 sherds resemble the transcultural ceramics from La Vega, and were probably manufactured at Concepción de la Vega. This might reflect the historically known connection between Concepción de la Vega and Cotuí (Deagan and Cruxent 2002). The impressed decorations and the addition of a yellowish slip on three red paste sherds are consistent with and provide evidence for West African continuity (Ernst 2016). Other expressions of transculturation within the Cotuí assemblage include the presence of sherds with the same multi-banded comb-dragged pattern. The manufacturing technique consists of coiling finished on the potter's wheel. The sherds are similar in style as the ones in Concepción de la Vega. Since only four sherds of this type were recovered from Cotuí, it is at this stage hard to tell if they were made at Cotuí. It might be suggested that their presence is a result of interactions between Cotuí and Concepción de la Vega.

The last indicator of change is the clay pipe found in the site. Clay tobacco pipes have been found in the precolonial Caribbean, however not of this type. The shape of this pipe suggests an African introduction without a clay stem where a piece of reed or a straw was inserted. The impressed decorations also resemble African traditions, but a similar placement of decorations is also found on European pipes (Ernst 2016; Sudbury and Gerth 2014).

5 Conclusion

Following the first encounters between Amerindians and Spanish in the Greater Antilles, material culture, notably ceramics, have been exchanged between the two societies. Many types of European ceramics have been recovered from indigenous sites throughout Hispaniola. Likewise, Amerindian ceramics have been found in Spanish sites. The functions of the ceramics in their respective settlement's context differ greatly. This is partially due to difference in the nature of interactions in indigenous towns and those in Spanish ones.

However, the nature of contact between the indigenous inhabitants and the Spaniards was also different in Playa Grande than in El Cabo. In the sixteenth century, the closest Spanish settlement to El Cabo was about 200 kilometers away. The first encounters in the region of El Cabo were rather late in the colonization of the island and mainly due to Spanish trade of manioc between Santo Domingo and Isla Saona (Samson 2010). On this basis, as well as the characterization of the European materials, it is suggested that the colonial assemblage of El Cabo is the result of a single instance of direct trade between the Spanish and the inhabitants of El Cabo, or more likely the result of indirect (down-the-line) trade within local exchange networks (Hofman et al. 2014; Samson 2010; Valcárcel Rojas et al. 2013). The nature of the discard patterns indicates that there was a shift in value from regular Spanish ceramic vessels (or even sherds) to those ceramics being valuable items for the indigenous communities. In contrast, Playa Grande was situated in a region where Spanish presence was stronger as many of the initial navigations by the Spanish occurred along the northern coast (Deagan and Cruxent 2002; Sauer 1966). The amount and diversity of European earthenwares found throughout the site suggests that direct contact occurred more often in Playa Grande than El Cabo. By the time the Spaniards arrived, Playa Grande belonged to the *cacicazgo* of Magua, under the leadership of *cacique* Guarionex who was known to pay tribute to the fort of La Vega (Sauer 1966). It is not known if inhabitants from Playa Grande were put to work in the Spanish towns. However, the importance of the region for the Spanish meant a substantial presence of Europeans in the vicinity of Playa Grande. The presence of European earthenwares may reflect direct trade between the indigenous people of Playa Grande and the Spanish, although down-the-line exchange cannot be excluded. Not enough is known regarding discard patterns of the European earthenwares to say much about the meaning of these artifacts to the inhabitants of the village. The finding of the abraded Isabela Polychrome sherd indicates the reuse or revaluing of this particular piece; in this artifact we see an outcome of transculturation and the coming together of cultural traditions.

The function of European earthenwares in indigenous sites like Playa Grande and El Cabo differs from the function of Amerindian ceramics in early Spanish towns. Within Amerindian settlements European earthenwares were regarded as items of high status. In Spanish towns local ceramics were brought in or made by Amerindian and African laborers, and were seen as utilitarian objects. The local manufacture of ceramics in Concepción de la Vega and Cotuí shows that the process of ceramic production is not as static as sometimes thought. Ceramics from the two Spanish colonial assemblages reveal that there are both continuities and changes in the manufacturing of ceramics in the early colonies. The forming of vessels by coiling is still the most common practice. However, the excavated potter's wheel from Concepción de la Vega and the presence of non-European wheel-finished ceramics show that wheel-made pottery was introduced to the colonies. Pastes, wall thicknesses, and specific vessel shapes from both sites correspond with precolonial ceramics. Continuity in the presence of Chicoid ceramics in Concepción de la Vega was more common than in Cotuí. However, new vessel shapes and decorative modes were introduced in both places, resembling Spanish, African, and Caribbean precolonial traditions. The assemblages of Cotuí and Concepción de la Vega reflect a process of transculturation in the *chaîne opératoire* of the ceramic repertoire, whereby multiple traditions came together in the creation of a new material set. Continuities and changes in the production of ceramics in the colonies provide insight into the formal processes of colonization and the inter-cultural dynamics that occurred. Although laborers, captives, and enslaved people often lived marginal lives in the colonies, their role in the transculturation process is evident in the ceramic assemblages. Caribbean, African, and Spanish influences are reflected in the transcultural ceramics. The evidence of these cultural groups' influences on ceramic manufacturing is the first visible indicator of Caribbean-Spanish-African congruence in Spanish towns. The presence of Spanish vessel shapes made with local techniques may reflect the intention of maintaining the Iberian lifestyle in the colonies, also evidenced by the presence of the potter's wheel. The decline in Amerindian decorative forms may be ascribed to the effects of Spanish domination, including Christianization and labor obligations. At the same time, the assemblages show that the Spanish did not fully reinforce Iberian life. Some of the locally made ceramics were used as utilitarian vessels (Pagán-Jiménez 2012). Local cuisine is a pivotal marker denoting social and cultural identity (Beaudry 2013; Hofman et al. 2018; Mintz and Price 1985; Rodríguez-Alegría 2005). Amerindian and African influences in the making of cooking vessels reveal some form of cultural preservation among the enslaved at least in the less visible spaces of colonial society. It is here that indigenous as well as African ceramic cultural aspects were maintained by the people who made and used them.

Both the incorporation of European earthenwares in the indigenous settlements and the transcultural ceramics present at Spanish colonial sites have shown that the outcomes of the first encounters between indigenous peoples of Hispaniola, Spanish colonizers, and in some cases African enslaved peoples, differed greatly. The exchanges of ceramics and ceramic technologies occurred on various levels in differing contexts. The ceramic materials presented in this chapter show us that the intensity of contact was one of the most important factors for why and how the material implications of these communities in contact differ from case to case. They reveal cases of incorporation as well as cultural preservation, of maintaining social memory, and of expressing social agency in the diverse intercultural situations in early colonial Hispaniola.

Acknowledgments

The research leading to these results has received funding from the European Research Council under the European Union's Seventh Framework Programme (FP7/2007–2013)/ERC-NEXUS1492 grant agreement no. 319209. We thank the Ministerio de la Cultura in Santo Domingo, the Museo del Hombre Dominicano, Adolfo López Belando, and the Oficina de Patrimonio Monumental for granting us permission to study the materials from El Cabo, Playa Grande, Concepción de la Vega and Cotuí, respectively. We are grateful to Jorge Ulloa, an anonymous reviewer, and Peter Siegel for their editorial comments.

References

Anderson-Córdova, Karen F. 1990. *Hispaniola and Puerto Rico: Indian Acculturation and Heterogeneity 1492–1550.* Ann Arbor: University Microfilms.

Anderson-Córdova, Karen F. 2017. *Surviving Spanish Conquest. Indian Fight, Flight, and Cultural Transformation in Hispaniola and Puerto Rico.* Alabama: University of Alabama Press.

Balfet, Hélène. 1984. "Methods of Formation and the Shape of Pottery." In *The Many Dimensions of Pottery, Ceramics in Archaeology and Anthropology,* edited by Sander van der Leeuw and Alison Pritchard, 171–202. Amsterdam: University of Amsterdam.

Bar-Yosef, Ofer and Philip van Peer. 2009. "The Chaine Operatoire Approach in Middle Paleolithic Archaeology." *Current Anthropology* 50 (1): 103–131.

Beaudry, Mary C. 2013. "Mixing Food, Mixing Cultures: Archaeological Perspectives." *Archaeological review from Cambridge* 28 (1): 285–297.

Bhabha, Homi K. 1994. *The Location of Culture.* London and New York: Routledge.

Card, Jebb J. 2007. *The Ceramics of Colonial Ciudad Vieja, El Salvador: Culture Contact and Social Change in Mesoamerica.* Tulane: Tulane University.

Cohen, Jeremy. 1997. *Preliminary Report on the 1996 Field Season at Concepción de la Vega. Project report submitted to the Dirección Nacional de Parques*. Gainesville: Florida Museum of Natural History.

Deagan, Kathleen A. 1987. *Artifacts of the Spanish Colonies of Florida and the Caribbean, 1500–1800, vol. 1: Ceramics, Glassware, and Beads*. Washington, D.C.: Smithsonian Institution Press.

Deagan, Kathleen A. 1988. "The Archaeology of the Spanish Contact Period in the Caribbean." *Journal of World Prehistory* 2 (2): 187–233.

Deagan, Kathleen A. 1996. "Colonial Transformation: Euro-American Cultural Genesis in the Early Spanish-American Colonies." *Journal of Anthropological Research* 52 (2): 135–160.

Deagan, Kathleen A. 1998. "Transculturation and Spanish American Ethnogenesis: The Archaeological Legacy of the Quincentenary." In *Studies in Culture Contact: Interaction, Culture Change, and Archaeology*, Occasional Paper Vol. 25, edited by James G. Cusick, 23–43. Carbondale: Center for Archaeological Investigations, Southern Illinois University.

Deagan, Kathleen A. 2002. *Artifacts of the Spanish Colonies of Florida and the Caribbean, 1500–1800, vol.2: Portable Personal Possessions*. Washington D.C.: Smithsonian Institution Press.

Deagan, Kathleen A. 2004. "Reconsidering Taino Social Dynamics after Spanish Conquest: Gender and Class in Culture Contact Studies." *American Antiquity* 69 (4): 597–626.

Deagan, Kathleen A. and Joseph M. Cruxent. 2002. *Archaeology at La Isabela: America's First European Town*. New Haven: Yale University Press.

DeCorse, Christopher R. and Mark W. Hauser. 2003. "Low-Fired Earthenwares in the African Diaspora: Problems and Prospects." *International Journal of Historical Archaeology* 7: 67–98.

Dommelen, Peter van. 1997. "Colonial constructs: colonialism and archaeology in the Mediterranean." *World Archaeology* 28 (3): 305–323.

Eliot, Charles W. 2001. *American Historical Documents, 1000–1904*, Vol XLIII. New York: P. F. Collier and Son.

Ernst, Marlieke. 2016. "(Ex)Changing the Potter's Process: Continuity and Change in the Non-European Ceramics of Cotuí, the First Colonial Mine in Hispaniola, after 1505." Research Master thesis, Leiden University.

Ernst, Marlieke and Corinne L. Hofman. 2015. "Shifting Values: A Study of Early European Trade Wares in the Amerindian Site of El Cabo, Eastern Dominican Republic." In *GlobalPottery 1. Historical Archaeology and Archaeometry for Societies in Contact*, edited by Jaime Buxeda i Garrigós, Marisol Madrid i Fernández, and Javier G. Iñañez, 195–204. Oxford: British Archaeological Reports, International Series 2761.

Farbstein, Rebecca. 2011. "Technologies of Art: A Critical Reassessment of Pavlovian Art and Society, Using Chaîne Opératoire Method and Theory." *Current Anthropology* Vol. 52 (3): 401–432.

Fontijn, David. 2002. *Sacrificial landscapes: cultural biographies of persons, objects and 'natural' places in the Bronze Age of the southern Netherlands, c. 2300–600 BC.* Leiden: Sidestone Press.

French, Neal. 2005. *Ceramics shapes and glazing guide.* Kerkdriel: Librero.

Giddens, Anthony. 1979. *Central Problems in Social Theory.* London: Macmillan.

Goggin, John M. 1960. *The Spanish Olive Jar: An Introductory Study*, Yale University Publications in Anthropology Vol. 62. New Haven: Department of Anthropology, Yale University.

Gosden, Chris. 2006. "Material culture and long-term change." In *Handbook of Material Culture*, edited by Chris Tilley, Webb Keane, Susanne Küchler, Mike Rowlands, and Patricia Spyer, 24–42. London: Sage Publications.

Gosselain, Olivier G. 2000. "Materializing identities: an African perspective." *Journal of Archaeological Method and Theory* 7 (3): 187–217.

Hannerz, Ulf. 1987. "The World in Creolization." *Africa* 57 (4): 546–559.

Hernández Sánchez, Gilda. 2011. *Ceramics and the Spanish Conquest, Response and Continuity of Indigenous Pottery Technology in Central Mexico.* Leiden: Brill.

Hofman, Corinne L. 1993. *In Search of the Native Population of Pre-Columbian Saba (400–1450 A.D.). Part One: Pottery Styles and their Interpretations.* PhD diss., Leiden University.

Hofman, Corinne L., Menno L.P. Hoogland, and Alice V.M. Samson. 2008. "Investigaciones arqueológicos en El Cabo, oriente de la República Dominicana: resultados preliminares de las campañas 2005 y 2006." *Boletín del Museo del Hombre Dominicano* 35 (42): 307–316.

Hofman, Corinne L., Angus A.A. Mol, Menno L.P. Hoogland, and Roberto Valcárcel Rojas. 2014. "Stage of Encounters: migration, mobility and interaction in the precolonial and early colonial Caribbean." *World Archaeology* 46 (4): 591–609.

Hofman, Corinne L., Jorge Ulloa Hung, Eduardo Herrera Malatesta, Joseph Sony Jean, and Menno L.P. Hoogland. 2018. "Indigenous Caribbean Perspectives: Archaeologies and legacies of the first colonized region in the New World." *Antiquity* 92 (361): 200–216.

Hoskins, Janet. 2006. "Agency, biography and objects." In *Handbook of Material Culture*, edited by Chris Tilley, Webb Keane, Susanne Küchler, Mike Rowlands, and Patricia Spyer, 74–84. London: Sage Publications.

Knight, Vernon J. 2010. "La Loma del Convento: Its centrality to current issues in Cuban archaeology." In *Beyond the Blockade: New Currents in Cuban Archaeology*, edited by Susan Kepecs, L. Antonio Curet, and Gabino La Rosa Corzo, 26–46. Tuscaloosa: University of Alabama Press.

Knippenberg, Sebastiaan. 2012. "Jadeitite axe manufacture in Hispaniola. A preliminary report on the lithics from the Playa Grande site, Northern Dominican Republic." In *El sitio arquelógico de Playa Grande Río San Juan, María Trinidad Sánchez, informe de las excavaciones arquelógicas campaña 2011–2012*, edited by Adolfo J. López Belando. Santo Domingo: Museo del Hombre Dominicano.

Kulstad, Pauline M. 2008. *Concepción de la Vega 1495–1564: A Preliminary Look at Lifeways in the Americas' First Boom Town*. Master thesis, University of Florida.

Las Casas, Bartolome de. (1527) 1974. *Historia de las Indias. vol I, II and III*. Santo Domingo: Editora Corripio, Sociedad Dominicana de Bibliófilos.

Leeuw, Sander E. van der. 1993. "Giving the potter a choice. Conceptual aspects of pottery techniques." In *Technological choices: transformation in material cultures since the Neolithic*, edited by Pierre Lemonnier, 23–28. London: Routledge.

Lemmonier, Pierre. 1986. "The study of material culture today: An anthropology of technical systems." *Journal of Anthropological Archaeology* 5 (2): 147–186.

Lemmonier, Pierre. 1992. *Elements for an Anthropology of Technology*. Anthropological Papers No. 88. Ann Arbor: Museum of Anthropology, University of Michigan.

Lemmonier, Pierre. 2002. *Transformation in material culture since the Neolithic*. London: Routledge.

Leroi-Gourhan, André. 1964. *La gueste et la parole. Paris: French ministry of culture.* Translated by Anna Bostok Berger in 1993. Massachusetts: Massachusetts Institute of Technology.

Löbert, Horst W. 1984. "Types of Potter's Wheels and the Spread of the Spindle-Wheel in Germany." In *The Many Dimensions of Pottery: Ceramics in Archaeology and Anthropology*, edited by Sander E. van der Leeuw, 202–230. Amsterdam: Van Giffen Instituut voor Pre-en Protohistorie.

López Belando, Adolfo J. 2012. *El sitio arquelógico de Playa Grande Río San Juan, María Trinidad Sánchez, informe de las excavaciones arquelógicas camaña 2011–2012*. Excavation report. Santo Domingo: Museo del Hombre Dominicano.

López Belando, Adolfo J. 2013. "Excavaciones arqueológicas en el poblado taíno de Playa Grande, República Dominicana." In *Proceedings of the 25th Congress of the International Association for Caribbean Archeology*, edited by Laura Del Olmo, 254– 280. San Juan: Instituto de Cultura Puertorriqueña, el Centro de Estudios Avanzados de Puerto Rico y el Caribe, y la Universidad de Puerto Rico, Recinto de Río Piedras.

Mártir Angleria, Pedro. 1964. *Décadas del Nuevo Mundo 1*. México DF: José Porrua e Hijos.

Mintz, Sidney W., and Sally Price. 1985. *Caribbean Contours*. Baltimore: Johns Hopkins University Press.

Moya-Pons, Frank. 1978. *La Española en el siglo XVI*. Santiago: Universidad Católica Madre y Maestra.

Nielsen, Axel. 1991. "Trampling the archaeological record: An experimental study." *American Antiquity* 56 (3): 483–508.

Ogundiran, Akinwumi. 2007. "Living in the Shadow of the Atlantic World: History and Material Life in a Yoruba-Edo Hinterland, ca. 1600–1750." In *Archaeology of Atlantic Africa and the African diaspora*, edited by Akinwumi Ogundiran and Toyin Falola, 77–99. Bloomington: Indian University Press.

Olsen Bogaert, Harrold G. 2004. "Sitio Arqueológico Playa Grande, Río San Juan, Provincia María Trinidad Sánchez." *Boletín del Museo del Hombre Dominicano* 37: 126–142.

Olsen Bogaert, Harrold G., Jorge Ulloa Hung, Victor Avila, and Frank Coste. 2011a. *Pueblo Viejo Dominicana Corporation, Projecto Pueblo Viejo. Sitio Arqueológico No. 11 Investigación Estructuras Coloniales*. Unpublished excavation report, on file at the Faculty of Archaeology, Leiden University.

Olsen Bogaert, Harrold G., Jorge Ulloa Hung, Victor Avila, and Frank Coste. 2011b. *Pueblo Viejo Dominicana Corporation, Projecto Pueblo Viejo, Sitio Arqueológico No. 11 2da fase de investigación*. Excavation report. Santo Domingo: Unpublished excavation report, on file at the Faculty of Archaeology, Leiden University.

Ortega, Elpidio and Carmen Fondeur. 1978. *Estudio de la Cerámica del Período Indo-Hispánico de la Antigua Concepción de La Vega*. Santo Domingo: Taller.

Ortiz, Fernando. (1940) 1995. *Cuban Counterpoint: Tobacco and Sugar*. Translated by Harriet de Onís. Durham: Duke University Press. Originally published in Spanish in 1940, Havana: Editorial de Ciencias Sociales.

Orton, Clive, Paul Tyers, and Alan Vince. 2005. *Pottery in Archaeology*. Cambridge: Cambridge University Press.

O'Shepard, Anna. 1963. *Ceramics for Archaeologists*. Washington DC: Carnegie institution of Washington.

Pagán-Jiménez, Jaime R. 2012. *Almidones antiguos recuperados en artefactos cerámicos atribuidos a la fase de cantacto indohispánica del Sito 11, Pueblo Viejo, república Dominicana*. Leiden: Internal report Leiden, Faculty of Archaeology.

Rice, Prudence M. 1984. "Change and Conservatism in Pottery-Producing Systems." In *The Many Dimensions of Pottery: Ceramics in Archaeology and Anthropology*, edited by Sander E. van der Leeuw, 231–293. Amsterdam: Van Giffen Instituut voor Pre-en Protohistorie.

Rice, Prudence M. 1987. *Pottery Analysis, A source book*. Chicago: University of Chicago Press.

Rivera-Pagán, Luis N. 2003. "Freedom and Servitude: Indigenous Slavery and the Spanish Conquest of the Caribbean." In *General History of the Caribbean Vol. 1: Autochthonous societies*, edited by Jalil Sued Badillo, 316–362. Paris: UNESCO Publishing.

Rodríguez-Alegría, Enrique. 2005. "Eating Like an Indian: Negotiating Social Relations in the Spanish Colonies." *Current Anthropology* 46 (4): 551–573.

Roe, Peter G. and Hernan Ortíz Montañez. 2014. "Small vessels, precious contents: miniature pots and ceramic discs from the Punta Mameyes site (DO-42), Dorado, Puerto Rico." *Proceedings of the 24th Congress of the International Association for*

Caribbean Archeology, edited by Benoit Bérard, 300–336. Martinique: Université des Antilles et de la Guyane.

Rouse, Irving. 1992. *The Taínos: Rise and Decline of the People who Greeted Columbus.* New Haven: Yale University Press.

Roux, Valentine. 2016. “Ceramic Manufacture: The Chaîne Opératoire Approach.” In *The Oxford Handbook of Archaeological Ceramic Analysis*, edited by Alice M.W. Hunt, 101–113. Oxford: Oxford University Press.

Samson, Alice V.M. 2010. *Renewing the House: Trajectories of social life in the yucayeque (community) of El Cabo, Higüey, Dominican Republic, AD 800 to 1504.* Leiden: Sidestone Press.

Sauer, Carl O. 1966. *The Early Spanish Main.* Berkeley and Los Angeles: University of California Press.

Schiffer, Michael B. 1983. “Towards the identification of formation processes.” *American Antiquity* 48 (4): 675–706.

Schreg, Rainer. 2010. “Panamanian coarse handmade earthenware – a melting pot of African, American and European traditions?” *Post-medieval Archaeology* 44 (1): 135–164.

Shephard, Herschel. 1997. *A Report: The Fort at Concepción de la Vega, Dominican Republic and Standards, Interpretation and Conservation. Consultant report, submitted to the Dirección Nacional de Parques, Santo Domingo, and U.S.A.I.D.* Gainesville: Florida Museum of Natural History.

Skibo, James M. and Micheal B. Schiffer 2008. *People and Things. A Behavioural Approach to Material Culture.* New York: Springer.

Sued Badillo, Jalil. 2001. *El Dorado Borincano: La Economía de la Conquista 1510–1550.* San Juan: Ediciones Puerto.

Sudbury, Byron, J. and Ellen Gerth. 2014. “Clay Tobacco Pipes from the Tortugas Shipwreck, Florida (1622).” *Odyssey Papers* 41: 1–15.

Ting, Carmen, Jorge Ulloa Hung, Corinne L. Hofman, and Patrick Degryse. 2018. “Indigenous Technologies and the Production of Early Colonial Ceramics in Dominican Republic.” *Journal of Archaeological Science: Reports* 17: 47–57.

Torres, Joshua M. and Lisbeth A. Carlson. 2014. “Spindle whorls and fiber production: evidence from two Late Ceramic Age sites in eastern Puerto Rico.” In *Proceedings of the 24th Congress of the International Association for Caribbean Archeology*, edited by Benoit Bérard, 180–188. Martinique: Université des Antilles et de la Guyane.

Valcárcel Rojas, Roberto. 2012. “Interacción colonial en un Pueblo de indios encomendados, El chorro de Maita. Cuba.” PhD diss., Leiden University.

Valcárcel Rojas, Roberto, Alice V.M. Samson, and Menno L.P. Hoogland. 2013. “Indo-Hispanic Dynamics: From Contact to Colonial Interaction in the Greater Antilles.” *International Journal of Historical Archaeology* 17 (1): 18–39.

Veloz Maggiolo, Marcio and Elpidio Ortega. 1980. "Nuevos Hallazgos Arqueológicos en la Costa Norte de Santo Domingo." *Boletín del Museo del Hombre Dominicano* 13: 11–48.

Voss, Barbara L. 2008. *The Archaeology of Ethnogenesis: Race and Sexuality in Colonial San Francisco*. Berkeley and Los Angeles: University of California Press.

Weaver, Brendan J.M. 2015. *"Fruit of the vine, work of human hands": An Archaeology and Ethnohistory of Slavery on the Jesuit Wine Haciendas of Nasca, Peru.* PhD diss., Vanderbilt University.

CHAPTER 7

Rancherías: Historical Archaeology of Early Colonial Campsites on Margarita and Coche Islands, Venezuela

Andrzej T. Antczak, Ma. Magdalena Antczak, Oliver Antczak, and Luis A. Lemoine Buffet

1 Introduction

The frantic nature of the contact period followed by the unrelenting forging of quotidian colonial realities brought dramatic changes to indigenous peoples across the Americas. Each of these phases assumed specific social expressions and pulsated with diverse regional tempos. Undeniably, some threads that interconnected indigenous populations of the late precolonial times were irreversibly severed during the contact period. Other links survived and underwent various processes of transformation. Along the Venezuelan coast and the parallel chain of Southeastern Caribbean islands, the arrival of the Europeans had categorical consequences. It cut off or thoroughly transformed traditional circuits of exchange and spheres of interaction which crossed boundaries of archaeologically defined precolonial cultures and united diverse – protohistorically known – linguistic and ethnic units (Amodio 1991; Antczak and Antczak 2006; Biord Castillo 1985; Biord Castillo and Arvelo 2007; Heinen and García-Castro 2000; Henley 1985; Perera 2000; Scaramelli and Tarble de Scaramelli 2005; Tiapa 2008). Although little is known about the nature of the social processes behind the scene, there is some persuasive archaeological evidence of their operation in the late precolonial and early colonial Venezuelan Caribbean (Antczak and Antczak 2015a, 2015b; Antczak et al. 2015; Rivas 2001).

Northeastern Venezuela is exceptionally well-suited for archaeological research into the encounter of differing material cultures and socio-cultural transformations in early colonial settings. Since the early sixteenth century, the Spanish were present in this area on the barren island of Cubagua surrounded by extensive pearl oyster beds. There, the town of Nueva Cádiz was officially founded in 1528. Ever since then, facts and fictions about this town and its inhabitants have played an important role in Venezuelan historiography and anthropology; they have also stirred the imagination of artists. The 'story'

DOI:10.1163/9789004273689_008

of Cubagua has played an important role in the (re)formulations of national identity in Venezuela (e.g., Armand 2017; IPC 2009; Suárez 2014).

However, for all its impact on present-day culture, the intricacies of early sixteenth-century Cubagua remained only superficially disclosed by archaeologists. When excavations were carried out on this island in the 1950s they left behind more questions than answers. The overarching goal of the research backing up this chapter is to thoroughly recontextualize the town of Nueva Cádiz de Cubagua. This is achieved based on new archaeological data from our fieldwork on the islands of Margarita and Coche (Figure 7.1). This data is further compared with the information obtained during the study of collections from Nueva Cádiz, currently held in Venezuelan and North American museums. All this data is also contextualized within the large body of information amassed during the long-term Venezuelan Islands Archaeology Project directed by the first two authors since 1983, allowing for the construction of a solid late-precolonial backdrop and further critical evaluation of early colonial disentanglements, inceptions and transformations (Antczak and Antczak 2015). These activities are part of the ERC-Synergy project NEXUS1492 led by the Faculty of Archaeology, Leiden University in collaboration with several Venezuelan institutions.

In this chapter, we critically merge the archaeological data, documentary sources, and the ecology of the pearl oyster in order to provide a novel understanding of the socionatural conditioning of the earliest colonial settlements on the islands of Margarita, Coche and Cubagua. This chapter opens with a historical contextualization of Nueva Cádiz reviewing previous investigations in the area. Next it discusses the results of the research conducted by the authors. Broad interpretations of early colonial campsites together with suggestions for future research conclude this chapter.

2 Nueva Cádiz de Cubagua: History and Research Antecedents

Cubagua's pearls achieved European notoriety after Christopher Columbus' third voyage to the New World in 1498. He was stunned by the natural abundance of pearls and their widespread use as indigenous body ornaments, and baptized northeastern Venezuelan region as the Gulf of Pearls (*Golfo de las Perlas*) (Colón [1498]1997, 79–80; de Las Casas 1981, 31). Columbus' letter to the Catholic Monarchs of Spain, Ferdinand and Isabella, spread the news across Europe (del Verde [1499]1989; Cantino [1501]1989) and attracted a throng of sailors to the region, also called the Coast or Island(s) of Pearls (Castellanos [1589]1987; de Las Casas 1997; del Verde [1499]1989; Fernández de Oviedo

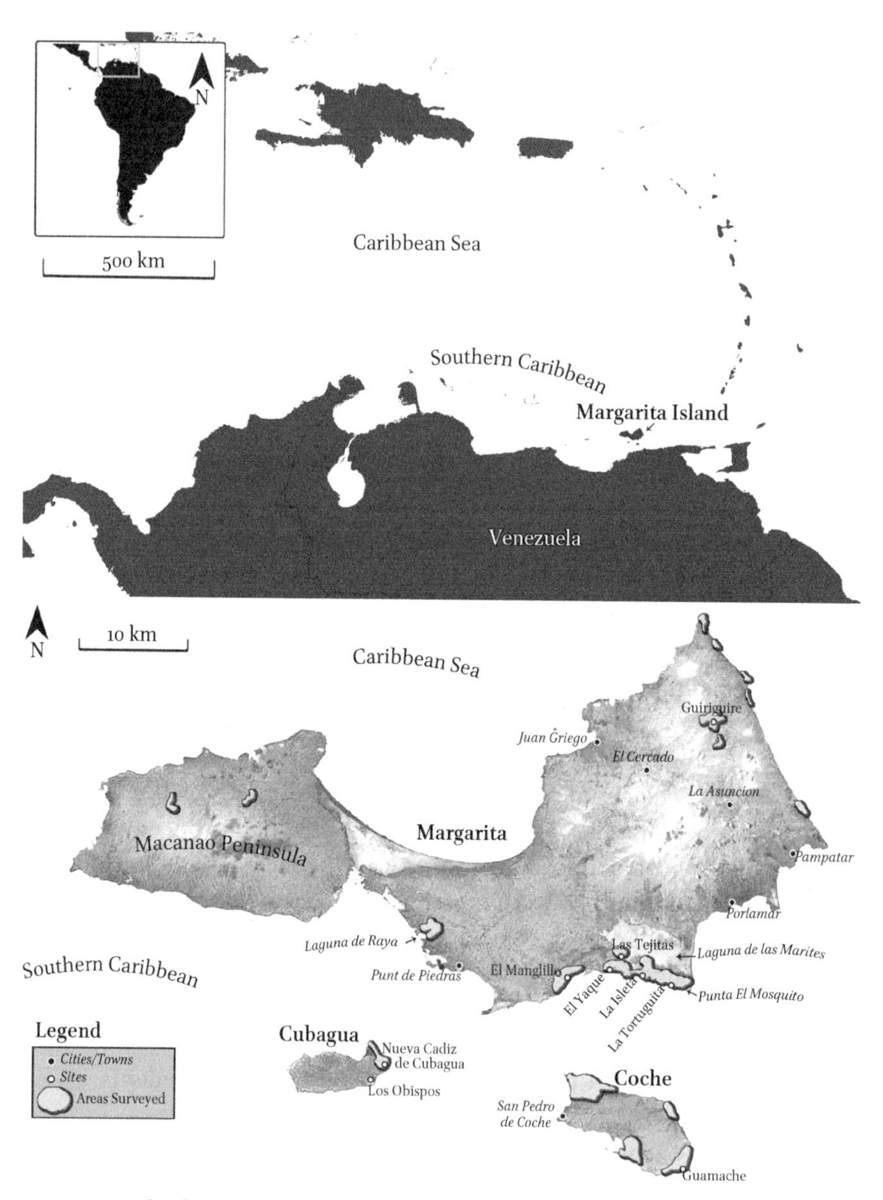

FIGURE 7.1 Islands and sites. Top: Margarita, Coche, and Cubagua Islands in the Caribbean; bottom: sites referred to in this chapter

[1535]1986; Mártir de Anglería [1530]1988; López de Gómara [1552]1988). Voyages as early as 1499 by Alonso de Ojeda, Amerigo Vespucci, Pedro Alonso Niño and Cristóbal Guerra procured detailed information about the feasibility of pearl 'businesses' (Donkin 1998, 314–315; Sauer 1966; Zubiri 2002). The exploitation of this resource began soon after.

In 1521, the previously existing cluster of temporary campsites (or *rancherías* as they were called in documentary sources) was converted into an *asiento* or administrative seat. In 1526, the *asiento* became the *villa* or town hall of Santiago de Cubagua (Cunnil Grau 2004, 60), and on the 12th of September 1528, the *Cédula Real* (Royal Decree) proclaimed the foundation of the *ciudad* (town/city) of Nueva Cádiz (Otte 1977, 87). After a decade of prosperity, the yields of Cubagua pearls dwindled and a frenetic, but ultimately unsuccessful, search for new oyster beds was performed on almost all of the Venezuelan islands (Arellano Moreno 1950, 180; Ramos 1976, 179–207). By 1538, a new campsite of pearl fishers from Cubagua was established in Cabo de La Vela in present-day Colombia, some 1000 km to the west (González 2002; Guerra Curvelo 1997; Vásquez 1989). Farther away still, pearls had also begun to be exploited in the Pearl Islands in Panama (Camargo 1983; Cipriani et al. 2008) and in Baja California (Gerhard 1956). By the early 1540s the town of Nueva Cádiz had been abandoned (Vila 1948; see also Vela Cossio and García Hermida 2014).

All in all, Amerindian slaves, pearls and gold, in addition to the quotidian 'bartering' or just forceful stealing of staple foods from the indigenous peoples (maize, cassava and other foodstuffs), triggered and sustained Spanish interest in the exploration and conquest of the northeastern part of South American *Tierra Firme* through the initial decades of the sixteenth century (Arcila Farias 1946, 1983; Arellano Moreno 1950; Jiménez 1986; Vila 1978).

This eventful period of Venezuelan history has generated much scholarly work in Venezuelan historiography and anthropology (Ayala Lafée-Wilbert and Wilbert 2011; Boulton 1961; Brito Figueroa 1966; Cervigón 1997, 1998a; Cunnil Grau 1993, 2004; Gabaldón Márquez 1988; Morón 1954; Otte 1961, 1977; Rodríguez Velásquez 2017; Rojas 2008; Sanoja Obediente and Vargas Arenas 1999; Velázquez 1956; Vila 1961, 1963, 1978). Topics related to the rowdy and boisterous social life of this early town nourished Venezuelan literature (Azócar de Campos 2009; Nuñez 1988; Pacheco et al. 2006) and cinematography (e.g. Arreaza-Camero 1993), and stirred up the imagination of sixteenth-century engravers (Champlain 1989; de Bry 1990; O'Brian 1996).

However, despite Cubagua's near omnipresence in Venezuelan narratives, these are characterized by three important drawbacks: (1) the heavy emphasis placed on Spanish deeds known from documentary sources; (2) the paucity of archaeologically grounded interpretations where Amerindian and African social actors may yet become visible and endowed with agency; and (3) the emphasis on early colonial urban spaces and lack of studies on the indigenous depopulation and desettlement and European resettlement of non-urban scapes on the islands of Margarita, Coche and Cubagua after 1498. These constraints contribute to the perpetuation of flawed, incomplete and unidirectional grand narratives.

Although several documents from the early sixteenth century mention the presence of Spanish huts and, later, buildings and urbanized spaces on Cubagua Island, the archaeological evidence of these phenomena was for a long time understudied. In Nueva Cádiz as in the surrounding region, due to the imposition of colonial regimes, indigenous peoples and European newcomers were involved in myriad new entanglements in which ethnic, linguistic, racial, gender and other social statuses were often violently redefined in everyday 'civilizing' practices (e.g., Deagan 2003; Gosden 2004; Silliman 2015; Voss and Casella 2012). The ever-changing nature, dynamics and intensities of these processes inspired international researchers to work on the topic, essentially drawing from rich documentary sources (Bénat Tachot 2015; Dawson 2006; Helmer 1962; Idyll 1965; Mosk 1938; Orche 2009; Perri 2009; Quiévreux 1900; Warsh 2010, 2018; Willis 1976, 1980; Woodruff Stone 2014). The pearl fishery was also approached by historical ecologists (Cipriani et al. 2010; MacKenzie et al. 2003; Romero 2003; Romero et al. 1999). But despite the evident remains of Nueva Cádiz lying in the sands of Cubagua (Rugil 1892), archaeology was long silent. At the beginning of the twentieth century, Leonard Dalton (1912, 183–184) summarized this situation well by arguing that on Cubagua '...diligent search and delving will reveal relicts of the fifteenth-century [this is an error; it should be sixteenth century] settlement [and yet] ... nothing seems to have been done in the way of archaeological excavation....'

The needed excavations were finally carried out in the second half of the 1950s by José Maria Cruxent and his collaborators; but these only half-opened a fascinating Pandora's Box (Cruxent 1955, 1964, 1969, 1972, 1980; Cruxent and Rolando 1961; Cruxent and Rouse 1958; Ferris 1991; Goggin 1960, 1968; Lister and Lister 1974; Rouse and Cruxent 1963; Vaz and Cruxent 1978; Vila 1961; Willis 1976, 1980; Wing 1961). According to Cruxent (1972), the cultural strata in the ruins of Nueva Cádiz lay only at a shallow level. We confirmed this shallowness by analyzing the annotations made by John Goggin who accompanied Cruxent in his initial excavation in December of 1954 as well as the data attached to the objects that are currently held in the Museum of Natural History in Gainesville and in the Peabody Museum at Yale University in New Haven. In general, the maximum depth of the excavation was 40 cm but approximately 98% of all excavated units reached only the depth of 15 cm. This resolution precludes any clear-cut spatial or temporal discrimination among the different categories of Amerindian materials from: (1) precolonial times (pre-1498); (2) post-Contact but pre-Nueva Cádiz times (1498–1528); (3) the time span of the town of Nueva Cádiz (1528–1542); and (4) post-1542 to the end of the sixteenth century. In addition, there is no doubt that Nueva Cádiz's boisterous lives taking place on the sandy surface of a semi-deserted island contributed to the daily intermingling of the archaeological materials entrapped in the shallow superficial strata. The postdepositional processes that could bias the original deposition include centuries of almost uninterrupted human transit across the ruins associated with trampling, building, transporting loads, preparing and repairing fishing gear, disposing of rubbish, goat roaming, and looting. These processes acted together with natural agents such as winds, tropical storms, salinity, high temperatures, and the bioturbation produced by land crabs (*Cardisoma guanhumi* and *Gecarcinus ruricola*), rabbits (*Sylvilagus floridanus margaritae*), burrowing owls (*Athene* sp.) and, probably, common rats (*Rattus* sp.). Finally, the archaeological field methods used in the 1950s also lacked the microstratigraphical and microcontextual approach, and fine-mesh sieving.

Despite the above weaknesses, some of the intricacies of the contact period on Cubagua began to be illuminated by Cruxent and Rouse (1958) within the perspective of the historical-cultural chronology of north-eastern Venezuela. Close to the La Aduana Archaic Age shellmidden (its beginnings date to 4150±80 BP), Cruxent and Rouse (1958, 1, 112) found a small surface scatter of potsherds painted with straight lines that seemed like the Playa Guacuco style from Margarita. This scatter was dated to between ca AD 1150 and 1500 and classified as a member of the (impoverished) Dabajuroid series from north-western Venezuela that 'traveled' from there through trade. Cruxent and Rouse found the Playa Guacuco pottery closely related to the early colonial pottery of Nueva Cádiz and to the Los Obispos styles, which purportedly developed subsequently *in situ* on Cubagua Island.

In 2008, Aníbal Carballo performed an archaeological survey on Cubagua aimed at reconstructing the changing cultural landscapes of the island. Although 36 new sites and 27 isolated features were added to the archaeological map of this island, the survey confirmed the scarcity of Saladoid pottery (Carballo 2014, 16, 79–80, Plano 13; 2017). This singularity was also confirmed by unsystematic surveys conducted on Cubagua by the authors in 2014. Carballo (2014, Planos 14, 15, p. 66, Lám. 25, p. 152) further reported only a small number of Playa Guacuco potsherds and found one potsherd assigned to the Krasky style from Los Roques Archipelago, pertaining to the Valencioid series from north-central Venezuela. Some new data related to the first religious sanctuary of Nueva Cádiz and its cemetery has also been yielded by excavations carried out between 2007 and 2008 by the archaeologist Jorge Armand (2017). Our research into the early colonial settlements on Margarita, Coche and Cubagua began to take shape based on the above-outlined scenario.

3 Surveys of Early Colonial Sites on Margarita and Coche Islands

Searching for the archaeological signatures of sixteenth-century campsites, in 2014 we carried out initial surveys of the heavily populated town of Punta de Piedras on Margarita Island as well as in the surroundings of the ruins of Nueva Cádiz de Cubagua. These attempts produced archaeologically unsatisfactory results. Therefore, our attention moved eastwards towards the southeastern coast of Margarita and to Coche Island where four pedestrian archaeological surveys were undertaken throughout 2015 and 2016. These included systematic monitoring of surface in a strip of ca 500 m along the coasts and the excavation of test pits and small trenches. In the next sections we focus on material remains recovered at two sixteenth-century sites: La Tortuguita on Margarita and El Guamache on Coche Island.

3.1 *Margarita Island*

3.1.1 Documentary Sources

The number of early historical sources related to the southeastern Margarita is very limited. Subero (1996, 132), drawing from unspecified documents, assures that the first dwellings were erected by European pearl seekers not only on Cubagua but also on the southeastern coast of Margarita as early as August of 1509. Some other sources put the origin of *rancherías* on Cubagua at the very beginning of the sixteenth century (Benzoni [1565]1991, 570). This early origin, however, was erroneously attributed to de Las Casas's time-independent statement (de Las Casas 1981, 25). Here, we use the date of 1516 as the documentarily supported origin of the early Spanish *rancherías* on Cubagua (Otte 1977, 87). Nonetheless, we are aware that the gargantuan profits offered by the bountiful pearl oyster beds drew European pearl seekers to these islands very soon after the year 1500.

3.1.2 Archaeological Evidence

The "Punta Mosquito" site on Margarita was mentioned by Theodoor de Booy (1916, 11–12) as "a favourite abode of the aborigines." We presume that he did not refer exactly to the site that currently bears that name on the most southeastern point of Margarita, because it shows a rather inhospitable environment to be an "abode." De Booy (1916, Figure 3) further stated that pearl oyster shell deposits found on the shore dunes at "Punta Mosquito" and interspersed by layers of ash could be interpreted as the remains "from early Spanish pearl fisheries." According to our surveys, these shell deposits had to be situated to the west of Punta Mosquito, somewhere between the El Manglillo and El Yaque sites (Figure 7.1b). Towards Punta Mosquito, the coast not only is barren and hardly accessible to canoes but farther to the east – especially around the La Tortuguita site – colonial remains are very abundant, visible on the ground surface and cannot go unnoticed.

Cruxent and Rouse (1958, 1, 117) mentioned "Punta Mosquito" and the archaeological site of "Los Mayas" on the southeastern coast of Margarita. This site, pointed out in the first volume of the above referenced publication, corresponds, most probably, to the "Las Maras" site that is marked on the map provided in the second volume (Cruxent and Rouse 1958, 2, Figure 11). There, the "Las Maras" site is adjacent to "Punta Mosquito," the southeasternmost tip of the island. We conclude that it is this site that coincides with the La Tortuguita site that was found during our surveys. Cruxent and Rouse (1958, I, 117) further stated that Las Maras yielded some European pottery from the sixteenth and seventeenth centuries and assigned it to the Obispo style (post-Nueva Cádiz dated to ca 1622–1640; but see Cruxent and Vaz 1978, 369). However, they recognized that scarce European materials found in this and other surface scatters across the Macanao Peninsula (the western part of Margarita), at the Güirigüire (eastern Margarita), and in Guamache on the island of Coche, all included by them in the Obispo style, may in fact be classified in future research as an independent style (Cruxent and Rouse 1958, 1, 117–118). The only decoration found on the associated (possibly) transformative ceramic materials consists of rectangular or tubular appendixes or false handles applied to the shoulders of the vessels (Cruxent 1980, 174; Cruxent and Rouse 1958, 1, 118)

Our surveys yielded new archaeological data from several colonial sites located on Margarita and Coche islands. But before turning to discuss in more detail the sites of La Tortuguita (Margarita) and Guamache (Coche), let us briefly introduce the related sites of El Manglillo and El Yaque (Margarita) whose locations were indicated to us by Mr. Luis Lemoine (Fundación Arca, Caracas) (Figure 7.1b). They were surveyed in 2015 with Luis Lemoine, Cecilia Ayala and Pedro Rivas (Fundación La Salle de Ciencias Naturales, Caracas), and Werner Wilbert (Instituto Venezolano de Investigaciones Científicas, Caracas). The El Manglillo site features a series of dune formations covered with xerophytic plants and is crosscut by deep ravines conducting rainwater to the sea. Fragments of Spanish olive jars and Columbia Plain dishes dated to the first half of the sixteenth century and sherds of coarse earthenware, probably of local transcultural production, were recovered on the dunes (Luis Lemoine, personal communication 2015). Test pits of 1x1 m were excavated but only one trench of 2x9 m situated on the slope of a large dune revealed a cultural layer at a depth between 20 and 65 cm. It yielded ceramic materials like those found on the surface as well as bivalve shells, fish vertebrae, and the features of four hearths. These fireplaces could easily be seen from the sea as well as from inland both day and night indicating that the camp was not a clandestine location. The environmental characteristics of the El Yaque site, located on the western border of Laguna de Las Marites, are similar to those of El Manglillo. A test pit of 2x1 m and a small trench of 2x5 m were excavated revealing a cultural layer of greyish sand at a depth between 15 and 35 cm. Hearths are absent. Ceramic materials are overtly like those found at El Manglillo. Small patches of bivalve shells show rather marginal dependence on locally available marine resources. The mandible of a non-local deer (*Odocoileus virginianus*) suggests that game was brought from the adjacent mainland. El Manglillo and El Yaque most probably did not host a well-established group of *rancherías*, but represented the remains of recurrent encampments under the open sky taking advantage of the benign climate and sandy seacoast. These camps may be a result of intense mobility on the part of largely 'amicable' indigenous peoples who interacted with the Spanish in pearl fishery provisioning. They indicate a rather 'safe' atmosphere that, according to Juan de Castellanos ([1589]1987, 120–123), a resident of Nueva Cádiz, characterized the Spanish emplacements on Margarita in the 1520s and 1530s.

The site of La Tortuguita is located between the modern settlement of La Isleta and Punta El Mosquito (Figure 7.1b). To the north, a sandy shore gives way to hypersaline lagoons that produce salt. Behind the saltpans, the terrain rises slightly and undulates with small hills cut by ravines serving as cones of ejection for sporadic but heavy rains. The vegetation is largely xerophytic (Campos and Guzmán 2002); firewood and permanent freshwater reservoirs are absent (Rojas 2010). The terrain is hardly apt for agriculture and except for rabbits and snakes, the land fauna is poor. We cautiously assume – until new, especially paleobotanical, evidence can test the matter – that similar semi-deserted conditions existed in this area during the sixteenth century. Several surface scatters of sixteenth-century pottery and other materials were found dispersed in patches across one square kilometer at this site (Figure 7.2d, e). Test pits showed that the material may be found to a depth of 15–25 cm. This fact, together with the discovery of one fragment of Spanish stucco or plaster, makes us confident that postholes and hearths of sixteenth-century *rancherías* and probably later *estancias*, or homesteads, are possible finds in future research.

The surveys at La Tortuguita yielded hundreds of archaeological remains. The quantity and variety of pottery shows a wide range of techniques and decorative motifs ascribed to European along with Amerindian manufacture (Figures 7.3–7.5). There are also possible transcultural or hybrid wares that seem to contain both European and Amerindian characteristics (Figure 7.5d–f). Future analytical scrutiny may also identify elements of African origin. The presence of more permanent dwelling structures may further be strengthened by the recovery of stone grinding tools (*manos* and *metates*), ceramic griddles,

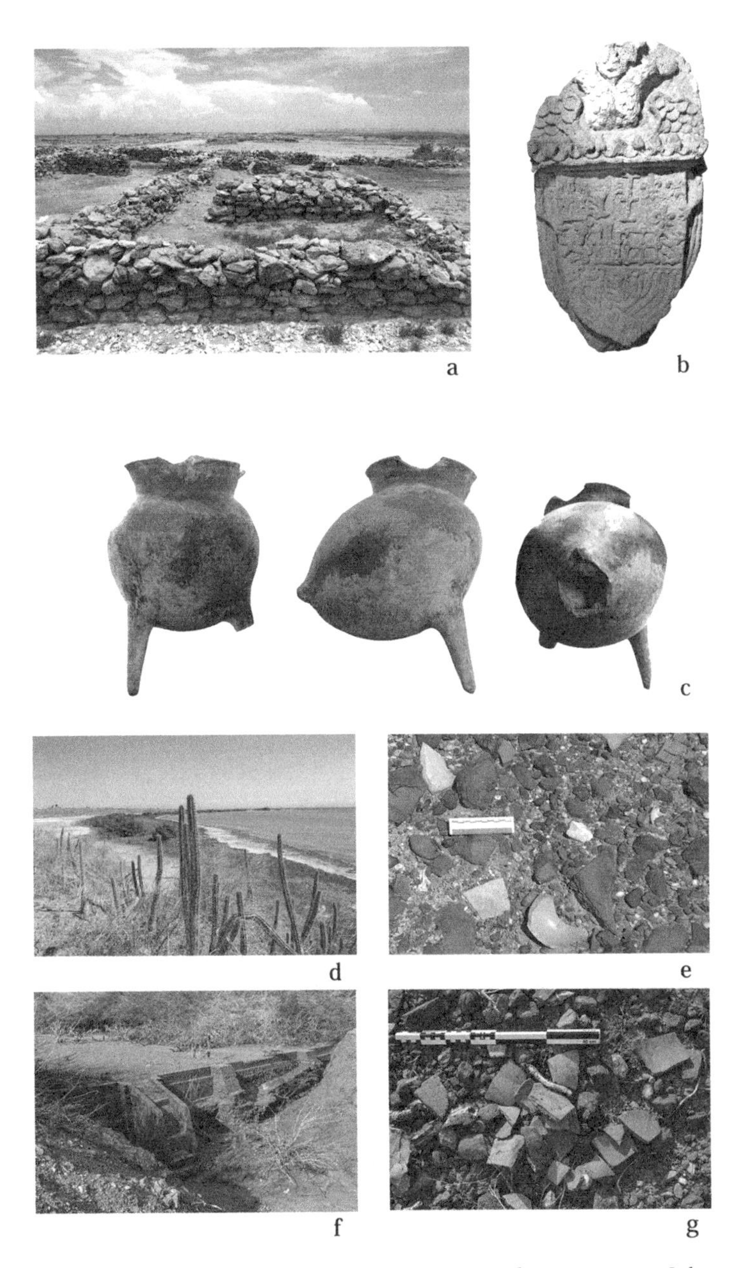

FIGURE 7.2 Sites and materials: (a) ruins of Nueva Cádiz town in 2014, Cubagua; (b) coat of arms of Nueva Cádiz, stone, Museo de Nueva Cádiz, Asunción, Margarita; (c) large tripod vessel with red-slipped upper body part, found in Nueva Cádiz in 1950s, probably Amerindian pottery but of undefined stylistic affiliation, height 28.5 cm, Museo Marino, Punta de Piedras, Margarita; (d) view from La Tortuguita westwards to Punta Mosquitos, Margarita; (e) scatter of Spanish pottery, La Tortuguita, Margarita; (f) abandoned dam with 'old' stony foundations, Guamache, Coche; (g) scatter of Spanish olive jar fragments, Guamache, Coche

FIGURE 7.3 Selection of primarily sixteenth-century European ceramics from Margarita, la Tortuguita, and Güiri-güire, Venezuela (surveys 2014–2016): (a) linear Blue Morisco Ware, *plato* (plate), Seville, prob. pre-1550 due to well-defined angle between the rim and the center of the interior surface; (b) Mottled Blue Morisco Ware, *jarro* (jug), Seville, ca. 1550–1625; (c) Decorated Blue Morisco Ware, *jarro*, Seville, ca. 1550–1625; (d), Decorated Blue Morisco Ware, *cuenco* (bowl), Seville, ca. 1550–1625; (e) Plain White Morisco Ware, *plato*, Seville, pre-1550 due to marked central boss and well-defined angle between the rim and the center of the interior surface; (f) Plain White Morisco Ware with Green Edge, *cuenco*, Seville, prob. pre-1550; (g) Decorated Blue Morisco Ware, *plato*, Seville, ca. 1550–1625; (h) Plain Blue Morisco Ware, *escudilla* or *albarello*, Seville, ca. 1500–1570; (i) Decorated Blue Morisco Ware, *cuenco*, Seville, ca. 1550–1625; (j) Plain White Morisco Ware, *escudilla*, Seville, prob. pre-1550; (k) Lead-glazed red earthenware, *plato*, prob. Seville, prob. sixteenth century; (l) Plain White Morisco Ware, plato, Seville, pre-1550, with possible owner's/user's mark on back; (m) Spanish *botija* (olive jar) with incised mark on shoulder; (n) Spanish *botija* rim with impressed mark, Seville, sixteenth to mid-seventeenth century; (o) Spanish green lead-glazed *cantimplora*, Seville, pre-1550; (p) Spanish *botija* rim with disk-shaped *botija* sherds probably used as stoppers, Seville, sixteenth to mid-seventeenth century; (q) perforated triangular Spanish *botija* fragment, Güiri-güire

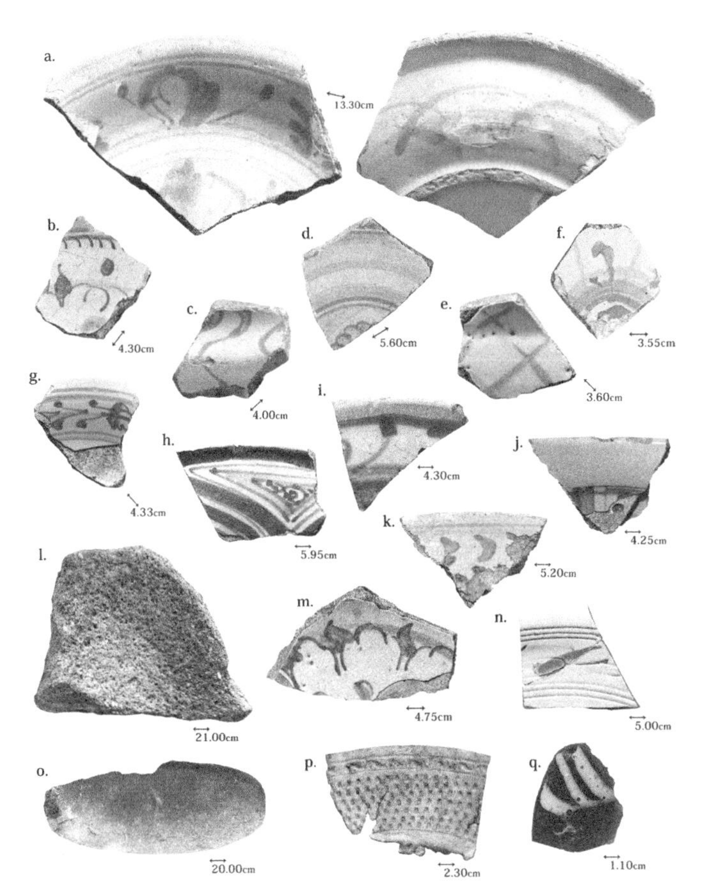

FIGURE 7.4 Selection of late sixteenth/seventeenth-century materials from Margarita, La Tortuguita site, Venezuela (surveys 2014–2016): (a) Ligurian *berettino* majolica, with *calligrafico a volute* design, bowl, late sixteenth to first quarter of seventeenth century; (b-f) Ligurian *berettino* majolica, plate, late sixteenth to first quarter of seventeenth century; (g) Ligurian berettino majolica, with *calligrafico a volute* design, bowl, late sixteenth to first quarter of seventeenth century; (h) possibly Portuguese Blue on White faience, plate, 1625–1650; (i) Seville Blue on White ware, *plato*, late sixteenth to mid-seventeenth century; (j) Ligurian *berettino* majolica, bowl, late sixteenth to first quarter of seventeenth century; (k) Ligurian Blue on White majolica, plate, second half of sixteenth to first half of seventeenth century; (l) fragment of lithic *metate*; (m) Ligurian Blue on White majolica, plate, second half of sixteenth to first half of seventeenth century; (n) polychrome lead-glazed *graffita tarda*, plate, Pisa, ca. 1550–1650; (o) lithic *mano* (grinding stone/percutor); (p) flattened copper-alloy thimble; (q) fragment of hand-blown blue glass with painted white enamel, prob. Venetian, sixteenth or seventeenth century

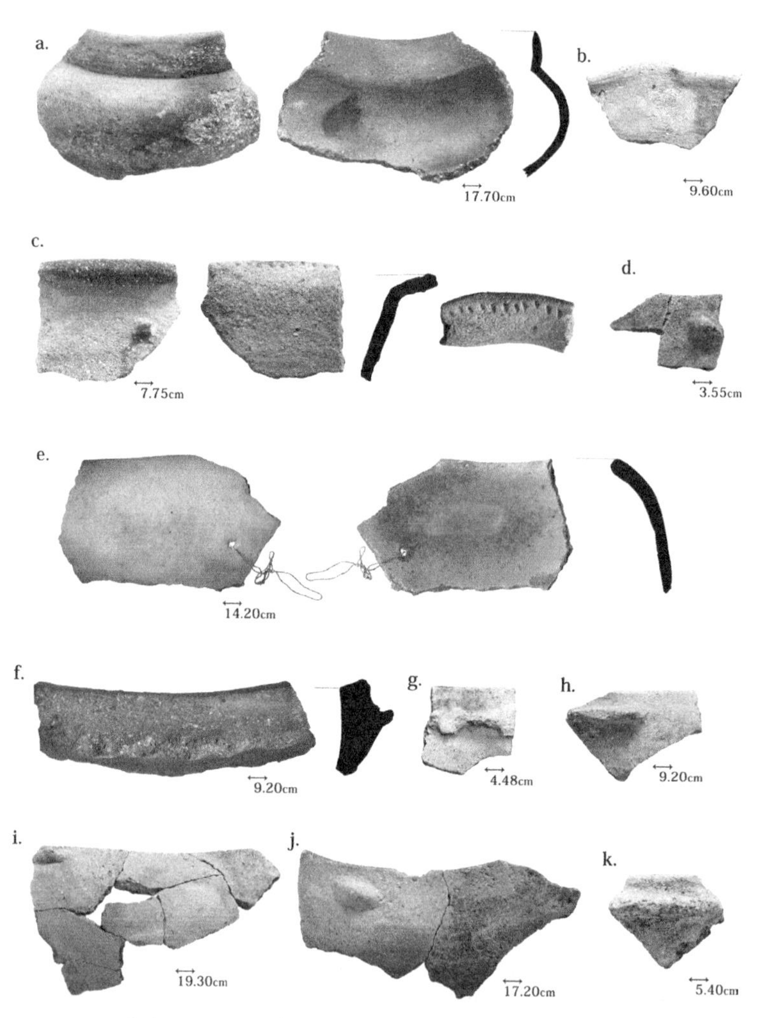

FIGURE 7.5 Indigenous transformative wares (sixteenth–seventeenth century) and criollo-ware (eighteenth–nineteenth century) from Margarita, Venezuela (for localities see Figure 7.1 [bottom]): (a) large indigenous cooking *olla*, La Tortuguita, prob. sixteenth or seventeenth century; (b) wavy-shaped and externally red-slipped rim, open bowl, Las Tejitas, prob. eighteenth–nineteenth century; (c) open bowl rim, incised on lip's internal side, a small knob or false handle on shoulder, La Tortuguita, prob. sixteenth or seventeenth century; (d) bulbous pellet applied to the shoulder of red-slipped indigenous bowl, La Tortuguita, prob. sixteenth or seventeenth century; (e) fragment of large restricted bowl, perforated and with copper wire attached to it, La Tortuguita, sixteenth–nineteenth century; (f) wavy strip or 'false' handle applied around the shoulder of open bowl, El Manglillo, prob. eighteenth–nineteenth century (similar decoration still produced by potters in El Cercado, Margarita); (g) wavy protuberance with digital impressions, Las Tejitas, prob. nineteenth century; (h, k) possible false handles, Las Tejitas, prob. nineteenth century; (i-j) bulbous pellets on shoulders of large restricted bowls, La Tortuguita, sixteenth–nineteenth century

and many spatially concentrated fragments of olive jars presumably reused to store freshwater. These artifacts accounting for food preparation and consumption rather than food procurement indicate that the staple food was brought rather than produced *in situ*. The absence of metal tools compared to the presence of multifunctional indigenous tools made of quartz indicates that Amerindian peoples inhabited the La Tortuguita site together with the Spanish (*peninsulares* and *criollos*), and probably also with enslaved Africans and *mestizos*

All these characteristics may have met the requirements of the early colonial pearl fishery *rancherías* but the occupation of La Tortuguita site lasted much longer. European pottery reveals that the occupation of this site could have started contemporaneously with Nueva Cádiz, i.e., somewhere in the early 1520s. It continued across the colonial times, far beyond the demise of Nueva Cádiz in the early 1540s. A map from 1661 drawn by the Spanish military engineer Juan Betin depicts some forms of habitation west of Punta El Mosquito that could coincide with the La Tortuguita site (Nectario Maria 1960, 131–136). Also, fragments of blue shell-edged whiteware, red clay pipes, and nineteenth-century case bottles collected during our surveys indicate that this site was still inhabited or temporarily frequented during the late eighteenth and nineteenth centuries (von Humboldt [1814–1825]1995). Fishery of the pearl oysters in front of Punta Mosquitos and on Coche was active between 1845 and 1850s (Quiévreux 1900, 446). Through time, the settlement at La Tortuguita would have gradually lost its initially purely temporal character, which was akin to some other encampments on Margarita before 1536 (Cunnil Grau 2004, 63). According to the documents its occupation would fluctuate according to the erratic pearl fishery. It would become more permanently inhabited as it transitioned into a multifunctional colonial ranch (*estancia*), or it could even have been temporarily abandoned. Remarkably, the persistent presence at almost all surveyed sites of hand-coiled coarse red earthenware indicates the continuous use of locally made pottery (*cerámica criolla*). Some of its forms and decoration derived from the town of Nueva Cádiz, persisted in the post-Nueva Cádiz *rancherías*, and seem to be still alive today in the locally made traditional pottery from El Cercado in Margarita (Acosta Saignes 1964; Gómez 2004, 160–162, 184; Ocanto and Baptista 1998) as well as from Manicuare, near Cumaná on the adjacent mainland coast (Ginés et al. 1946).

The elements that would have been deemed attractive and, therefore, could have pulled sixteenth-century settlers into the La Tortuguita site, were the proximity to (1) harbors and sandy shores; (2) pearl oyster beds; (3) fishing and mollusk-gathering grounds; (4) saltpans; and (5) pasture for goats. Also, the visibility from the hilly terrain towards both the sea and the interior of Margarita would have permitted the monitoring of the seascape traffic between Margarita, Coche, Cubagua and the mainland. However, fresh provisions such as fruits, vegetables, meat and water had to be brought by canoes or accessed from inland localities of Margarita.

3.2 *Coche Island*

3.2.1 Documentary Sources

The Island of Coche appears in Spanish documents beginning in mid-1520. On the 27th of August, 1520, the Spanish Crown granted a license to Juan de Cárdenas to "establish the seat of the barter of pearls [*asiento de rescate*] in Cubagua, [and] to get pearls on Coche [Island] and in Punta de Araya [on the adjacent mainland]" (Subero 1996, 143). On July 28th, 1526, the island was entrusted (*encomendada*) to Juan López de Archuleta (Subero 1996, 127), a member of the elite in Nueva Cádiz and the owner of two pearl fishing canoes (*señor de dos canoas*) (Otte 1977, 50). The *Cédula Real* of this assignment describes an island called *Conche*, a name derived from the Spanish name *concha* (shell) and not, as suggested by Bartolomé de Las Casas (1981, 32), from a small local deer still abundant on this island during late colonial times (von Humboldt [1814–1825]1995, 74). The island was "[F]rom 2 to 3 leagues around, uninhabited and without being previously entrusted to any other person, and because he [Archuleta] wanted to have in it his cultivation, animal and cattle breeding and other husbandry in order to provide provisions for his farmhouse and [for the benefit] of that island" (Otte 1961, 12–13). In 1799, Alexander von Humboldt ([1814–1825]1995, 47) was informed by local indigenous informants that Coche "...had never been inhabited [by the indigenous peoples]."

In 1529, according to the letter that Diego Caballero (resident of Nueva Cádiz) sent to King Carlos v, rich pearl oyster beds were discovered close to Coche (Morón 1954; Pinto 1967; Subero 1996, 46). During solely the month of January 1529, more than 1,500 marks (12,000 ounces) of pearls, the equivalent of over 17 million carats, were gathered from the pearl beds of Coche (Galtsoff 1950, 4). By this year, other Spaniards already had *granjerías* (husbandry operations, in this case dedicated to pearl fishing) on Coche and some royal officials were already established there (Otte 1961, 108–110). In June of 1529, the Queen wrote that Hernando Carmona, the bailiff (*Alguacil Mayor*) of Nueva Cádiz, had informed her that because of the discovery of rich oyster beds, many people who lived in Nueva Cádiz town had moved to Coche. Carmona even asked the Court for permission to use the title of *Alguacil Mayor* of Cubagua on Coche (Subero 1996, 104). A month later, probably as the result of the burgeoning pearl boom, the Court unveiled immediate plans for the establishment of a "village of Christians" (*pueblo de cristianos*) on Coche including the services of a clerk and a village council (Subero 1996, 112). The new oyster beds were large, and their discovery was not a short-lived boom. Still by 1573, the bishop Fray Pedro de Agreda was informing the Crown that Coche was benefitted by its "generous" pearl oyster beds (Agreda [1581]1964; Subero 1996, 199).

Drawing from these documentary sources, it becomes clear that specific sectors of Coche Island were visited by Europeans from the early 1520s, if not before. Later, these sites were populated by the owners of the canoes (*señores de las canoas*) plus their Amerindian slaves and household servants (*naborías*). As Nueva Cádiz de Cubagua was the only Spanish town in northeastern Venezuela during the first decades of the sixteenth century, the earliest settlements on Coche Island most probably took the form of *rancherías* of pearl fishers.

3.2.2 Archaeological Evidence

During the survey of Coche Island, we located a few Archaic through Ceramic Age sites and materials which are currently under analysis. The only sixteenth-century site was found in Guamache. Overly, the archaeological evidence on Coche is very scarce compared to other Venezuelan islands (Antczak and Antczak 2006).

The Guamache site, briefly mentioned by Cruxent and Rouse (1958, vol. 1, 117, vol. 2, Figure 7), yielded some European pottery assigned to the sixteenth and seventeenth centuries and classified as belonging to the Obispo style. It is located on the southeastern shore and close to the largest contemporary artisanal fishery port on this island (Figure 7.1b). Similar to La Tortuguita, the terrain is hilly, covered by xerophytic vegetation and cut by deep ravines. The location of the port is privileged: boats can be safely beached there, contrary to the rather cliffy windward coast to the northeast. One of the site's great assets is that the deep ravines could be easily connected and closed by a simple dam fashioned to conserve the freshwater (Figure 7.2d). Today abandoned, the dam still presents its stony structure with bases that may date to colonial times when settlers could also have stored up freshwater in this manner. However, an unquestionable signature of the existence of such a dam in early colonial times did not emerge in our survey even if the dam lies only a few hundred meters from the seashore and from the hilltop where the archaeological remains were found. The southern slope of the hill closest to the port contained scatters of Spanish olive jars and majolica (Figure 7.2e). Three test pits of 1x1m excavated on this slope provided no evidence whatsoever of a cultural layer under the surface. Yet we continue to presume that the hilltop may still contain some structural remains (e.g. postholes, remains of adobe walls) of earlier *rancherías* which could have been established here in or near the 1520s. Some cultivation of land and grazing (mainly goats) could have been carried out in the past on the slopes of the hills. Their tops could, at the same time, have provided a fresher climate while offering ample strategic views towards the surrounding seascape and the mainland coast. All these locational advantages might have been attractive to early colonial settlers.

4 Pearl Oyster Ecology Comes to the Fore

The marine conditions between the islands of Margarita, Coche and Cubagua and the mainland peninsulas of Araya and Paria, conditions featuring seasonal upwellings bringing deep cold waters rich in nutrients to the surface (Villamizar and Cervigón 2017), have been indicated as highly beneficial for the thriving of mollusk communities – especially large pearl oysters (*Pinctada radiata*) (Capelo and Buitrago 1998). The abundance of these mollusks in the local waters shines important light on the study of human-environment interactions some of which might have been anthropogenically manipulated during the contact period (MacKenzie et al. 2003). In fact, ecologists have documented the existence of large pearl oyster beds along the coasts of Cubagua where an early sixteenth-century pearl fishery flourished (Carballo 2014, Lám. 47, p. 165; Romero 2003; Romero et al. 1999; Tagliafico et al. 2012). Large oyster populations were also indicated off the southeastern coasts of Margarita and Coche (Romero 2003, Figure 2). Statistical information collected during the 1930s and 1940s (Donkin 1998, Map 38; Galstoff 1950, Figure 1 and Table 2) documented pearl oyster beds facing the sites of Los Mosquitos, La Isleta, and El Yaque on the southeastern coast of Margarita, and fronting the southeastern coast of Coche. Salaya and Salazar (1972, Tables 1, 12, 17, Maps 1, 3, 4, 6, 8–10, 13) compiled pearl oyster fishery data collected between 1946 and 1969. These data show the persistent importance of the above-mentioned oyster beds. It is very important for this research to emphasize that the current location of rich pearl oyster beds coincides with the data provided by the early colonial documentary sources. This indicates the persistence of these beds in question across five centuries. Our research underlines that their current location is adjacent to the archaeological sites identified during the recent archaeological surveys. Thus, the material culture recovered at the sites of La Tortuguita on Margarita and Guamache on Coche (Figures 7.2–7.5), confirms the existence of sixteenth-century campsites at these sites corresponding to the historically documented pearl fishery. These sites might have endured through time, changing their character from temporary to permanent and from monofunctional to multifunctional. The adjacent oyster beds could also have undergone qualitative transformations. This has already been noted for the oyster beds at Cubagua. There, once-enormous aggregations of pearl oyster have been largely replaced by Turkey Wing (*Arca zebra*), the species currently more profitable for its protein content (Casas et al. 2015; Cervigón 1997, 1998a, 1998b; Hernández-Ávila et al. 2013).

5 Concluding Remarks and Future Research

The material remains discussed in this chapter were recovered during pedestrian surveys and surface collection offering little information about connections between scattered artifacts. Aware of these constraints, we confronted the recovered materials with pertinent documentary data and environmental and ecological variables. In this way we opened new avenues for further inquiry rather than provided final interpretations into the specific forms and dynamics of interaction between indigenous, African and European peoples and their material culture.

Drawing from the above insights we consider that further research in the field should adopt new methodological characteristics. Surveys should look to systematic large-scale aerial excavations to discover contextually bounded signatures of housing structures and delineate settlement layouts, in their stratigraphic sequences. Understanding the palimpsestic character of the archaeological deposits may prove especially fruitful at sites such as La Tortuguita and Guamache. This will allow us to bring the sixteenth-century *rancherías* to life and help us understand how, during the twilight of the early town of Nueva Cádiz de Cubagua, Spanish power was also expanding over the adjacent islands of Margarita and Coche.

Comparisons of the spatial organization of the above discussed sites with other similarly excavated colonial contexts in the region may permit the evaluation of transformations in the organizational principles and interactions of the pluralistic collectives that inhabited them. Understanding the past vibrant lives of what today are potsherds, lithics, glass and metal fragments, animal remains, and hearths scattered across the islandscape is crucial to this intellectual undertaking. However, in this chapter we have only begun to explore how – from the archaeological perspective – European and natives co-participated in the creation and transformation of larger social trends on the early colonial islandscapes of Margarita, Coche, and Cubagua. Further tasks require stretching the 'social imagination,' which must be in agreement with soundly recovered, identified and dated material culture that is itself in constant juxtaposition with documentary sources and environmental variables.

Accurate identification and dating of the recovered materials is essential. This chapter places special emphasis on determining the origin of European wares due to the necessity of dating the occupation of identified campsites and discussing their co-existence, or lack thereof, with the town of Nueva Cádiz. While analyzing the materials found at the colonial site of Maurica (near the modern town of Barcelona) currently at the Museum of Natural History in Gainesville (Rouse and Cruxent 1963, 132, 138–139) we realized that they were initially dated to 1620–1640 (Cruxent and Rouse 1958, vol. 1, 202). However, some of them may date to the sixteenth century (see also Cruxent and Vaz 1978) and indicate the expansion of Spanish power to the adjacent mainland coast during the time of existence of Nueva Cádiz.

Activities such as food provisioning and preparing, freshwater storing, firewood hauling and chopping, fireplace preparing and maintaining, shade and rain shelter erecting, post hammering, mollusk crushing, maize and coarse salt grinding, fishing net mending, and bartering with close and more distant neighbors all may have been largely intended human actions. But people themselves "are not always engaged in activities that have a clear objective" (Thomas 2017, 281). Artifact use at a specific site can be highly ambiguous, and often not the one intended by the producer (Silliman 2009, 213–214). We should also consider activities which were not merely labor-oriented, those which contributed vibrancy and lifelike texture to past human lives. Although such activities do not often leave material signatures, nevertheless, talking, praying, joking, smoking, resting, gaming, admiring, as well as those activities related to sex and hygiene, have formed an important part of *ranchería* life. None of these activities may be considered exclusively related to pearl fishery, but gave life to the early colonial *rancherías* and contributed to their endurance and transformation observable through the passing of colonial times.

We have discussed how the early colonial power's rapacious devouring of human and other-than-human resources was operationalized through the *ranchería* campsites and their material culture. The Spaniards were engulfing the different groups of indigenous peoples in myriad swiftly shifting scenarios of interaction and, at the same time, accumulating the native space in their non-native hands. In, around and through the *rancherías*, the New World was being forged by and for both the autochthons and the newcomers. But the temporary or seasonal early colonial *ranchería* as a specific unit of work and life has deep precolonial antecedents that precede the symbolic date of 1498 which marks the Third Voyage of Columbus. Therefore, a sound understanding of the late precolonial backdrop is necessary to study the early colonial disentanglements, transformations and enduring elements concomitant with European arrival.

With the passing of colonial time, some *rancherías*, such as those at La Tortuguita, may have lost their earliest, largely unifunctional character which derived from the activities related to pearl gathering that permeated interactions among Spaniards, Amerindians and Africans. The human components of these interactions also lost their initially relatively clear-cut ethnic, cultural or linguistic distinctions. Human relations then became largely structured by the colonially created racial distinctions (e.g. Mira Caballos 1997). Remarkably, in northeastern Venezuela, the very first generation of post-Contact Americans included such historically renowned *mestizos* as the conqueror Francisco Fajardo who was the son of Doña Isabel – a Guaiquerí Indian woman from Margarita – and the Spanish lieutenant Francisco Fajardo (Ayala Lafée-Wilbert and Wilbert 2011; McCorkle 1965). The category of social actors embodied by Fajardo should be considered when studying intercultural interactions in *rancherías*, especially in view of the fact that these Spanish-induced settlements of precolonial origin played an important though barely understood role in the early political-administrative structure of colonial northeastern Venezuela (Castillo Hidalgo 2005, 216).

In fact, we argue that precisely in this region it is in *rancherías* where the origin of several colonial transformations should be sought. They were par excellence "transculturation sites" *sensu* Domínguez (1978). To date, we are still unable to say much about the specificities of these interactions and transformations but diverse activities (e.g., pearling and goat grazing) could have occurred within the same islandscape in multifunctional *rancherías*, according to diversified task distribution and task timing. The extractive colonization so characteristic to early colonial encounters exploited native labor and local resources and influenced new habits and thoughts of peoples who clearly coexisted spatially and often shared the *materiality* referred to in this chapter. Acknowledging the latter fact is not to deny the ubiquitous presence of unequal power that endorsed further divisions and marginality (Silliman 2014, 68).

Finally, crucial is the critical interweaving of independent lines of evidence stemming from archaeology, archaeometry, archival research, ethnohistory, ethnography, experimental research, linguistics, descendant perspectives, and oral lore to examine possible convergences and contradictions. Both, the Nueva Cádiz town and the *rancherías* were an integral part of colonial scapes and their transformations were influenced by a fluctuating combination of socioenvironmental factors (e.g., Sluyter 2001, 411). In the case of Nueva Cádiz town, further research is necessary to examine natural factors such as reduced rates of accretion associated with low seawater temperatures and increased salinities in the Caribbean during the Little Ice Age that could have played an important role in the demise of its pearl fishery (Cipriani et al. 2010). Also, our research discussed in this chapter suggests that the ultimate collapse of the town may be related to the great versatility of temporary *rancherías*. The latter could be a more flexible and convenient spatial means to expand Spanish colonial influence beyond the stony walls of Nueva Cádiz' urban structure that embodied immobility and rigid centrality. In *rancherías* shared histories were being forged (Harrison 2014) and activities-centered life unfolded (Ingold 2007). Being the most popular form of settlement persistently utilized across the centuries, *rancherías* continue to thrive along the coasts and on the islands of the Southeastern Caribbean to the present day.

Acknowledgments

We thank Dan Bailey for valuable comments and Konrad A. Antczak for great insights and the identification of European pottery. Many thanks go to Cecilia Ayala Lafée-Wilbert and Werner Wilbert, as well as to Pedro Rivas, who actively participated in the early fieldworks and contributed with their knowledge, local contacts and experience. We also thank Fernando Cervigón (†), Bladimir Rodríguez, Teobaldo Castañeda, and Pablo Rodríguez from the Museo Marino de Margarita, and Grecia Salazar from Museo de Nueva Cádiz in Asunción. We express gratitude to our friends from Margarita: Régulo Briceño, José Voglar, Pablo Segundo Mata, Pavel Nuñez (†), Emanuel Narváez (Nelo), and Fernando Fernández. We acknowledge the companionship of the Venezuelan ONG Provita group. The Instituto del Patrimonio Cultural issued the respective permissions and the Guardia Nacional protected us during the fieldwork. This project has received funding from the European Research Council under the European Union's Seventh Framework Programme (FP7/2007–2013) ERC Grant agreement No. 319209, under the direction of Prof. dr. C.L. Hofman.

References

Acosta Saignes, Miguel. 1964. "Cerámica de El Cercado, Margarita." *Anuario* 1: 1–20.

Agreda, Pedro de. (1581)1964. "Relación de la Provincia de Venezuela, hecha por el Obispo de Coro, año 1581." In *Relaciones Geográficas de Venezuela*, compiled by Antonio Arellano Moreno, 221–231. Caracas: Biblioteca de la Academia Nacional de la Historia.

Amodio, Emanuele. 1991. "Relaciones Interétnicas en el Caribe Indígena: Una Reconstrucción a Partir de los Primeros Testimonios Europeos." *Revista de Indias* 51 (193): 571–606.

Antczak, Andrzej T. and Maria M. Antczak. 2015a. "Revisiting the Early 16th-Century Town of Nueva Cádiz de Cubagua, Venezuela." Paper presented at the *1st EAA-SAA Joint Meeting* Connecting Continents, *Curaçao, 24–27 November.*

Antczak, Maria M. and Andrzej T. Antczak. 2006. *Los Ídolos de las Islas Prometidas: Arqueología Prehispánica del Archipiélago de Los Roques*. Caracas: Editorial Equinoccio.

Antczak, Maria M. and Andrzej T. Antczak. 2015b. "Late Pre-Colonial and Early Colonial Archaeology of the Las Aves Archipelagos, Venezuela." *Contributions in New World Archaeology* 8: 1–37.

Antczak, Andrzej T., Maria M. Antczak, Roberto Valcárcel Rojas, and Andrej Sýkora. 2015. "Rethinking guanín: The role of northern Venezuela in circulation and valuation of Indigenous metal objects in the Circum-Caribbean macroregion." Paper presented at the *26th Congress of the International Association for Caribbean Archaeology, Sint Maarten, 19–25 July*.

Arcila Farías, Eduardo. 1946. *Economía Colonial de Venezuela*. México: Fondo de Cultura Económica.

Arcila Farías, Eduardo. 1983. *Hacienda y comercio de Venezuela en el siglo XVI*. Caracass: Banco Central de Venezuela.

Arellano Moreno, Antonio. 1950. *Fuentes para la Historia Económica de Venezuela (siglo XVI)*. Caracas: Tercera Conferencia Interamericana de Agricultura.

Armand, Jorge N. 2017. *Nueva Cádiz de Cubagua (1502–1545): Parábola de la Venezuela Pospetrolera*. Balti: Editorial Académica Española.

Arreaza-Camero, Emperatriz. 1993. "Cubagua, or the search for Venezuelan national identity." *Iowa Journal of Cultural Studies* 12: 108–131.

Ayala Lafée-Wilbert, Cecilia and Werner Wilbert. 2011. *Memoria Histórica de los Resguardos Guayquerίes: Propiedad y Territorialidad Tradicional*. Caracas: Ediciones IVIC.

Azócar de Campos, Elba. 2009. *Cubagua: Mas Allá del Signo*. Caracas: Universidad Simón Bolívar.

Bénat Tachot, Louise. 2015. "L'île de Cubagua. Réflexions sur les mécanismes et les enjeux de la société coloniale des Caraibes au XVIe siècle." In *A la recherche du Caraïbe Perdu*, edited by Bernard Grunberg, 285–304. Paris: L'Harmattan.

Benzoni, Girolamo. (1565) 1991. "Historia del Mondo Nuovo [1565]." In *Nuovo Mondo. Gli Italiani, 1492–1565*, edited by Paolo Collo and Pier L. Crovetto, 547–589. Torino: Giulio Einaudi Editore.

Biord Castillo, Horacio. 1985. "El contexto multilingüe del sistema de interdependencia regional del Orinoco." *Antropológica* 63/64: 83–103.

Biord Castillo, Horacio and Lilliam Arvelo. 2007. "Conexiones Interétnicas entre el Orinoco y el Mar Caribe en el siglo xvi: La región centro-norte de Venezuela." In *Lecturas Antropológicas de Venezuela*, edited by Lino Meneses Pacheco, Gladys Gordones, and Jacqueline Clarac de Briceño, 239–245. Mérida: Editorial Venezolana.

Booy, Theodoor de. 1916. *Notes on the Archaeology of Margarita Island, Venezuela*. New York: The Museum of the American Indian.

Boulton, Alfredo. 1961. *La Margarita*. Caracas: Ediciones Macanao.

Bry, Theodor de. 1990. *America de Bry*. Berlin and New York: Casablanca Verlag.

Brito Figueroa, Federico. 1966. *Historia económica y social de Venezuela*. Caracas: Universidad Central de Venezuela.

Camargo, R. Marcela. 1983. "Las pesquerías de perlas y conchas madreperla en Panamá." *Lotería* 326: 32–76.

Campos, S. Corina and G. Oswaldo Guzmán. 2002. "Estratigrafía Secuencial y Sedimentología de las Facies Turbidíticas del Flysch Eoceno de la Isla de Margarita, Edo. Nueva Esparta, Venezuela." PhD diss., Universidad Central de Venezuela.

Cantino, Alberto. (1501) 1989. "Carta de Alberto Cantino al Excmo. Principe Hercules I Duque de Ferrara, Lisboa, 17 de Octubre de 1501." In *El Mar de los Descubridores*, edited by Marisa Vannini de Gerulewicz, 41–43. Caracas: Fundación de Promoción Cultural de Venezuela.

Capelo, Juan C. and Joaquín Buitrago. 1998. "Distribución Geográfica de los Moluscos Marinos en el Oriente de Venezuela." *Memoria Sociedad de Ciencias Naturales La Salle* 58 (150): 109–160.

Carballo, Álvarez L. 2014. "Cambios de Los Paisajes Ancestrales de la Isla de Cubagua (4000 A.C.–1955 D.C.), Arqueología y Etnohistoria." Master thesis, Instituto Venezolano de Investigaciones Científicas (IVIC).

Carballo, Álvarez L. 2017. "Paisajes Ancestrales de la Isla de Cubagua (4000 A.C.–1500 D.C.)." *Boletín Antropológico* 35 (93): 7–31.

Casas, Paola, William Villalba, and Roberta Crescini. 2015. "Producción Específica de la Pepitona *Arca zebra* (Swainson, 1833) en la Bahía de Charagato, Isla de Cubagua, Venezuela." *Saber* 27 (4): 659–667.

Castellanos, Juan de. (1589) 1987. *Elegías de Varones Ilustres de Indias*. Caracas: Biblioteca de la Academia Nacional de la Historia.

Castillo Hidalgo, Ricardo I. 2005. *Asentamiento Español y Articulación Interétnica en Cumaná (1560–1620)*. Caracas: Biblioteca de la Academia Nacional de La Historia.

Cervigón, Fernando M. 1997. *Cubagua: 500 Años*. Margarita: Fundación Museo del Mar.

Cervigón, Fernando M. 1998a. *Las Perlas en la Historia de Venezuela*. Margarita: Fundación Museo del Mar.

Cervigón, Fernando M. 1998b. *La Perla*. Margarita: Fondo para el Desarrollo del Estado Nueva Esparta.

Champlain, Samuel de. 1989. "Breve Discurso sobre las Cosas más Notables que Samuel Champlain de Brouages ha Encontrado en las Indias Occidentales en el Viaje que Hizo a Ellas en el Año de 1599 y en el Año de 1601 como Sigue." In *El Mar de los Descubridores*, edited by Marisa Vannini de Gerulewicz, 201–238. Caracas: Fundación de Promoción Cultural de Venezuela.

Cipriani, Roberto, Héctor M. Guzmán, and Melina López. 2008. "Harvest History and Current Densities of the Pearl Oyster *Pinctada mazatlanica* (Bivalvia: Pteriidae) in

Las Perlas and Coiba Archipelagos, Panama." *Journal of Shellfish Research* 27 (4): 691–700.

Cipriani, Roberto, César Lodeiros, and Andrzej T. Antczak. 2010. "The exploitation of the Atlantic Pearl Oyster during early 1500's in the eastern coast of Venezuela." Paper presented at the *140th Annual Meeting of the American Fisheries Society, Pittsburgh, Pennsylvania, 12–16 September.*

Colón, Cristóbal. (1498) 1997. "La historia del viaje que el Almirante Don Cristóbal Colón hizo la tercera vez que vino a las Indias cuando descubrió la Tierra Firme, como la envió a los Reyes desde la Isla Española." In *Margarita y Cubagua en el Paraíso de Colón*, edited by Alí E. López Bohórquez, 71–83. Mérida: Universidad de Los Andes.

Cruxent, José M. 1955. "Nueva Cádiz: Testimonio de Piedra." *El Farol* 159 (17): 2–5.

Cruxent, José M. 1964. "Cuentas de collar de vidrio de sección estrellada." *Boletín Informativo IVIC* 3: 10–12.

Cruxent, José M. 1969. "Plano de Nueva Cádiz, Isla de Cubagua, Venezuela." *Boletín Informativo IVIC* 6: 6.

Cruxent, José M. 1972. "Algunas Noticias sobre Nueva Cádiz (Isla de Cubagua), Venezuela." *Memorias VI Conferencia Geológica del Caribe, Isla de Margarita, 6–14 July*: 33–35. Caracas.

Cruxent, José M. 1980. *Notas ceramología: Algunas sugerencias sobre la práctica de la descripción de las cerámicas arqueológicas de la época Indo-Hispana*, Cuaderno Falconiano 3. Coro: Ediciones UNEFM.

Cruxent, José M. and J. Eduardo Vaz. 1978. "Provenience studies of majolica pottery: Type Ichtucknee Blue on Blue." In *Archaeological essays in honor of Irving B. Rouse*, edited by Robert C. Dunnell and Edwin S. Hall, 343–374. The Hague: Mouton.

Cruxent, José M. and Maruja Rolando. 1961. "Tipología Morfológica de Tres Piezas de Cerámica; Nueva Cádiz, Isla de Cubagua." *Boletín Informativo IVIC* 2: 7–19.

Cruxent, José M. and Irving Rouse. 1958. *Arqueología Cronológica de Venezuela* (2 vols.). Caracas: Ernesto Armitano.

Cunnil Grau, Pedro. 1993. "Geografía y Poblamiento de Venezuela Hispánica." In *Los Tres Primeros Siglos de Venezuela, 1498–1810*, edited by Pedro Grases, 3–94. Caracas: Grijalbo.

Cunnil Grau, Pedro. 2004. "Biodiversidad y recursos naturales venezolanos para la sensibilidad euroamericana: sus paisajes geohistóricos siglos XV–XIX." *Boletín de la Academia Nacional de la Historia (Venezuela)* 87 (346): 25–156.

Dalton, Leonard V. 1912. *Venezuela*. New York: Charles Scribner's Sons.

Dawson, Kevin. 2006. "Enslaved swimmers and divers in the Atlantic World." *The Journal of American History* 92 (4): 1327–1355.

Deagan, Kathleen A. 2003. "Colonial Origins and Colonial Transformations in Spanish America." *Historical Archaeology* 37 (4): 3–13.

Domínguez, Lourdes. 1978. "La transculturación en Cuba (s. XVI–XVII)." *Cuba Arqueológica* 1: 33–50.

Donkin, Robin A. 1998. *Beyond Price. Pearls and Pearl Fishing: Origins to the Age of Discoveries.* Philadelphia: American Philosophical Society.

Fernández de Oviedo y Valdéz, Gonzalo. (1535) 1986. *Historia General y Natural de las Indias, Islas y Tierra Firme del Mar Océano.* Caracas: Biblioteca de la Academia Nacional de la Historia.

Ferris, Carmen L. 1991. "Nueva Cádiz de Cubagua. Aspectos Históricos y Arqueológicos." *Tierra Firme* 9 (9): 185–197.

Gabaldón Márquez, Joaquín (ed). 1988. *Descubrimiento y Conquista de Venezuela, Vol. 2: Cubagua y la Empresa de los Belzares.* Caracas: Biblioteca de la Academia Nacional de la Historia.

Galtsoff, Paul S. 1950. "The Pearl Fishery of Venezuela." *Special Scientific Report – Fisheries* 26: 1–26.

Gerhard, Peter. 1956. "Pearl Diving in Lower California, 1533–1830." *Pacific Historical Review* 25 (3): 239–249.

Ginés, Hno., Fr. Cayetano de Carrocera, José M. Cruxent, and Jesús M. Rísquez. 1946. "Manicuare." *Memorias de la Sociedad de Ciencias Naturales La Salle* 16: 157–200.

Goggin, John M. 1960. *The Spanish Olive Jar. An Introductory Study.* New Haven: Yale University.

Goggin, John M. 1968. *Spanish majolica in the New World: types of the sixteenth to eighteenth centuries.* New Haven: Yale University.

Gómez, Ángel Félix. 2004. *Margarita 1757; Censo del Gobernador Alonso del Río y Castro* Juan Griego: Alcaldía de Marcano.

González, Taliana L. 2002. "Pesquería de perlas durante la colonia en Nuestra Señora de los Remedios del Cabo de la Vela al Río de La Hacha (1538–1545)." *Jangwa Pana* 2: 26–34.

Gosden, Chris. 2004. *Archaeology and Colonialism: Cultural Contact from 5000 BC to the Present.* Cambridge: Cambridge University Press.

Guerra Curvelo, Weildler. 1997. "La ranchería de las perlas del Cabo de la Vela (1538–1550)." *Huellas* 49/50: 33–51.

Harrison, Rodney. 2014. "Shared Histories: Rethinking 'Colonized' and 'Colonizer' in the Archaeology of Colonialism." In *Rethinking Colonial Pasts through Archaeology*, edited by Neal Ferris, Rodney Harrison, and Michael V. Wilcox, 37–57. Oxford: Oxford University Press.

Heinen, H. Dieter and Alvaro García-Castro. 2000. "The Multiethnic Network of the Lower Orinoco in Early Colonial Times." *Ethnohistory* 47 (3/4): 361–379.

Helmer, Marie. 1962. "Cubagua, l'ile des perles." *Annales* 17: 751–760.

Henley, Paul. 1985. "Reconstructing Chaima and Cumanagoto kinship categories: An exercise in 'tracking down ethnohistorical connections.'" *Antropológica* 63/64: 151–197.

Hernández-Ávila, Iván, Alejandro Tagliafico, and Nestor Rago. 2013. "Composition and structure of the macrofauna associated with beds of two bivalve species in Cubagua Island, Venezuela." *Revista de Biología Tropical* 61 (2): 669–682.

Humboldt, Alexander von. (1814–1825) 1995. *Personal Narrative of a Journey to the Equinoctial Regions of the New Continent.* London: Penguin Books.

Ingold, Timothy. 2007. *Lines: A Brief History*. Abingdon: Routledge.

Idyll, Clarence P. 1965. "The Pearls of Margarita." *Sea Frontier* 11 (5): 268–280.

Instituto del Patrimonio Cultural. 2009. *Cubagua: Parque Arqueológico, Paleontológico y Geológico*. Caracas: Instituto del Patrimonio Cultural.

Jiménez, G. Morella A. 1986. *La Esclavitud Indígena en Venezuela (Siglo XVI)*. Caracas: Fuentes para la Historia Colonial de Venezuela.

Las Casas, Bartolomé de. 1981. *Historia de las Indias, por Fray Bartolomé de Las Casas, Libro II*. México: Fondo de Cultura Económica.

Las Casas, Bartolomé de. 1997. "Historia de las Indias." In *Margarita y Cubagua en el Paraíso de Colón*, edited by Alí E. López Bohórquez, 119–182. Mérida: Universidad de Los Andes.

Lister, Florence C. and Robert H. Lister. 1974. "Maiolica in Colonial Spanish America." *Historical Archaeology* 8 (1): 17–52.

López de Gómara, Francisco. (1552) 1988. "Historia General de Las Indias." In *Cronistas y Primitivos Historiadores de la Tierra Firme, Vol. 1*, edited by Horacio J. Becco, 161–184. Caracas: Fundación de Promoción Cultural de Venezuela.

MacKenzie, Jr., Clyde L., Luis Troccoli, and León B. León. 2003. "History of the Atlantic Pearl-Oyster, *Pinctata imbricata*, industry in Venezuela and Colombia, with biological and ecological observations." *Marine Fisheries Review* 65 (1): 1–20.

Mártir de Anglería, Pedro. (1530) 1988. "Décadas del Nuevo Mundo." In *Cronistas y Primitivos Historiadores de la Tierra Firme, Vol. 1*, edited by Horacio J. Becco, 47–84. Caracas: Fundación de Promoción Cultural de Venezuela.

McCorkle, Thomas. 1965. *Fajardo's people: Cultural Adjustment in Venezuela and the Little Community in Latin American and North American Contexts*. Los Angeles: University of California.

Mira Caballos, Esteban. 1997. *El Indio Antillano: repartimiento, encomienda y esclavitud (1492–1524)*. Sevilla-Bogotá: Muñoz Moya Editor.

Morón, Guillermo. 1954. *Los orígenes históricos de Venezuela. Introducción al Siglo XVI*. Madrid: Consejo Superior de Investigaciones Científicas.

Mosk, Sanford A. 1938. "Spanish pearl-fishing operations on the Pearl Coast in the Sixteenth Century." *The Hispanic American Historical Society Review* 18: 392–400.

Nectario Maria, Hno. 1960. *El Gran Santuario de Venezuela: La Virgen del Valle de Margarita*. Madrid: Imprenta Juan Bravo.

Nuñez, Enrique B. 1988. *Cubagua*. Caracas: Monte Ávila.

O'Brian, Patrick, ed. 1996. *The Drake Manuscript*. London: André Deutsch Limited.

Ocanto, David and Félix Baptista. 1998. *Presagios de tierra y fuego: Análisis de la cerámica popular en El Cercado, Margarita*. Caracas: Consejo Nacional de la Cultura.

Orche, Enrique G. 2009. "Exploitation of Pearl Fisheries in the Spanish American Colonies." *De Re Metallica* 13: 19–33.

Otte, Enrique. 1961, *Cedulario de la Monarquía Española Relativo a la Isla de Cubagua (1523–1550), Vol. 1, 1523–1534*. Caracas: Fundación John Boulton y la Fundación Eugenio Mendoza.

Otte, Enrique. 1977. *Las Perlas del Caribe: Nueva Cádiz de Cubagua*. Caracas: Fundación John Boulton.

Pacheco, Carlos, Luis Barrera Linares, and Beatriz González Stephan, eds. 2006. *Nación y Literatura: Itinerarios de la palabra escrita en la cultura Venezolana*. Caracas: Fundación Bigott.

Perera, Miguel A. 2000. *Oro y hambre: Guyana siglo XVI: Ecología cultural y Antropología Histórica de un Malentendido 1498–1597*. Caracas: Universidad Central de Venezuela.

Perri, Michael H. 2009. "'Ruined and Lost': Spanish Destruction of the Pearl Coast in the Early Sixteenth Century." *Environment and History* 15 (2): 129–161.

Pinto, Manuel C. 1967. *Visión Documental de Margarita*. La Asunción: Ejecutivo del Estado Nueva Esparta.

Quiévreux, Henry. 1900. "La pêche des perles au Venezuela." *La Revue Maritime* 146: 444–448.

Ramos, Demetrio. 1976. *Estudio de Historia Venezolana*. Caracas: Fuentes para la Historia Colonial de Venezuela.

Rivas, Pedro. 2001. "Arqueología de los procesos de etnogénesis y ocupación territorial en la región norcentral de Venezuela." In *La arqueología venezolana en el nuevo milenio*, edited by Lino Meneses and Gladys R. Gordones, 211–273. Mérida: Universidad de Los Andes.

Rodríguez Velásquez, Fidel. 2017. "Representación e historiografía: miradas múltiples al pasado de la Isla de Cubagua (1892–2014)." *Ouro Preto* 23: 28–42.

Rojas, Weimber J. 2010. "Incidencia de la agricultura en el contexto económico de la isla de Margarita durante la segunda mitad del siglo XVIII." *Tiempo y Espacio* 30 (53): 40–66.

Rojas, Arístides. 2008. *Orígenes Venezolanos: Historia, Tradiciones, Crónicas y Leyendas*. Caracas: Fundación Biblioteca Ayacucho.

Romero, Aldemaro. 2003. "Death and Taxes: the Case of the Depletion of Pearl Oyster Beds in Sixteenth-Century Venezuela." *Conservation Biology* 17 (4): 1013–1023.

Romero, Aldemaro, Susanna Gilbert, and M.G. Eisenhart. 1999. "Cubagua's Pearl-Oyster Beds: The First Depletion of a Natural Resource Caused by Europeans in the American Continent." *Journal of Political Ecology* 6: 57–78.

Rouse, Irving and José M. Cruxent. 1963. *Venezuelan archaeology*. New Haven: Yale University Press.

Rugil. 1892. "Historia Patria: El Escudo de Cubagua." *Cojo Ilustrado* 1(6).

Salaya, Juan J. and Luis Salazar. 1972. *Exploraciones y Explotaciones de la Ostra Perla (Pinctada imbricata) en Venezuela, 1946–1969*, Informe técnico 44. Caracas: Ministerio de Agricultura y Cría.

Sanoja, Obediente M. and Iraida Vargas Arenas. 1999. *Orígenes de Venezuela: Regiones Geohistóricas Aborígenes hasta 1500 d.c.* Caracas: Comisión Presidencial v Centenario de Venezuela.

Sauer, Carl O. 1966. *The Early Spanish Main*. Berkeley and Los Angeles: University of California Press.

Scaramelli, Franz and Kay Tarble de Scaramelli 2005. "The Roles of Material Culture in the Colonization of the Orinoco, Venezuela." *Journal of Social Archaeology* 5 (1): 135–168.

Silliman, Stephen W. 2009. "Change and Continuity, Practice and Memory: Native American Persistence in Colonial New England." *American Antiquity* 74 (2): 211–230.

Silliman, Stephen W. 2014. "Archaeologies of Indigenous Survivance and Residence: Navigating Colonial and Scholarly Dualities." In *Rethinking Colonial Pasts through Archaeology*, edited by Neal Ferris, Rodney Harrison, and Michael V. Wilcox, 57–75. Oxford: Oxford University Press.

Silliman, Stephen W. 2015. "A requiem for hybridity? The problem with Frankensteins, purées, and mules." *Journal of Social Archaeology* 15 (3): 277–298.

Sluyter, Andrew. 2001. "Colonialism and Landscape in the Americas: Material/Conceptual Transformations and Continuing Consequences." *Annals of the Association of American Geographers* 91 (2): 410–429.

Suárez, Carlos. 2014. "Arqueología, Representación y Patrimonio: Las 'otras historias' de Cubagua y Nueva Cádiz." *Nuestro Sur* 5 (8): 41–57.

Subero, Jesús M. 1996. *Pespunteo en la Cronología Histórica de Margarita*. Porlamar: Editorial Benavente and Martínez.

Tagliafico, Alejandro, María S. Rangel, and Néstor Rago. 2012. "Distribución, Densidad y Estructura de Tallas del Genero *Strombus* de la Isla Cubagua, Venezuela." *Interciencia* 37 (5): 381–389.

Thomas, Julian. 2017. "Landscape, Taskscape, Life." In *Forms of Dwelling: 20 years of Taskscapes in archaeology*, edited by U. Rajala and P. Mills, pp. 268–279. Oxford: Oxbow Books.

Tiapa, Francisco. 2008. "Resistencia indígena e identidades fronterizas en la colonización del Oriente de Venezuela, siglos XVI–XVIII." *Antropológica* 109: 69–112.

Vásquez, Socorro. 1989. "Pesquerías de Perlas del Cabo de La Vela." *Boletín de Antropología* 4 (4): 45–48.

Vaz, J. Eduardo and José M. Cruxent. 1978. "Gamma-Ray induced thermoluminescence of majolica pottery as an indicator of its provenience." *Revista Española de Antropología Americana* 8: 49–54.

Vela Cossio, Fernando and Alejandro García Hermida, eds. 2014. *Arqueología de los primeros asentamientos urbanos españoles en la America Central y Meridional.* Madrid: Edita Mairea Libros.

Velásquez, Justo S. 1956. "Petróleo y Perlas de Cubagua." *Revista Shell* (5): 44–52.

Vila, Pau. 1948. "La destrucción de Nueva Cádiz. ¿Terremoto o huracán?" *Boletín Academia Nacional de la Historia* 31 (123): 213–219.

Vila, Pau. 1961. "Cubagua y el Poblamiento Oriental de Venezuela en los Comienzos de la Colonia." In *Miscellània Fontserè*, edited by Gustavo Gili, 435–443. Barcelona: Gustavo Gili.

Vila, Pau. 1963. "Las Actividades Perlíferas y sus Vicisitudes en Venezuela." *Revista de Historia* 3 (17): 13–37.

Vila, Marco-Aurelio. 1978. *La Geoeconomía de la Venezuela del siglo XVI*. Caracas: Universidad Central de Venezuela.

Villamizar, G., Estrella Y. and Fernando Cervigón. 2017. "Variability and sustainability of the Southern Subarea of the Caribbean Sea large marine ecosystem." *Environmental Development* 22: 30–41.

Verde, Simone del. (1499) 1989. "Carta de Simone del Verde a Mateo Cini, Cádiz, 2 de Enero del 1499." In *El Mar de los Descubridores*, edited by Marisa Vannini de Gerulewicz, 38–39. Caracas: Fundación de Promoción Cultural de Venezuela.

Voss, Barbara L. and Eleanor C. Casella, eds. 2012. *The Archaeology of Colonialism: Intimate Encounters and Sexual Effects*. Cambridge: Cambridge University Press.

Warsh, Molly A. 2010. "Enslaved Pearl Divers in the Sixteenth Century Caribbean." *Slavery and Abolition* 31 (3): 345–362.

Warsh, Molly A. 2018. *American Baroque: Pearls and the Nature of Empire, 1492–1700*. Chapel Hill: The University of North Carolina Press.

Willis, Raymond F. 1976. "The Archaeology of 16th Century Nueva Cádiz." Master thesis, University of Florida.

Willis, Raymond F. 1980. "Nueva Cádiz." In *Spanish Colonial Frontier Research,* edited by Henry F. Dobyns, 27–40. Albuquerque: Center for Anthropological Studies.

Wing, Elizabeth S. 1961. "Animal Remains Excavated at the Spanish Site of Nueva Cádiz on Cubagua Island, Venezuela." *Nieve West-Indische Gids* 41 (2): 162–165.

Woodruff Stone, Erin. 2014. "Indian Harvest: The Rise of the Indigenous Slave Trade and Diaspora from Española to the Circum-Caribbean, 1492–1542." PhD diss., Vanderbilt University.

Zubiri, Maria T. 2002. "Gobierno y Perleros en la Costa Venezolana a Inicios del siglo XVI." In *Conflicto y Violencia en América: VIII Encuentro-Debate América Latina ayer y hoy*, edited by Gabriela Dalla Corte, Pedro García Jordán, Miquel Izard, Javier Laviña, Ricardo Piqueras, Meritxell Tous, and Maria T. Zubiri, 69–73. Barcelona: Universidad de Barcelona.

CHAPTER 8

Santa María de la Antigua del Darién: the Aftermath of Colonial Settlement

Alberto Sarcina

1 Introduction: from Center to Periphery (from Glory to Obscurity)

Chance played a great role in the entire first part of the Iberian conquest of the continent nowadays known as America. Among the many plays of destiny, the first and crucial one was that Columbus by chance (and by mistake of calculations) found the Antilles, while he was sailing towards Cathay and Cipango. The encounter with what was perceived more and more clearly as a new land of considerable dimensions confronted the Spanish rulers with a totally new situation, which they began to face with strategies that were sometimes contradictory, but always following from the political and military experience they had gained in the phase of European expansion and consolidation. In Late Medieval Europe, the concept of empire was linked to an idealized line of succession dating back to the Holy Roman Empire. However, it was precisely the fall of the Byzantine or Eastern Roman Empire, with the conquest of Constantinople by Mehmed II in 1453, that elicited the need to open new roads from the West to the Indies.

The kings of Spain were inspired by the Roman imperial model when they had to face the abyss of the unknown. The governors and governances of the New World colonies were the equivalents of the Roman governors in the imperial provinces. Likewise, the main base of territorial domination was the founding of cities, which acted as military as well as symbolic bastions of the nascent Spanish imperial expansion. The cities of the new colonies were built with inspiration in the ideal model of the orthogonal Greek-Roman city, in a new Renaissance version that placed the cathedral church and the Plaza Mayor at the center of the urban grid. However, the models of Spanish imperial domination and the ideal plans of the cities to be founded in the New World, so clearly conceived in theory, were reshaped and transformed when confronted with the reality of the new lands, that is, with the indigenous peoples who inhabited it and with the environment so different from that of Europe.

Santa María de la Antigua del Darién is a paradigmatic case since it is the first Castilian city founded on the American continent. We do not know the

DOI:10.1163/9789004273689_009

plan of this city, but we have the historical record on how it was planned to be built. We possess the King's instructions given to the governor Pedro Arias de Ávila on how and where to found new cities. Also, we possess some detailed information about the incredible campaign financed by the Spanish Crown that consisted of a fleet of twenty ships loaded with all the elements supposed to be essential in a sixteenth-century Castilian city and with two thousand passengers representative of all kinds of trades, from peasants to the bishop. These documents are the testimony of the ideal of a city that the Spanish wished to build in the middle of an unknown continent. The ambitious plans failed, and the ideal had to be molded and transformed into something probably less sublime, but surely novel.

In 1510, an armed and desperate group of Spaniards landed on the banks of the Darién River, a few kilometers inland from the western side of the Gulf of Urabá, in what is today Colombia. Over the past six months, the Spaniards had resisted the attacks of the indigenous Urabaes in their fort of San Sebastián de Urabá, on the opposite (eastern) shore of the Gulf. They had suffered great losses and of the three hundred men that had arrived, only some fifty had remained. The rest had died of malnutrition, illnesses, or infections caused by indigenous arrows poisoned with curare (Oviedo, Historia, Vol. 2, Book XXVII, Chap. IV). The purpose of the Spaniards had been the founding of a settlement that was intended to be the first component and future capital of the new province of Nueva Andalucía, under the direction of governor Alonso de Ojeda. Ojeda himself, meanwhile, had been lost at sea in an attempt to reach Hispaniola and collect the reinforcements that had been gathered there by lieutenant-captain Martín Fernández de Enciso. Now, having reached the Darién River, the survivors' only goal was to stay alive.

What they found 6 km upriver from the coast was a Cueva-speaking indigenous village called Darién, like the river that passed alongside it (Oviedo, Historia, Vol. 2, Book XXVII, Chap. IV), and some five hundred indigenous men ready to defend their land, commanded by a *cacique* named Cemaco. The Spanish eventually managed to win the conflict and they settled in the village, which they named Santa María de la Antigua del Darién, in honor of the image of the Virgin that is kept in the Cathedral of Seville, a primary site of devotion for seafarers heading for the West Indies. The settlement, arisen more by reason of hunger and chance than through planning, soon became the first outpost of the conquest – the first Spanish enclave in an indigenous universe. With the discovery of the South Seas in 1513 and the subsequent possibility of the Spanish to continue on course towards the East Indies, as well as upon receiving the title of city and diocese, Santa María de la Antigua del Darién became the center of the new empire (or rather the hopes of a new empire) for the Spanish. This is the only moment in (modern) history in which western Urabá was the center of any state interest.

Today, the area where Santa María de la Antigua del Darién was founded lies in the north of the Chocó Department of Colombia, some five hundred meters from the Tanela River, a tributary of the Atrato by way of the Cienaga de Marriaga. This area of low hills belonging to the foothills of the Serranía del Darién mountain range is now characterized by extensive cattle ranches. These ranches are evidence of territorial occupation by a few large landowners, and in some cases represent the direct or indirect outcome of the conflicts and violence in this part of the country between the end of the 1990s and the first years of the current century (Grupo de Memoria Histórica 2013). Colombian state authority is minimal here, and in fact the area is controlled by armed groups named BaCrim ('Criminal Bands'), which represent a direct continuation of the paramilitary groups that preceded them. This situation of lack of state control essentially began with the abandonment of Santa María de la Antigua.

In 2013, the Colombian Ministry of Culture and the Colombian Institute of Anthropology and History initiated an archaeological project in this area under the direction of the author. The goal of the project was to identify, delimit, and scientifically study the site of the first Spanish city founded on the American continent. This project, which resulted in the creation of an archaeological park and the construction of a Patrimonial House, has been the first step towards state presence in the area, based on a cultural approach and one of social inclusion.

In the present chapter, I wish to reflect on the relationship between the indigenous population that inhabited the city of Santa María (or, rather, frequented it) and the *conquistadores* during and after the city's existence, using chronicles and archaeological data recovered during five years of investigations (2013–2017). Firstly, a spatial proposal of the city will be made as maps and detailed written descriptions are lacking. Next, I shall analyze some data from the stratigraphic excavations of 2015 and, subsequently, the final moments of the city by comparing documentary sources and archaeological data. Finally, the years immediately following the abandonment of the city, on which there is practically no evidence, will be reflected upon.

2 Appearance of the City: an Approach to the Nonexistent Map of Santa María de la Antigua del Darién

In 1514, when Pedro Arias de Ávila, the new governor appointed by the Spanish Crown, arrived at Santa María de la Antigua directly from Seville, together with more than two thousand people, he was greeted by Vasco Núñez de Balboa, who received him

> ... with five hundred and fifteen men who were living there and had built more than one hundred houses or buhíos [*bohíos*: indigenous houses]: and the population was very kind, and a beautiful river passed alongside the houses of the city, of very good water and with many good fish. This is the river of Darién, and not the one that in book XXVII the bachelor Vadillo called the river of Darién, and this one comes from the eastern part, and the one that he mentions is a branch of the river Sanct Johan, that enters the posterior part of the Gulf of Urabá, as history has already said. ...
>
> There were among those first settlers more than one thousand five hundred indigenous men and women *naborías* [servants] who served the Christians in their *haciendas* and homes.
>
> OVIEDO, HISTORIA, Vol. 3, Book XXIX, Chap. VIII

Gonzalo Fernández de Oviedo is the principal source for the reconstruction of Santa María de la Antigua's history. As an observer of the founding and later mayor of the city, he offers relevant information about the first Spanish fort, which was established within the indigenous settlement conquered in 1510. When Fernández de Oviedo arrived in 1514 with the fleet of Pedrarias Dávila, approximately two thousand people lived in Darién: five hundred Spaniards and one thousand five hundred indigenous peoples; the latter being servants (*naborías*) of the first. The town had more than one hundred houses and the river ran alongside it. However, another chronicler, named Pascual de Andagoya, says: "Effectively, la Antigua, formed by some two hundred houses of indigenous style, and inhabited by the Spaniards from Balboa and their indigenous servants, could not comfortably accommodate the 1,500 new inhabitants who arrived with Pedrarias" (Andagoya, 1986). Obviously, here we face our first problem: were there one hundred or two hundred houses? Whom do we believe, Oviedo or Andagoya? Both arrived there at the same time in the *armada* of Pedrarias Dávila, so that both are direct, and theoretically, reliable witnesses. Nonetheless, it is hard to imagine that two thousand people would fit in the one hundred houses mentioned by Oviedo, unless he referred to only Spanish houses.

The issue leads to the question how the first Spanish cities founded on the new continent were spatially organized? In this respect the instructions King Ferdinand II (the Catholic) gave to Pedro Arias de Ávila on August 4, 1513 are a frequently cited and studied source. Among instructions and recommendations about various issues, they address the new foundations and their layout:

> ... you will distribute the *solares* [plots, pieces of land] of the place to build houses, and these are to be distributed according to the qualities of the people and are to be orderly from the beginning. Once the *solares* are made, the town is to look ordered: in the place destined for the *plaza*, where the church is to be, and by the orderliness of the streets. Since in the places that are newly built and where there is orderliness from the beginning, without any work or cost they remain orderly, and the others never become orderly [by themselves]. ... The distribution is to be such, that everyone is to receive part of the good [land], part of the mediocre [land], and part of the less good [land].[1]

The King specified the dimensions of the farmland and the *solares* in a subsequent royal decree on August 9, stating that the *solares* would be 100×80 paces, some 56×44.8 m.[2] According to Aprile-Gniset (1991, 186–215), originally the construction of the colonial American cities was not developed following the regulations established by the Crown. Indeed, many of these cities were developed prior to the *Instrucciones y reglas para poblar* issued between 1523 and 1529, and some, as Santa María de la Antigua, were constructed even prior to the *Instrucciones a Pedrarias Dávila* (1513). *Manzanas* (blocks), city blocks, and squares, according to the author, were developed in a quadrangular rather than rectangular shape as was planned. A city block, consisting of four *manzanas*, would then measure 85×95 m (Aprile-Gniset 1991, 202), as the *Plaza Mayor*. The sides of the *solares* would measure between 42.5 and 47.5 m. Aprile-Gniset (1991, 198), after analyzing twenty centers founded by the Spanish in the first phases of conquest, summarizes their layouts as three different shapes of an orthogonal "grid": (1) a completely orthogonal reticle; (2) atypical cases formed by the seascape or fluvial conditions; and (3) non-conventional layouts of "spontaneous character." Neither Santo Domingo nor Panamá la Vieja seem to have had a strict grid. Instead, it seems that the foundations had to be adapted and their layout molded following the topography of the chosen places. According to Panama's street plan of 1586 by Juan Bautista Antonelli, the city blocks do not seem to have had a fixed measure and, as in Santo Domingo, there existed a certain polycentrism since the city had several squares besides the *Plaza Mayor*, and the *Casas Reales* were not on it (Tejeira Davis 1996, 57). Additionally, both cities were far from having *manzanas* of the dimensions

1 Published in Manuel Serrano y Sanz, *Origenes de la dominación española en América*, Madrid, Libreria General de Victoriano Suarez, 1918, vol. I, CCLXXIX–CCLXXXVI.

2 Following Tejeira Davis (1996, 45), in the times of Pedrarias one pace corresponded to 2 *pies* (Castillian feet = *c.* 28 cm).

suggested by the Crown (100.8×100.8 m): they had *manzanas* of much smaller size, especially Panamá la Vieja.

The layout of Santa María de la Antigua del Darién must have been even more atypical. It has to be considered in relation to cities founded next to a river, such as Mompox, since as Oviedo mentions, the Darién River ran alongside the houses. But there is another important characteristic: the first fort was founded within the indigenous settlement, using its houses and architecture. In fact, the Spaniards arriving with Balboa and Enciso in 1510, beat the indigenous peoples commanded by the *cacique* Cemaco, entered the town "and there these people fortified themselves, and so they settled" (Oviedo, Historia, Vol. 2, Book XVII, Chap. IV). Of course, the succeeding settlement must have been different with respect to the indigenous town, because after the conquest of the town by Balboa and Enciso (and before the arrival of Pedrarias),

> ... the captain, Rodrigo de Colmenares, had gone with a vessel, and landed at Gaira, below Sancta Marta, and the Carib Indians killed more than thirty of his men because of not being cautious; and from there he went to the Darién with those who remained, who were more than a hundred. Then the Captain Cristóbal Serrano went and took with him more than two hundred people, among which there were one hundred and fifty fighting men, and in other vessels others went. So, that small town was more populated. With which the first *conquistadores*, before Colmenares and Serrano went, were joined by those that remained from the *armada* of Captain Diego de Nicuesa, as has been said ... .
>
> OVIEDO, HISTORIA, Vol. 3, Book XXIX, Chap. II

The original town surely expanded, and probably partially changed its layout. However, it must have preserved most of its original orientation, especially the area where the first chapel was founded (Anglería 1989, 104), at the house of the defeated *cacique* himself.

Five years of archaeological research have shown that the region where the city of Santa María de la Antigua del Darién developed consists of two different and adjacent sites, the first corresponding to the indigenous town of Darién conquered in 1510, and the second to the town founded by governor Pedro Arias de Ávila in 1514 (Sarcina 2017). In order to distinguish them archaeologically, these sites have been named Darién and Santa María de la Antigua (Figure 8.1). The archeological evidence behind this distinction consists of a larger presence of Spanish material among the finds encountered during the test pit survey at the second site (Sarcina 2017) and the distinct stratigraphic sequences of both sites. The stratigraphic excavations of 2014 and 2015 at the site of Santa María de la Antigua showed post-abandonment, abandonment

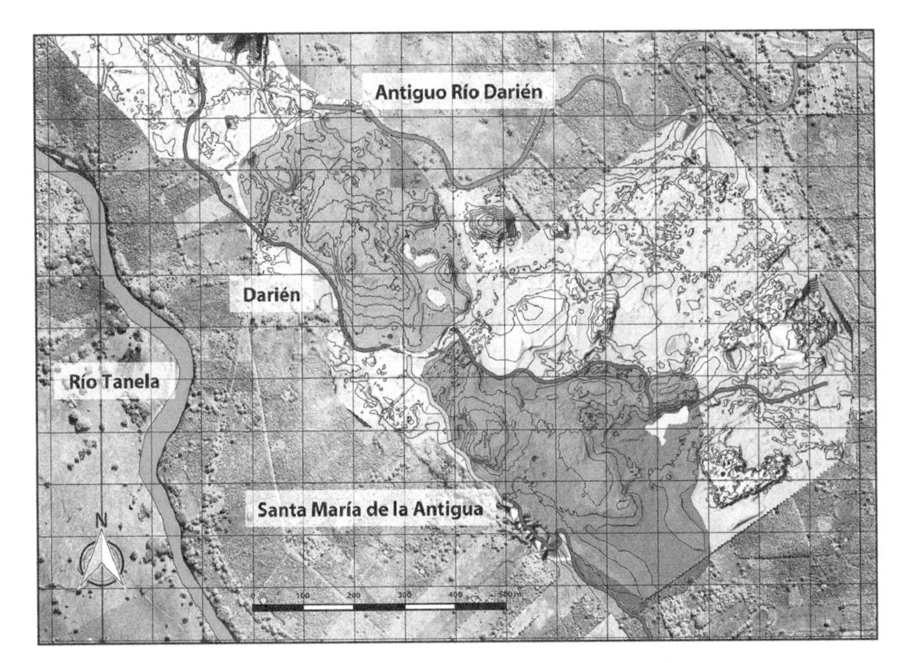

FIGURE 8.1 The area of Santa María de la Antigua del Darién, formed by its two sections, Darién and Santa María de la Antigua, Colombia, according to the results of the 2013 test pit survey, and the 2014–2016 stratigraphic excavations. To the left, the current Tanela River, and next to the city, the reconstruction of the old course of the Darién River and its two branches

and colonial-contact phases in the stratigraphy. Below these phases, there is no evidence of human occupation. In Darién, to the contrary, there is a relevant pre-Hispanic phase, characterized by an occupational model of house and garden, which is currently being dated (Sarcina 2018).

Another matter that was clarified is the old course of the Tanela River (Old Darién) that today runs 500 m from the archaeological sites. As we can see in the reconstruction in Figure 8.1, we now know that a branch of the river ran immediately to the city's west and another one cut the city in half, while the main course ran towards the northeast of the ancient indigenous town.

3 An Indigenous *Naboría* House in Darién

The 2015 investigations mainly consisted of the excavation of a 23×23 m excavation unit (excavation unit F) in the area of Darién. Satellite image analysis suggested the existence of two rectangular anomalies, one within another, probably related to a *solar* (or *manzana*) dating from the Spanish founding.

This hypothesis was proven by the results of the archaeological investigations that saw a spatial distribution of European findings concentrated in the portion of the excavation unit corresponding to these anomalies (Sarcina 2018).

If these anomalies actually do correspond to a *solar* or *manzana* from the old city, they would have dimensions of approximately 39×54 m (2106 m^2) in the case of the "inner" and darker anomaly and of 48×63 m (3024 m^2), in the case of the lighter one. This land would have been adjacent to the old branch of the Darién River, and not much deviant from the rules on *solares* dictated by the Crown. Nonetheless, when comparing the situation to the plan of Panamá la Vieja, and observing how in these first foundations the *solares* and *manzanas* had dimensions much smaller than the ones dictated by the royal rules, it is likely that in this case it could have been a *manzana*.

The area of dispersion of the archaeological material at Darién is about 12.3 ha. This would give us the space (excluding streets and possible squares) for 58 *manzanas* in the first case and 40 *manzanas* in the latter; that is 224 or 160 *solares*. This comes closer to Andagoya's assertions, although these *solares* of small dimensions would surely be distributed according to the importance of the *vecino* (resident) or the institution. The Cathedral, for example, had four *solares* as did the Monastery of San Francisco, while Balboa had two *solares* and two houses. The excavation unit F investigations in 2015, which yielded part of these possible *solares*, revealed the presence of a small structure inhabited by indigenous peoples, from which we found postholes, two hearths, and a trash midden. All of the material encountered in this space in the northeastern part of the excavation unit was of indigenous manufacture. European material appeared abruptly at the southwestern limit of the excavation unit (Figure 8.2). The geophysical investigations, which were carried out in the area immediately southwest of the excavation unit, showed a rectangular anomaly. This allows us to hypothesize that what was found may have been the house of an indigenous family of *naborías* (servants) in the service of a Spanish family whose house (the anomaly detected) was erected on the adjacent *solar* to the southwest (Sarcina 2018). This would imply a physical co-presence of structures and people, including Spaniards and indigenous peoples, at least in the part of the city that was conquered by the Spaniards in 1510.

The trash midden of the indigenous house, excavated in 2015 and 2016, appeared to had been formed inside an old canal running northeast towards a shoal where today there is a wetland. Thus, it is possible that this midden had a drainage function as well. Among the large pottery sample recovered from the midden, one indigenous ceramic fragment stands out. This distinct piece was painted externally with a European-inspired apparently vegetal decorative motif (Figure 8.3). What is interesting about this discovery is that it was found in an indigenous context that contained only indigenous-made material, even while situated adjacent to a possible *solar* occupied by Spaniards. This piece,

FIGURE 8.2 Rectangular anomalies at the site of Darién (4 and 2), and another anomaly identified in the area. To the left, the old course of one of the branches of the Darién River. In white, the area of excavation unit F and the density of Spanish findings at the site, focused in the southwestern part

FIGURE 8.3 Indigenous-made pottery with vegetal motif of Spanish inspiration, found in the midden of excavation unit F, Darién

ultimately, represents the fruit of experiments by local potters that were inspired by the decoration of imported ceramics belonging to the invaders. This is an interesting case of syncretism where the change in taste of the indigenous peoples is evidenced, and where the *naborías* themselves reproduced the ornamental forms of their masters.

4 "They Burned Down Most of That City"

After the arrival of governor Pedro Arias de Ávila and more than two thousand men and women in 1514, the Spaniards had to establish a new part of the city to the southeastern end of the first foundation, following the course of the Darién River branch (Sarcina 2018). The area of dispersion of the archaeological material in this second foundation is about 18 ha, corresponding to 85 or 59 *manzanas*, measured according to the biggest or smallest of the anomalies studied in 2015. Santa María, formed by these two entities, would officially have the status of city in 1515, with a coat of arms and episcopal see. Nevertheless, it would be abandoned for reasons that go beyond the scope of this chapter in 1524. Its population would disperse mainly to Panamá, Acla and Nombre de Dios, which were settlements founded (or refounded as in the case of Nombre de Dios) by Pedrarias Dávila himself and his captains between 1515 and 1519.

Once again thanks to Gonzalo Fernández de Oviedo, we have a description of the last days of Santa María de la Antigua del Darién. He speaks of how between 1521 and 1523 the population gradually diminished, "[...] every day the *vecinos* left, because the governor promised and gave them *indios de repartimiento* [the natives given to an *encomendero*] and other advantages to those that left the city" (Oviedo, Historia, Vol. 3, Book XXIX, Chap. XIV). In his military action against *cacique* Bea, Oviedo himself faced troubles gathering enough armed people, "[...] because in the city there were few people, since every day we were less, because the governor, to all those who went where he was, would flatter them and give them repartimientos there, and they would not return to Darién [...]" (Oviedo, Historia, Vol. 3, Book XXIX, Chap. XV). The description of the last days is quite dramatic:

> Two to three months later, Darién was depopulated, in the month of September in the year fifteen twenty-four. And once the *vecinos* of the city left, among those who remained was Diego Rivero, who as was mentioned in Chapter II of Book XXV, went or rebelled against the governor Diego de Nicuesa with the vessel, and left him lost on the island of Escudo. And his own *indios*, of this Diego de Rivero, together with others who joined them, killed him. And they killed one of his sons, who was from eight to ten years old, they hanged him from the rafter of his own hut, and they killed the mother of this child and another three or four ill Christians, and they burned down most of that city, and among the other houses, mine ....
>
> OVIEDO, HISTORIA, Vol. 3, Book XXIX, Chap. XXII

However, Oviedo did not witness these occurrences since he departed to Spain on July 3, 1523. Upon reaching Spain, he complained before the court of the governor's behavior, who after moving the episcopal see and all major political positions to Panamá, continued to promote a policy of depopulating Santa María in favor of new foundations. How much truth would there then be in his relation?

The stratigraphic excavations of 2014 through 2016 allowed us to identify strata with signs of burning in almost all levels belonging to the contact period. In 2014, the small excavation unit D (2×2 m), excavated in an area of the Darién site adjacent to that of Santa María de la Antigua, yielded a dirt floor with evident traces of burning and parts of a moved cobble pavement. In excavation unit F, excavated in the central part of the Darién site in 2015, 76% of the postholes showed signs of burning. There were also signs of burning related to the rafters and other parts of the house (Sarcina 2018). Besides, the portion of the excavation unit with the greatest presence of Spanish material yielded things that apparently were broken in situ, such as an Isabela Polychrome majolica plate, a pair of scissors, and a sword blade that along with its hilt was shattered into four pieces. Hypothetically, this appears to be a site that went through an attack and was burned down intentionally. In 2016, excavation unit H, which is still under excavation at the site of Santa María de la Antigua, also presented clear evidence of burning in the levels corresponding to the contact period. There are three stains of rectangular shape, measuring 6.30×3.30 m, 4.30×3.10 m, and 4.00×2.90 m, and (as far as the current excavation has gone) a fourth stain from which only a corner has been identified, also with clear evidence of burning (Figure 8.4). All of them have the same SW-NE orientation. At the moment, this evidence has been interpreted as the burned floors of small wooden structures, probably related to each other. Furthermore, virtually intact Spanish material, left in situ, has been found at the site, in particular, a *botija* (olive jar) and three glazed vials. All this evidence may belong to different fires at different moments throughout the existence of the city, but at least in the cases of excavation units F and H, we would hardly find material of common use abandoned or broken at the site. At the moment, the most consistent interpretation seems to be a confirmation of Oviedo's words.

5 Post-abandonment "Ritual Caches"

Not much is known about what happened in the area of Santa María de la Antigua del Darién after it was abandoned in 1524. The sources reveal very little, although there are some interesting clues, as will be seen in the following

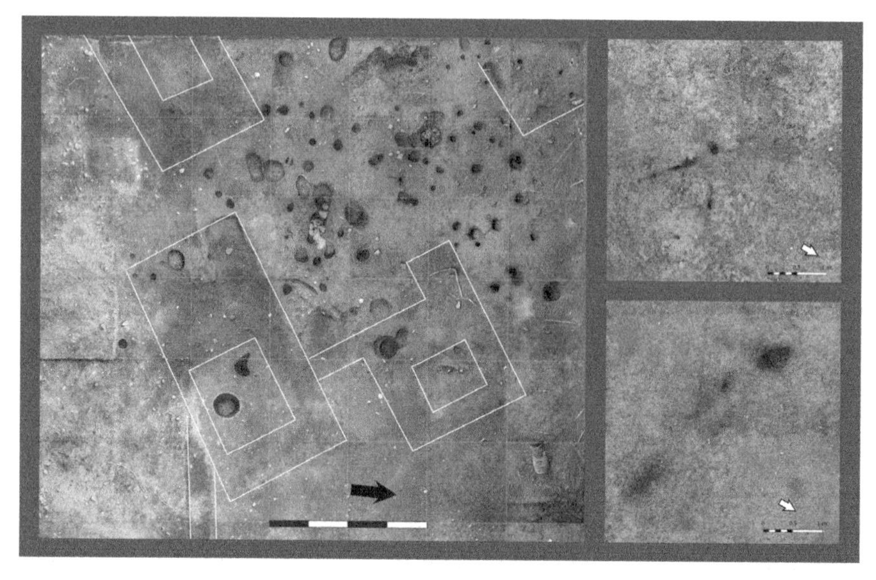

FIGURE 8.4 Santa María de la Antigua. On the left, four rectangular stains (in white lines) possibly related to burned cabin floors in excavation unit H (2016). On the right, signs of burning related to the rafters in excavation unit F (2015)

paragraph. With regard to the results of the stratigraphic excavations of the past few years, the stratigraphy related to this phase is very poor, and there is no evidence of a stable reoccupation of the site until modern times.[3]

However, there is clear evidence that this area was being frequented immediately following the Spanish abandonment. The principal indications of these visits are a series of offering activities which we have termed "ritual caches." These are, stratigraphically speaking, cuts made into the layers immediately above the phase of contact, that is, at a moment soon after abandonment. These cuts, found in excavation units A and H at the site of Santa María de la Antigua, were made in order to deposit offerings. Excavation unit A (2014) yielded two floors of cobbles and packed earth, possibly related to two internal patios of a large wooden construction, which, in turn, prompted the discovery of the "ritual caches," found some two meters north of the Spanish

3 Today there is a small village, called Santuario, that was built nearly exactly on top of the indigenous settlement of Darién (and the first Spanish foundation of 1510) in the 1980s. It is inhabited by colonists originating mostly in the coastal region of Cordoba.

floors (Sarcina 2014; Rivera 2014). Three groups of offerings were found: the first constituted five small vessels of 8–10 cm in height, while the second contained four vessels with similar dimensions, a small 5×3 cm axe, and two spindle whorls (Figure 8.5, left). The third offering held a semi-cylindrical *mano* of 30 cm in length. The vessels were roughly manufactured, many of them with irregular rims. To the contrary, the small axe and the *mano* were excellently made. In excavation unit H (2016), excavated 65 m north of excavation unit A, two more caches were found. One of these consisted of a large vessel with inside a small axe (4.5×3 cm). The second cache (UE 77) discovered a few meters away, was composed of a large globular container, which held a medium-sized vessel and a small other one, 9 cm high, with zoomorphic decoration (Figure 8.5, right).

If, as is suggested by the archaeological investigations, there was actually no true reoccupation of the site, what is the meaning of these offerings? The Cuevas did not return to the site where they lived previously, because of fear or because the physical and metaphysical environment had become somehow "contaminated" by too many years of violence and slavery? In this case the offerings, these "ritual caches," perhaps had the function of purifying the area from all the negative things that had occurred in the previous fifteen years, restoring some form of balance to this space. Hopefully, future (archaeological and ethnoarchaeological) investigations will help us resolve this enigma.

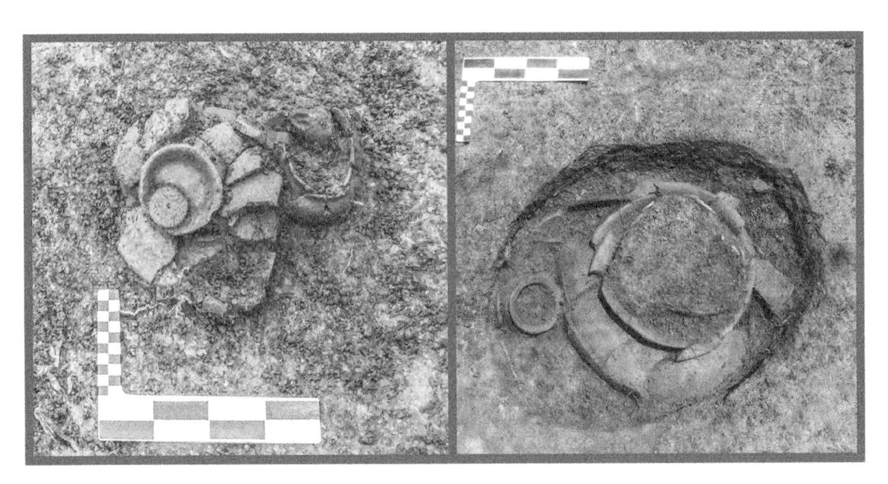

FIGURE 8.5 Ritual caches in the post-abandonment levels of Santa María de la Antigua. To the right, UE 77 from excavation unit H 2016, and, to the left, UE 30–31 from excavation unit A 2014

6 ...Seven Years Later

> In the Culata, of the gulf of Urabá, it was a blessing to discover certain rescate [barter] and engagement with the *indios* from there, in a very good manner. That is, a certain navy that stations there, and with the royal official of Your Majesty in this land, among the *indios* he took, brought an indigenous woman who said to be of the *cacique* Cemaco, who is the one who was of Darién, who is there settled. And with her went a *vecino* from Acla, servant and steward of *licenciado* Corral, to whom the said *cacique* was given in *encomiendo*. The *vecino* went with her and with other persons he had from said *cacique*, and they spoke to him and his *indios*, and they gave him six hundred or seven hundred pesos of gold, and he [the *cacique*] and other *caciques* from that Culata remained very peaceful.
>
> *Carta del licenciado Espinosa*, Panamá, August 15, 1532, doc. 392 in FRIEDE, 1955–1960, 286

This fragment from the letter of *licenciado* Espinosa to the King, written in Panama seven years after the abandonment and burning of Santa María de la Antigua del Darién, offers us some important insights. First of all, it is pertinent to note that the memory of the *cacique* Cemaco still lingered, the same *cacique* who had to receive the group of Spaniards headed by Nuñez de Balboa and Fernández de Enciso in 1510, 22 years previously. It is possible that what is being discussed here is Cemaco's family, and that the indigenous woman was part of this group. It is also mentioned that this group was related to the first inhabitants of Darién, which "is there settled," meaning in the zone of La Culata (the part to the south of the Gulf).

The "*vecino* from Acla" is Julian Gutiérrez, who had lived in Santa María de la Antigua in the service of bachelor Diego del Corral (Oviedo, Historia, Vol. 3, Book XXIX, Chap. XIX). The latter was one of the first settlers of the city, who arrived before the *armada* of Pedrarias, with Rodrigo de Colmenares. It is worthwhile to pause a moment on the figure of bachelor Del Corral, one of the most powerful and influential vecinos of Santa María, who owned a house in the principal part of the city and an *estancia* (country house) half a league outside the city (Oviedo, ibid., Chap. XX). Espinosa tells us in this letter, that the *cacique* Cemaco and his people were given in *encomienda* to him, although in Oviedo we read that it was the *cacique* Corobarí (Oviedo, ibid., Chap. XV), a detail confirmed by the words of governor De la Gama (Friede, doc. 499, 41). From the pages of Oviedo we can infer that in the houses of the bachelor Del Corral there was a level of familiarity with some indigenous families: the *cacique* Bea "was a close relative of one indigenous woman that the bachelor had as a concubine, with whom he had children" (Oviedo, ibid.); the *cacique* Corobarí had in the house of the bachelor "his mother, wife, and children" (Oviedo, ibid.); and "there was understanding between them, and they spoke in his *estancia* outside the city" (Oviedo, ibid., Chap. XX). Thus, Del Corral had close relations with these *caciques*, also in part because of his concubine/wife Elvira and her son Perico, as well as their relatives. So much so, that in 1521 the bachelor tried detaining an expedition organized by Fernández de Oviedo (at the time the mayor of the city) in order to bring to submission the *caciques* Bea, Corobarí, and Guaturo who had rebelled. He was then detained and sent to Spain, charged with publicly having an indigenous concubine (Oviedo, ibid., Chap. XV and XVI).

A very interesting fragment of the history of the first phase of the colonization is the story of Julian Gutiérrez and the india Isabel. She was a lengua or translator, related to some of the principal *caciques* of the lower Urabá, and later became wife of the same Gutiérrez. The narrative is found largely in the collection of the Patronato at the General Archive of the Indies in Seville, and the great majority of the folios have been transcribed in the "*Documentos inéditos para la historia de Colombia*," edited by Juan Friede (Friede, 1955–1960). In 1532, the territory that belonged to the governorate of Castilla del Oro was divided into various new governorates and provinces: Panamá, Santa Marta (from 1524), Nicaragua (from 1528), and Cartagena (precisely from 1532). The boundary line between the governorates of Panamá and Cartagena was the Gulf of Urabá, and the San Juan River (today the Atrato). There, however, was no clear agreement and this would lead the governorates to first confront each other legally and later (around 1534–1535) physically and militarily. This eventually resulted in Cartagena and its governor Pedro de Heredia emerging as winners.

At the time the letter by *licenciado* Espinosa was written, the new governor of Panamá, the *licenciado* Antonio de la Gama, decided to change the politics with respect to the indigenous groups that still remained in the lower part of the Gulf of Urabá. This change, reminiscent of the policy of Nuñez de Balboa, consisted of more peaceful relations with the natives based on barter and exchange. It is possible that this alteration arose as a result of the political situation created by the adjacent formation of a new and powerful governorate. The last violent "entry" of pillaging and plundering, during which the Spanish devastated various indigenous villages and took numerous prisoners, was undertaken by the royal official Miguel Juan de Ribas and his captain Esteban Milanés, who died during the expedition (Friede, doc. 393, 287–296). The decision of the governor was to return the prisoners to their respective settlements and make peace with all of the *caciques* of "la Culata del Urabá." This was done by using Julián Guitiérrez, a man who already had a relationship with an indigenous woman,[4] who was a relative of some of the *caciques* of the area. Gutiérrez was raised in Santa María de la Antigua in the house of bachelor Del Corral, in an environment where there were many relations with local *caciques* and their families. On August 25, 1532 the following statement was written:

> The *licenciado* Mr. Antonio de la Gama, governor of this Kingdom of Castilla del Oro for His Majesty, commands that any person who has an indigenous man or woman from among those brought by the *armada* of the royal official Miguel Juan de Ribas, of which Esteban Milanes, deceased, was the captain, would come and show them to him in the next three days, with a penalty of ten pesos of gold for each person who does not show up.
>
> FRIEDE, doc. 393, 292

This change in policy is quite impressive, especially when compared with the customs in force during the government of Pedro Arias de Ávila. In fact, it is probably the first time that we can observe the restitution of indigenous slaves to their lands. The peace with these groups, which evidently was still kept in the territory close to where Santa María de la Antigua once stood, had necessarily to result in new behaviors. Without doubt, this was imperative because this specific territory was in dispute between the governors and having the indigenous peoples allied was a point of strength. This part of the Gulf was frequented by various groups of Spaniards, some of whom probably belonged to the people of Pedro de Heredia, since Julián Gutiérrez asked the governor for a letter to give to the *caciques*: "so that if any vessel or people arrive there showing interest, do not hurt the said *indios*" (Friede, doc. 393, 295). The governor in effect wrote the letter, directed to

4 According to a letter sent by governor De la Gama from Panamá to the King on May 15, 1533, Isabel Corral was a "principal indigenous woman, very understood and a good Christian. Since a child, she had been raised with the said Julian Gutiérrez in the house of the said *licenciado* (Del Corral) who was in charge of his *hacienda* and house. She is the relative of the principal *caciques* and *indios* with whom friendships were settled, and this indigenous woman was the translator in the peace accords, and the principal part in their occurring. And having seen this, I spoke to the said Julian Gutiérrez and begged him that he marry her, because by doing this, apart from serving Our Lord and His Majesty, the good work she had done in the peacemaking would be rewarded" (Friede, doc. 499, 42).

> … all the captains, masters, pilots, majors, and whichever persons from whatever vessel, caravel, brigantine, ship or any other that arrive at the Culata and Gulf of Urabá, which is in this governorate, that because with the *caciques* peace and amity are beginning to settle, that none be reckless to do them wrong or injury, nor take anything from them against their will … under penalty of death and the loss of all his property.
>
> FRIEDE, doc. 393, 296

Julián Gutiérrez embarked on at least three voyages, always accompanied by the translator, Isabel, who became his wife in 1533. The first voyage is cited in the letter of *licenciado* Espinosa, which seemed to have been an initiative of Gutiérrez[5] himself. It resulted in such a success that once the facts were known, Espinosa and governor De la Gama "thought that this rescate should be preserved, and that it should not be done by hands other than the ones of this steward" (Friede, doc. 392, 286). The second voyage started on August 29, 1532 and is perfectly documented and described by the scribe and overseer Fernando Gallego. Gutiérrez arrived in a place "at the mouth of the Urabá river, where said *caciques* are"(Friede, doc. 396, 299), which could have been the southernmost mouth of the Atrato (known in eighteenth-century maps as the mouth of Urabá), a place somewhere further south, or even on the eastern coast of the gulf, on the cape of Urabá, near where San Sebastián de Urabá was founded. There he found 14 *caciques*, three of them "from the great river of Dabaive." Everyone, apparently, spoke the same language since the translator was always Isabel and they refer to their leaders and heads with the words *tiba* and *saco*, which are words in the Cueva language. This is an interesting issue, especially if the place for meetings was the east coast of the Gulf, because Gutiérrez said at some point: "I came to the culata and gulf of Urabá, until the hill of the *águila*, to pacify the principal *caciques* and *indios* of such coast and gulf" (Friede, doc. 396, 312). This part of the Gulf, far beyond the limits of the governorate of Panamá,[6] was theoretically

5 As can be inferred from the words of Espinosa, and even more so from those of governor De la Gama to the King: "[…] and if you think that good will be done by the hand of said Julián Gutiérrez, as it was he who began it."

6 The motivations of this policy of peace seem to become clearer. In these months Pedro de Heredia was discovering the famous tombs of the Zenú, some leagues from the eastern shore of the Gulf of Urabá, that would fill the personal coffers of Heredia and those of the nascent governorate of Cartagena with gold. The governor of Panamá was evidently overcoming the boundary between the two governorates with a clear objective, since in July of 1535 he pretended that Pedro de Heredia and his people "leave from the said provinces of the Cenú and from all the others that are and belong to this governorate" (Friede, doc. 752, 303). The final objective was the gold of Zenú that clearly pertained as a region to Cartagena.

inhabited by different peoples – the so-called *hurabaes*, archers that must have been linked to the Zenú people. But according to the accounts of Julián Gutiérrez' voyages, it appears that they understood each other and were related[7] to an indigenous woman from the family of the *cacique* Corobarí, on the other side of the Gulf, that must have spoken the Cueva language.

Another interesting fact is the exchange between the Spaniards and indigenous peoples, who by then already had a good understanding of the Europeans and their goods. Gutiérrez donated gifts such as knives, combs, needles, hooks, and certain shirts from Holland with their "deep red hoods plastered with blue velvet." The *caciques*, however, only wanted axes and iron knives, which they surely appreciated for their productive efficiency. On his second voyage, Gutiérrez sold about two hundred items consisting of axes and iron daggers in exchange for gold.

During the third voyage of Julián Gutiérrez, from September to October 1532, something particularly interesting happened. On his trip to the town of the *cacique* Everaba, the brigantine of Gutiérrez made a stop "at the port of Darién," where the indigenous peoples and some Spaniards traveling with him decided to go "to the town of Darién" in order to fish and hunt iguanas. Here they found "some tracks of negros" (Friede, doc. 401, 334). Seven years after its abandonment, the port and city of Santa María continued to be a geographical point of reference. On his way back, Gutiérrez stopped again in the port of Darién, accompanied by the *cacique* Everaba and 26 natives from his group, determined to kill the *cimarrones* (maroons).

> We arrived at the said port Friday afternoon. Another day after eating we went upriver from Darién by the toldo [warehouse], and others upriver, and went to where the town used to be ... and said Julián Gutiérrez went in pursuit and I went with him upriver one league, following the track of two or three negros. ... and we followed the track until night came and there said Julián Gutiérrez stopped. ... and after a while said *indios* returned to where said Julián Gutiérrez was and his companions to let them know how said Gonzalo had encountered a hut and in it there were some certain negros roasting meat over a fire ... and said Julián Gutiérrez asked said Gonzalo if with the negros he had seen any *indios*, and said Gonzalo said no, only the negros who were singing. ... And we stopped there until they fell asleep. ... And then said Julián

7 One of her sisters was the wife of the principal *cacique* of the area, Everaba, sometimes known as Hurava, and the other was the wife of the son of said *cacique*.

> Gutiérrez set off with his companions upriver, and the *cacique* with his *indios* through the thick vegetation, and when we arrived we encountered them. And said *cacique* and *indios* were frightened by them and we were left alone with said negros, and said *cacique* Everaba shot a poisoned arrow at one who was fleeing, hitting him in the side and said negro fell in the river. And since nobody went to him, since the *indios* did not dare to, said negro got up and left so that of the three negros we caught one and the other two escaped, one poisoned and the other with many cuts, and I think none would have escaped [far] and we gave them for dead.
>
> FRIEDE, doc. 401, 340–341

This excerpt was narrated by the scribe and observer Gil de Morales, who accompanied Gutiérrez on this third voyage. It contains some interesting information. First, the two ways of reaching the place where the city was located are described. The first one was a trail that evidently began where "the toldo" was, that is the construction that functioned as a warehouse on the coast. The second, going up the Darién River, was longer. Once they arrived at the site of Santa María, the group followed "the tracks of two or three negros," a small group of *cimarrones*, one league (5/5.5 km) upriver. According to the way Gutiérrez reacts to the news of the *cimarrones*, this does not seem to be a novelty to him. It would be interesting to know what these "tracks of negros" were, immediately recognizable as those of fugitive slaves, perhaps the remains of a campsite or the ash of hearths. Anyway, it is obvious that at the site where Santa María de la Antigua del Darién had been, there was no stable indigenous presence any longer. The presence of a small group of African fugitives that settled here in "a hut," probably in a transitory fashion, is interesting because it represents a very early occurrence and bears witness to a phenomenon that was probably much more widespread in a territory abandoned by the Spanish which was very difficult to control. We cannot speak of a palenque, but the image of a palisaded hamlet in which a group of Africans cooked meat and sang their songs is very evocative and attests to early, albeit ephemeral, settlements.

The fear of the natives when faced with the Africans is interesting: none of them, and they are 26 against 3, managed to get close to them. Although this would seem to attest to very few contacts between these two ethnic groups, Gutiérrez' question to his second translator Gonzalo that "if with the negros he had seen any *indios*" makes one think that the reception of such fugitives by indigenous groups must not have been rare.

7 Conclusions

Attempting to reconstruct the common life and social relations that took place at Santa María de la Antigua del Darién also means trying to reconstruct the layout of the city and the dynamics which developed here, where there have not remained any visible structures, plans, or clear descriptions. We are still at the beginning of this journey, but we can already place the city next to the Darién River, see it leaning out in front of one of the branches, and being bisected by another one. Comparing the data from documentary sources to those from the field investigations, we have been able to understand something about the relation between *solares* and houses. Furthermore, the stratigraphic excavations clearly demonstrate that indigenous servants were not relegated to the outskirts of the city, but lived in small wooden houses next to Spanish properties. The discovery of a fragment of indigenous-made pottery with European motifs in the midden of one of these houses is proof of very strong dynamics of cultural syncretism, in both directions. Not only did the Spanish apply indigenous techniques and materials in the construction of their houses and made daily use of local ceramics (Sarcina 2017), also the indigenous peoples adopted European motifs for decorating some of their vessels. Likewise, the indigenous peoples quickly began to appreciate metal objects, especially those made of iron, which were more efficient than lithic tools, as we can infer from the travel accounts of Julián Gutiérrez.

The dramatic ending of the city of Santa María de la Antigua del Darién, as recorded by historians, seems to be confirmed by the archaeological data from the stratigraphic excavations of 2015 and 2016 (Sarcina 2016, 2018). After its abandonment, it seems the city area was never permanently occupied again, but only frequented by indigenous groups with ritual purposes, possibly of "cleansing," by placing offerings of small ceramic pieces or worked stone items (Sarcina 2014, 2016). This same area was also frequented during the immediate years following the abandonment of the city by small groups of fugitive African slaves that set up temporary camps, at least in one case reported in the travel accounts of Julián Gutiérrez.

Santa María de la Antigua del Darién is the "missing link" in the history of early urban planning in the colonial Americas, filling an empty temporal space between Santo Domingo and Panama. Its study is a fundamental step to understand how the strategies of Spanish colonization, through the founding of cities, evolved and transformed with the passage of time and with increased control of territory. The results of archaeological research (with stratigraphic and geophysical methods and the study of soils, plant remains, etc.) compared to the study of primary sources and the plans of other contemporary Spanish cities already offer a glimpse into how real needs, geomorphology, and the relationship with local indigenous populations determined, in this first continental settlement, a city layout that was very distant from the Renaissance ideal derived from the original Greek-Roman model.

Acknowledgments

I want to thank the Colombian Institute of Anthropology and History (ICANH) and the Ministry of Culture of Colombia for having promoted and supported the research in Santa María de la Antigua from 2013 until today. In particular, I would like to thank the director of ICANH, Ernesto Montenegro, for his constant support. I also wish to thank Corinne Hofman and Floris Keehnen for inviting me to the symposium "Material Encounters and Indigenous Transformations in the Early Colonial Americas," held during the SAA 82nd Annual Meeting in Vancouver, which forms the basis of this volume.

References

Andagoya, Pascual de. 1986. *Relación y Documentos. Crónicas de América.* Madrid: Edición de Adrián Blázquez, Historia 16.

Anglería, Pedro Mártir de. 1989. *Decadas del nuevo mundo.* Madrid: Ediciones Polifemo.

Aprile-Gniset, Jacques. 1991. *La Ciudad Colombiana. Prehispanica, de conquista e indiana, Vol. 1.* Bogotá: Banco Popular.

Friede, Juan, ed. 1955–1960. *Documentos inéditos para la historia de Colombia, Vol. 1.* Bogotá: Academia Colombiana de Historia.

Grupo de Memoria Histórica. 2013. *¡Basta ya!. Colombia: Memorias de guerra y dignidad.* Bogotá: Imprenta Nacional.

Oviedo y Valdés, Gonzalo Fernández de. [1851–1855] 1535. *Historia general y natural de las Indias,islas y tierra firme del mar Océano, 4 vols.* Madrid: Imprenta de la Real Academia de la Historia.

Rivera, Javier. 2014. *En Plan Especial de Manejo y Protección Santa María de la Antigua del Darién, Inédito.* Bogotá: ICANH y Ministerio de Cultura de Colombia.

Sarcina, Alberto. 2014. "En Plan Especial de Manejo y Protección Santa María de la Antigua del Darién." Unpublished report. Bogotá: ICANH y Ministerio de Cultura de Colombia.

Sarcina, Alberto. 2015. "Santa María de la Antigua del Darién. Misión arqueológica 2015. Final report." Unpublished report. Bogotá: ICANH.

Sarcina, Alberto. 2016. "Santa María de la Antigua del Darién. Misión 2016. Final report." Unpublished report. Bogotá: ICANH.

Sarcina, Alberto. 2017. "Santa María de la Antigua del Darién, la primera ciudad española en Tierra Firme: una prospección arqueológica sistemática." *Revista Colombiana de Antropología.* 53(1): 269–300.

Sarcina, Alberto. 2018. "Santa María de la Antigua y Darién. Los dos caras de la primera ciudad europea en tierra firme." *INDIANA.* 35(2): 243–269.

Serrano y Sanz, Manuel D. 1918. *Origenes de la dominación española en América, Vol. 1.* Madrid: Casa Editorial Bailly Bailliere.

Tejeira Davis, Eduardo. 1996. "Pedrarias Dávila y sus fundaciones en Tierra Firme, 1513–1522." In *Anales del Instituto de Investigaciones Estéticas Vol XVIII,* 41–77. Mexico City: National Autonomous University of Mexico.

CHAPTER 9

Material Encounters and Indigenous Transformations in Early Colonial El Salvador

William R. Fowler and Jeb J. Card

In this chapter we explore the material encounters and indigenous transformations that took place in two different times and places in the Central American Republic of El Salvador, which in the early colonial period formed part of the *audiencia* of Guatemala. The cases consist of (1) the early Spanish colonial *villa* (town) of San Salvador – now the archaeological site of Ciudad Vieja – during the second quarter of the sixteenth century, and (2) the indigenous Pipil town of Caluco of the Izalcos region in western El Salvador during the second half of the same century. The two cases together cover a time span from about AD 1525 to 1600.

Any archaeological study of sixteenth-century Spanish America should confront the "haunts" of modernity enumerated by Charles Orser (1996, 2004, 2014): colonialism, mercantilism/capitalism, Eurocentrism, and racialization. These are structurally complex, interconnected forces or metaprocesses that operate on a global scale through simultaneous vertical and horizontal networks as a "unified ... system of activity, practice, and procedure" that came into existence after AD 1500 (Orser 2014, 27). While the haunts are massive global forces that change through time, we may address them in specific archaeological cases by contextualizing issues of power (e.g., gender, status, ethnicity, and identity) through material culture studies at the local, regional, and global scale and in varying time spans. The scale at which pattern can be comprehended or meaning attributed is referred to as the "effective scale" (Marquardt 1989, 7; 1992, 107; following Crumley 1979, 166). We view these issues from the perspective of the effective scales (segments of space and time) in multiscalar, dialectical analysis of archaeological landscapes and material culture (Marquardt 1989, 1992; Orser 1996, 184–190; 2014, 2–4, 66–69). Considering the nexus of everyday practice as seen through a multiscalar, dialectical analysis of the materiality of interaction and change, it becomes possible to plot the courses of relationships and intersections of different groups.

The issues of power came into play at the scales of the community and the household at Ciudad Vieja and Caluco as distinct social groups made efforts to preserve their own practices and traditions as they lived alongside each other

 | DOI:10.1163/9789004273689_010

and entered into social relations in new urban or semi-urban communities. We use the term "traditions" here in the sense discussed by Timothy R. Pauketat (2001a, 2001b), not as conservative structures that impede or constrain change, but rather as dynamic media of negotiations that generate cultural change. By this interpretation, tradition-making or cultural construction is embodied or represented through practice in material culture (objects and landscapes). Material culture "as a dimension of practice, is itself causal. Its production – while contingent on histories of actions and representations – is an enactment or an embodiment of people's dispositions – a social negotiation – that brings about changes in meanings, dispositions, identities, and traditions" (Pauketat 2001b, 88). Tradition in this sense bears a strong similarity to Gramsci's (1971) cultural hegemony and Bourdieu's (1977) doxa. The specific social groups included Spanish and indigenous *conquistadores* and colonists and their families, local indigenous populations, and (in Caluco) large numbers of mestizos and African slaves. Unequal relations of power among these groups led to the dialectical relationship of domination and resistance, the latter expressed in subtle forms. Attempts to assert, protect, and maintain identities succeeded partially or failed completely, leading to new social forms and classifications derived from a conjunction of sociocultural practices which was, in turn, a product of close proximity and permeable social boundaries between groups.

1 Historical Background

Dispatched from their base of operations in Olintepeque, Guatemala, under the command of Diego de Alvarado, a small group of Spaniards, accompanied by a much greater number of *conquistadores mejicanos* from Mexico and points south, pitched camp and founded the first *villa* of San Salvador in April 1525, possibly on the same site as the later 1528 settlement (Escalante Arce 2014, 54). If this was the location of the first Spanish colonial town in El Salvador, San Salvador was built in the valley of La Bermuda, a small pocket of land to the north of Cuscatlan Pipil territory which had little or no indigenous settlement at the time of the conquest (Fowler and Earnest 1985). Excavations beneath a sixteenth-century structure, likely a Christian church used by indigenous settlers, as well as isolated artifact finds across the site suggest a modest Terminal Classic/Early Postclassic occupation (ca. AD 800–1000) at this location but a lack of subsequent artifact or architectural evidence suggests it was abandonded until the arrival of the Spaniards. Despite the isolation of La Bermuda, San Salvador was still prone to attack, especially from the west and the south. The Pipils rebelled and drove out the Spaniards and their allies sometime in

1526 (Barón Castro 1996, 39–44), forcing them back to their base in Guatemala. A subsequent Spanish military push, aided by much larger numbers of *conquistadores mejicanos,* overcame Pipil resistance in early 1528, allowing the Spaniards to return and found a permanent settlement (Matthew 2012, 85–92). This was the second founding of the *villa* of San Salvador on 1 April 1528 by 73 Spanish *conquistadores* and their Mexican allies under the command of Diego de Alvarado. All of the Spaniards declared themselves residents of the town; many of the Mexican allies remained to settle there as well (Barón Castro 1996, 87–91, 197–202; Fowler 2011e, 28–29). Some 17 years later, in 1545, residents began to relocate to the modern location of San Salvador, on the left bank of the Acelhuate river, although our analysis suggests that occupation persisted at La Bermuda until about 1560.

Two major indigenous Nahuat-speaking Pipil polities dominated the territory of modern El Salvador at the time of the conquest and the founding of San Salvador: the relatively small Izalcos kingdom in the west and the larger Cuscatlan kingdom in the central region (Fowler 1989, 60–64). The Cuscatlan polity was centered on the *altepetl* ("mountain/water place," ethnic state) of Cuscatlan (modern Antiguo Cuscatlan), located just west of modern San Salvador. Cuscatlan had probably conquered a number of smaller Pipil polities such as Nonoalcos and Cojutepeque during the late pre-conquest period (Fowler 1989, 191, 208). The Spaniards effectively dominated the Pipils by the time of the second founding of San Salvador in 1528, although some armed resistance still occurred in the region until the end of the 1540s. The eastern portion of the country, between the Lempa River and the Gulf of Fonseca, was held primarily by the Lencas whose distribution extended from eastern El Salvador north into central and western Honduras (Fowler 1989, 64–65). The Lencas were still engaged in active resistance against colonization at the time of the second founding of San Salvador (Escalante Arce 2014, 77–84; Lardé y Larín 2000, 143–147).

The Izalcos polity of the Nahuatl-speaking Pipils, whose heartland lay in the Río Ceniza valley of western El Salvador, was a thriving economic and political power at the time of the first Spanish *entrada* in 1524 (Fowler 1987, 1991, 1993, 1994, 1995, 2006, 2009; MacLeod 1973, 80–95). The prodigious cacao production of the Izalcos region, recognized long ago by Millon (1955) and Bergmann (1969), provided the basis of this power, both before and after the conquest. Soon after the conquest, Spanish entrepreneurs, *vecinos* of Santiago de Guatemala (not San Salvador), recognized the potential value of cacao as a cash crop, and by 1535 they began to export small amounts of cacao from Izalcos to Mexico (Fowler 1987, 145; MacLeod 1973, 80). By the 1540s the Izalcos cacao plantations had expanded to such an extent that the region was hailed as a veritable jewel in the Spanish Crown. The 1548–1551 *Tasaciones de tributos* (tribute assessments) conducted by Alonso López de Cerrato and his associate justices of the audiencia of Guatemala (Fowler 1989, 26–27) indicate that Izalco, Caluco, Nahulingo, and Tacuscalco were the major cacao-producing towns of the region in the mid-sixteenth century, and they were among the principal cacao producers of Mesoamerica (Bergmann 1969, 92–93; Fowler 1989, 164; 1991, 1993). The audiencia established a Spanish town in the region, La Trinidad de Sonsonate, founded in 1553, ostensibly to control the activities of Spanish and *mestizo* merchants (Escalante Arce 1992, 1, 28, 34, 55–60, 109; Fowler 1995, 52–53; 2006; MacLeod 1973, 82).

Cacao production, which remained largely in indigenous hands while the Spaniards controlled trade, intensified in the region from the mid- to the late sixteenth century despite a serious decline in native population (Fowler 1995, 1997, 2006). By the end of the sixteenth century, the Izalcos towns had only a fraction of the indigenous populations that were recorded at mid-century by Cerrato, and this demographic collapse, combined with plant disease, soil exhaustion, and outside competition combined to destroy the once-lucrative cacao production and trade (Fowler 1987, 147; MacLeod 1973, 116–117).

2 Landscapes of Transformation

Discussion of landscapes of transformation requires first a brief consideration of early Spanish colonial urbanism. The relevant studies rely heavily on the trope of the Spanish colonial grid-plan city: A systematic implantation of Eurocentric spatial concepts and values ordained and structured as a strategic element in conquest and colonization throughout Spanish America, explicitly intended by the Crown, conquerors, and colonists as a vehicle for the spread of commercial capitalism (Kinsbruner 2005, 64–65). Indeed, Christopher N. Matthews (2010, 91) regards the American urban grid as "one of the largest materializations of the capitalist ideology ever constructed." Problematization of the Spanish colonial grid plan reveals the dialectic between the haunts of modernity and the processes of identity, gender, and hybridity, expressed through relations of interaction such as domination and resistance, accomodation, and appropriation. An effective way to decenter the grid plan with respect to these issues is through landscape analysis of the urban environment of conquest and a perspective from practice theory (Bourdieu 1977) which views the cultural production of the grid plan as the material and spatial embodiment of relationships and networks of power (Fowler and Zavaleta Lemus 2016; cf. Pauketat 2009, 245).

As J.H. Elliott (1963, 55) noted, the word *conquista* implied to the sixteenth-century Castilian the establishment of Spanish "presence." And presence was established through the formal founding and building of towns (Brewer Carías 2008, 15; Domínguez Compañy 1981, 39–43). Urban historian Richard L. Kagan (2000, 26–28) emphasizes that the grid-plan town represented an ideological statement by *conquistadores* acting on behalf of the Crown to impose moral, legal, and religious order, or *policía* on indigenous populations. Such ideological statements played a crucial role in the formation of community and the definition of structures of power. As the material and spatial embodiment of policía, the grid-plan city symbolized Hispanic civilization itself and evoked a propagandistic statement concerning the power of the empire that translated into very specific notions of spatial patterning derived from the social structure of the conquest. The symbolism of the siting of the government buildings and church on the central *plaza* underscored the power of the sword and the cross in imposing Castilian imperial will locally. Thus, in a very real spatial sense, the conquest emanated outward from the plaza and the grid to the surrounding countryside (Elliott 1963, 68). As Richard Morse (1962, 473), quoting Erwin Walter Palm (1951, 258), put it, the "geometric lines of force radiated out to the vast and often loosely settled surrounding space."

Conquistadors endeavored to found cities in the New World almost immediately after entering a region – in some cases even before violently subduing local indigenous societies or while they were still at war with rival Spanish factions. These cities, with rare exceptions, were built on a grid plan for reasons of both practicality and cultural tradition. Cultural tradition refers to the "genealogy of practices" (Pauketat 2001b, 80) that resulted in the grid-plan layout forming the arena for urban interaction throughout colonial Spanish America. The main thrust of the argument that we develop here is that Spanish urban traditions and institutions with very deep historical roots provided the long-term structural framework for conquest, colonialism, and colonial hegemony, yet Spanish and indigenous social actors dwelling in and near early Spanish colonial cities interacted within this spatial context from the perspective of their own traditions. Through the exercise of local agency, both Spaniards and Indians negotiated, made, and constantly remade the cultural traditions that would be passed down to future generations.

3 Ciudad Vieja

The best preserved example of a conquest-period, grid-plan town in Central America may be found at Ciudad Vieja. The rigid and orthogonal plan of San Salvador is a paragon of the Spanish American grid-plan, although potentially echoing highland Mesoamerican grid-plan cities including the Aztec capital of Tenochtitlan, especially with regard to the expansive use of space in these new colonial centers (Fowler 2008, 2009, 434–438; 2011e, 29–33; 2014a; Hardoy 1978, 228–229; Low 1995; McAndrew 1965, 110; Wagner et al. 2013, 41–45).

Located in a rural area 10 km south of the modern town of Suchitoto, the site is readily accessible and easily observed from the surface. Archaeological research at Ciudad Vieja conducted from 1996 to 2005 (Fowler 2008, 2009, 2011e, 2014a, 2014b, 2015; Fowler and Zavaleta Lemus 2016) and 2013–15 (Fowler and López Rodríguez 2016) has established that the urban core of the grid-plan town covered a core area of 45 ha (about 111 acres) (Figure 9.1), virtually all of which was artificially leveled and filled with various types of constructions. Visitors to the city, Spanish and indigenous alike, in its prime of occupation and expansion, surely would have been impressed with the strict, orderly layout of the townscape. They would have admired the spacious *plaza* at the center of the town, the church to the east of the *plaza*, the *casa de cabildo* (town hall) and other municipal buildings on the north, stores and shops on the west, and a tavern and blacksmith's shop on the south (Card 2007, 482–499; Card and Fowler 2012). Among the 20 structures and activity areas so far excavated, we have identified Spanish and indigenous residences (Gallardo Mejía 2004, 2011; Hamilton 2009, 2011b), ritual spaces (Fowler and López Rodríguez 2016), civic/administrative buildings (Fowler 2011c), food preparation areas (Card 2007, 2011), commercial and industrial buildings (Fowler 2011a, 2011b, 2011d), warehouses and storage areas, terraces, streets, ramps, and defensive features. The latter consist of guard houses, sentry stations, and a steep defensible cliff sloping away from the south and east sides of the site (Hamilton 2009, 247, 2011a).

Long, straight streets run from (or into) the four corners of the *plaza* in the cardinal directions. The one exception to the straightness of the streets is the street running south from the platform on the southeast corner of the *plaza mayor* which curves around a valuable natural spring. Other streets run parallel and perpendicular to the *plaza*, bisecting each other to form large blocks. The *plaza*, including surrounding streets on the four edges of the *plaza*, measures approximately 100 m on a side. The square city blocks measure approximately 80 m (roughly 100 Spanish *varas*, 1 vara = 83.5 cm) on a side, bounded by streets about 8 m (10 varas) in width. Most of the city blocks were probably subdivided into four *solares* (see Bayle 1952, 80–82) or house lots of approximately 40 x 40 m (50 x 50 varas).

The overall town plan shows strong adherence to the grid-plan layout, but our excavations and geophysical remote sensing data show that no prevailing norm governed the locations of structures within house lots, the orientation of architecture, or the size, shape, and internal arrangement of rooms within structures (Fowler et al. 2007). Spanish-style buildings are distinguished by

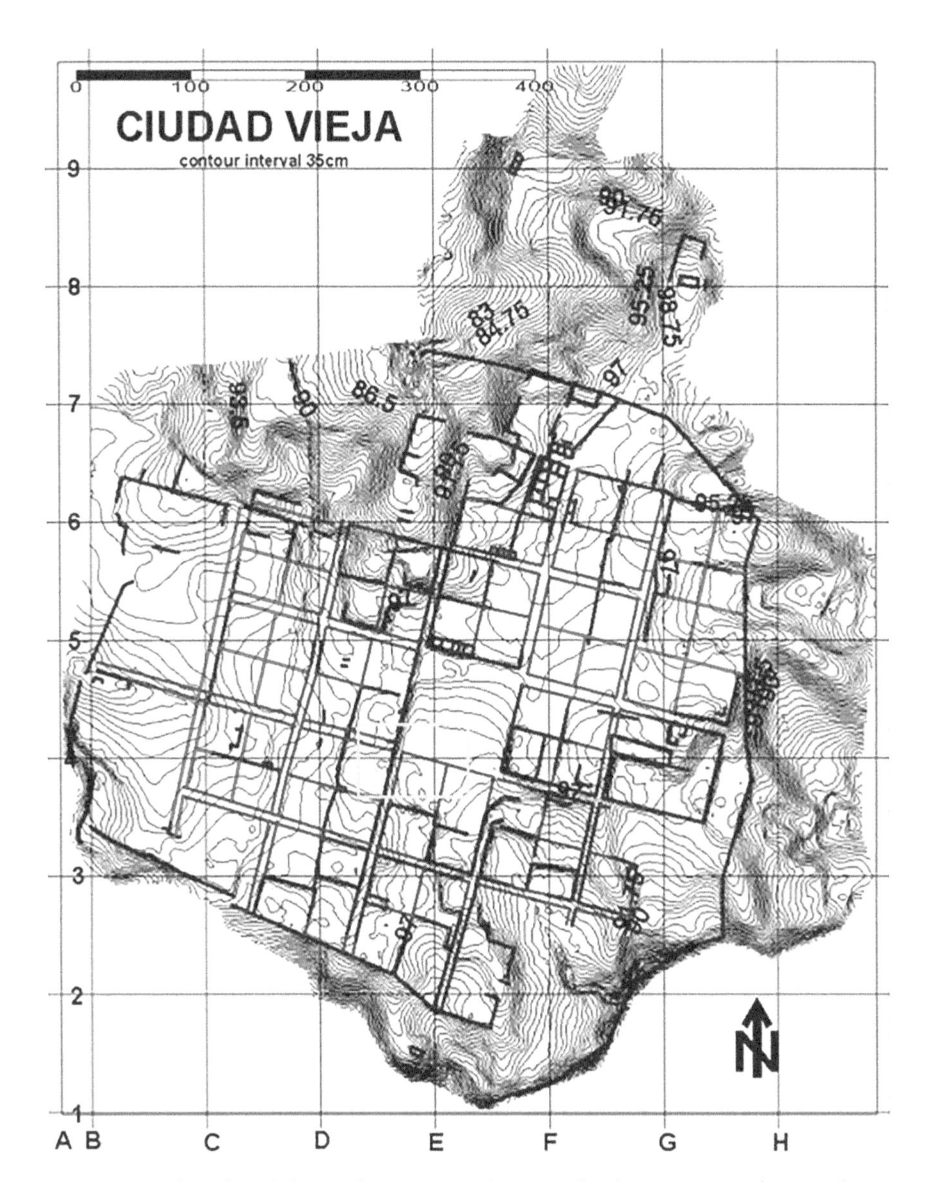

FIGURE 9.1 Plan of Ciudad Vieja showing natural topography, the reconstructed sixteenth-century urban grid plan, and the archaeological site grid
PLAN BY CONARD C. HAMILTON, 1998–99

multiroom floor plans, substantial stone foundations measuring 83–84 cm (1 vara) or greater in width, brick tile floors, ceramic roof tiles, and the use of iron nails and other hardware to secure structural elements. But variation occurs in Spanish-style architecture with wall foundations ranging above and below the standard *vara*. The structures are generally oriented to the overall site alignment of 12°, but exceptions do occur.

A good case for an indigenous-led household can be found in Structure 2F1, located on the southeastern edge of the site. This structure consisted of one room outlined by narrow foundations. It did not employ roof or floor tile, indicating packed earth floors and thatch roofing (Hamilton 2009, 258, 2011b). It appears to have had wattle-and-daub walls, with only a handful of Spanish-technology metal tacks involved in its construction (Hamilton 2009, 363). Trash middens associated with Structure 2F1 show similarities to those found in other households at the site in regards to ceramics as well as the use of obsidian blades (Card 2007, 476–479). Several spindle whorls suggest indigenous textile production, and the artifact assemblage as a whole indicates a household rather than a non-residential out-structure. The orientation of such buildings does not follow the overall site grid. Thus, while the overall urban layout appears quite rigid at first glance, a considerable amount of flexibility and variability in internal distribution and use of space characterizes the town.

4 Caluco

In contrast to the Spanish-directed founding of the *villa* of San Salvador which included many indigenous residents, the sixteenth-century town of Caluco was a reorganized, resettled indigenous community with few if any Spanish residents. The pre-Columbian Izalcos Pipil polity consisted of four principal towns: Izalco, Caluco, Tacuscalco, and Nahuilingo, referred to collectively during the early colonial period as "los Izalcos." In terms of native politics, these four towns formed a quadripartite altepetl (Lockhart 1992, 19) divided into the paired towns of lzalco/Caluco and Nahulingo/Tacuscalco, with the native lord of Tecpan Izalco controlling cacao production and exercising political domain at the time of the conquest (Fowler 1989, 195; 1991, 189). Lockhart (1992, 436–438) emphasized the cellular-modular nature of the altepetl and its tendency to create larger units by the aggregation of parts that remain relatively separate and independent. Archaeologically, this principle was manifest by a tendency toward dispersed settlement with strong nucleation or centralization occurring only in complex urban landscapes (Lockhart 1992, 19; cf. Smith 2008, 73).

Documentary data indicate that the rectilinear pattern of the modern Izalcos towns is a Spanish imposition. A visitor to the Izalcos region at the time of the conquest would have observed a rural landscape dotted by dispersed settlements with households dispersed evenly across the landscape and surrounded by *milpas* and cacao orchards crisscrossed by irrigation canals (Sampeck 2007). Beginning in 1529, a stream of royal decrees mandated that indigenous settlements be "reduced" and brought into Spanish-style nucleated towns or villages with streets running east-west and north-south departing from a central *plaza* in which was located the church and the *cabildo* (town hall) with a jail. House lots were parceled out to indigenous families by priests and *caciques* (indigenous political leaders), and common lands were assigned for cultivation. The new formalized settlements were probably in place by about 1553 in Izalco, Caluco, Nahulingo, and Tacuscalco, since by this time the towns had their own priests and Spanish-style indigenous governments (Escalante Arce 1992:1, 26). In addition, before enforced nucleation occurred, Izalco and Caluco formed a single dispersed, rural settlement that was broken up in 1532 to assign *encomiendas* to two Spaniards. At that time, Tecpan Izalco (the northern sector) became known simply as Izalco, and Caluco Izalco (the southern sector) became Caluco (Escalante Arce 1992:1, 218). Contemporary observers often referred to the four Izalcos settlements as "*dos pueblos hechos quatro*" (two towns made into four), a clear reference to *reducción*. Verhagen (1997, 235–238) found, however, that the earliest archaeological materials in Caluco date to the 1580s. Therefore, full, practical implementation of forced nucleation of the Izalcos Pipil towns appears not to have occurred until this decade.

Sampeck (2007, 2010) has presented the results of an intensive regional survey in the Río Ceniza valley, as well as excavations, lithic analysis, and ceramic analysis in order to assess key elements of the Izalcos political economy before and after the conquest. Her data show that the Pipils were central actors in Late Postclassic regional integration, which positioned the Izalcos region within the world genesis of capitalism and structured Spanish colonialism locally in the sixteenth century. Sampeck (2007, 232–257, 2010) found in her Izalcos regional settlement survey slight nucleation balanced by dispersion in late pre-conquest times, following the principles of Nahua cellular-modular organization, contrasted with a tendency toward stronger nucleation in post-conquest times, associated with Spanish concepts of urbanization and the legal requirements of *reducción* (Escalante Arce 1992:1, 22–23; Fowler 1995, 40; MacLeod 1973, 122). In addition to the role of Nahua cognitive principles in determining patterns of settlement before the conquest, Sampeck (2007, 232–233) also attributes a significant role in pre-conquest settlement dispersion to the requirements of tending cacao orchards at the household level of production (see Fowler 2006, 2009, 431–432).

Fowler (1995) and Verhagen (1997) describe Caluco in the late sixteenth century as a dense, nucleated settlement consisting of a series of artificial terraces,

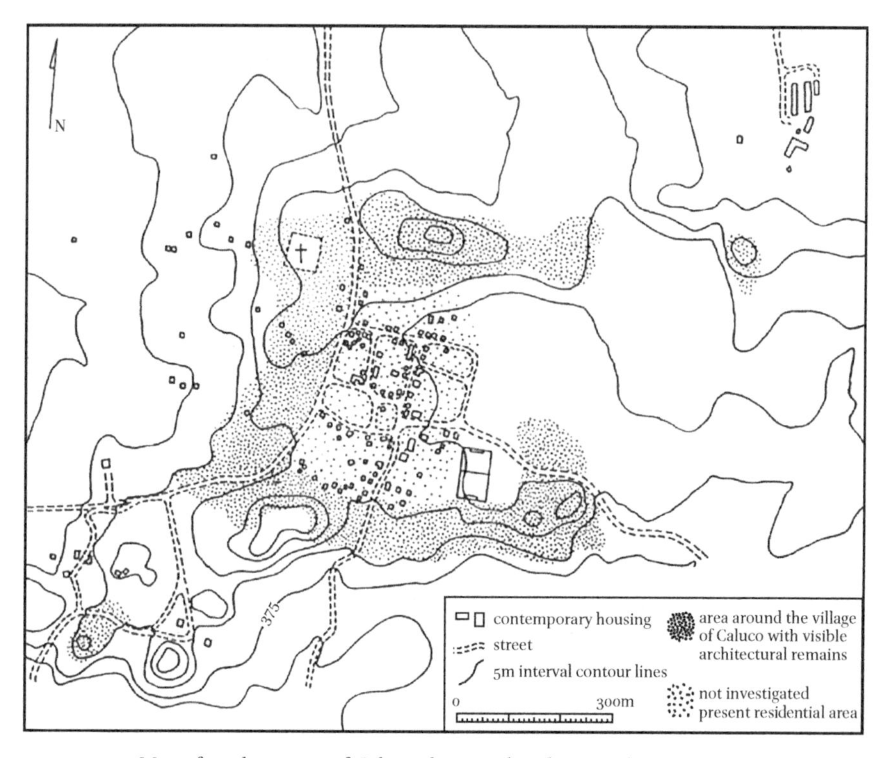

FIGURE 9.2 Map of modern town of Caluco showing distribution of sixteenth-century materials
FOWLER 1995, FIGURE 4

large platforms, a stone-paved street, house foundations, and check dams on gently undulating terrain covering an area of approximately 80 ha. (Figure 9.2). Also during this time, the once-sumptuous church of San Pablo y San Pedro, an exquisite example of Spanish American *mudejar* architecture, was built in Caluco, probably beginning around 1567–68 with construction completed in the early seventeenth century (Fowler 1995; Verhagen 1997, 147–175).

5 Local Production and Indigenous Material Transformation

5.1 *Indigenous Production*

While Spanish presence at Ciudad Vieja is clearly indicated by the townscape, the architecture, and certain industries and classes of material culture such as iron, brick, and glass, we see a strong indigenous presence at San Salvador also reflected in the material culture of Ciudad Vieja. Indigenous residents of the town would have included indigenous Pipils as well as Mexican (Nahua and Mixtec) and Kaqchikel groups allied with the Spaniards during the conquest.

Of the more than 44,000 potsherds from excavation and surface collection, the handful of majolica and a few hundred fragments of *botija* transport jars gives Ciudad Vieja one of the lowest proportions of Spanish ceramics found at any Spanish colonial site in the Americas and is more comparable with rural indigenous villages of the colonial period than with a Spanish capital (Card 2007, 514–519). The architecture and iron working of Ciudad Vieja are that of a Spanish town but the pottery is that of an indigenous one.

Obsidian artifacts also occur in great numbers on the surface and have been recovered from all excavated loci at Ciudad Vieja. *Manos* and *metates* for maize processing occur in domestic contexts and on the surface. Ceramic spindle whorls for spinning thread speak to gendered production practices. Polished ceramic ear flares and jade objects (elite bodily adornment) are significant in the inventory of indigenous-associated items and maintenance of indigenous identity. In sum, the archaeological evidence indicates that this Spanish American town had a indigenous Mesoamerican population of significant proportion and that at least some members of this population were engaged in maintenance or construction of indigenous identity. At the same time some men and women entered into relationships and engaged in productive activities as part of the new order.

The complex interplay of Spaniards, *conquistadores mejicanos*, and Pipils is particularly visible in the hybrid brimmed plates that make up the primary serving vessel at Ciudad Vieja (Figure 9.3). The form of these vessels closely mirrors variations in Spanish *morisco* and Italianate majolica, but the surface designs and potting technology are clearly part of the Late Postclassic Pipil ceramic tradition (Card 2007, 2013). Unlike the situation at most other Spanish colonial settlements, this new class of vessel is more associated with indigenous use than with Spanish use, and they are especially rare in the wealthiest Spanish households. This pattern of use is found in other examples of forced indigenous displacement in early colonial Spanish America (Card 2007, 276–299, 2013, 120). While San Salvador is within Pipil country, as there was no immediate precontact settlement on the site, all inhabitants would have been newcomers to the site by choice or by force.

Placing Ciudad Vieja locally produced earthenware brimmed plate forms within a seriation of European majolica finds that most of the plates agree with the historically documented occupation span of the site (1525/28–1545), but some of the forms post-date 1545, suggesting a protracted abandonment

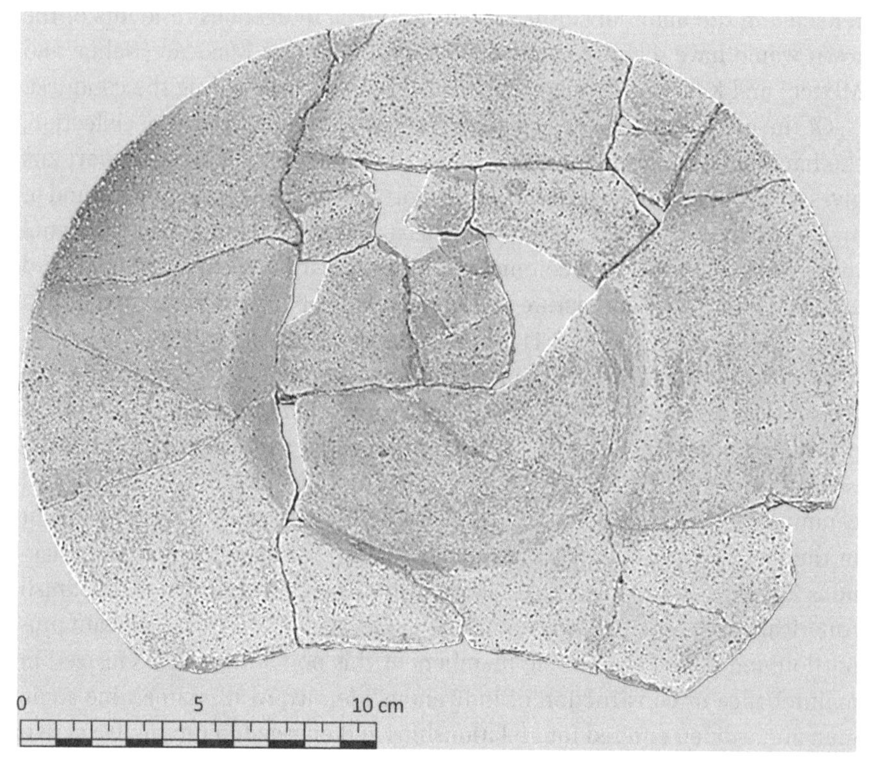

FIGURE 9.3 Hybrid brimmed plate, a very popular ceramic form at Ciudad Vieja. Provenience: Structure 4C1, unit 2002-3.7, level 3
PHOTO BY JEB J. CARD

lasting until possibly 1560. From an archaeological perspective, we can no longer privilege the record of the written text; we must revise the chronology of the abandonment of the town.

Analysis of vessel form and function from eight excavated contexts of Ciudad Vieja shows similar activities in Spanish and indigenous households, as well as the identification of a tavern or other commercial food and beverage vendor in the center of town. Three microstyles – possibly individual potters' "hands" – crosscut other classificatory categories suggesting localized or household production and distribution of ceramics.

These microstyles become less common during the later years of the occupation, suggesting the emergence of new community networks and production practices that tended to swamp the initial localized variability. The potters were likely Pipil women from different communities who initially brought minor differences on a regional style with them as well as adopting new pottery forms and styles. The interaction of their children as members of the new community left its mark in the more homogenous ceramics they produced. This transformation suggests that a model of creolization and ethnogenesis would be useful for understanding other interactions in early colonial Mesoamerica.

In contrast with the new concentration of power and economy around San Salvador, an essential element for understanding Izalcos Pipil settlement structure involves their engagement in the expansive trade network that reached into central Mexico in both pre- and post-conquest times. Even though the evidence from Caluco is decades after that of Ciudad Vieja, the importance of the Izalcos region in long-distance trade was undiminished after the Spanish conquest. Pipil concepts, institutions, and boundaries structured Spanish political and economic activities, while the Spanish rationalized the landscape as wage labor emerged in the late sixteenth century. This new local market of production, consumption, and speculation, however, did not respond entirely to the demands or mandates of colonial power or even the world-system. Characteristic ceramic markers of the late sixteenth-century town at Caluco consist of Ming porcelain and imported majolicas of Guatemalan, Mexican, and Panamanian origin (see Lister and Lister 1987; Sampeck 2015). Obsidian, groundstone objects, glass, and metal artifacts are also abundant in the Caluco collections.The Manila galleon trade was a prime catalyst for Mexico to consolidate power by moving the route of New World trade across the isthmus, but the dispersal of porcelain and majolica in the Izalcos region suggests that contraband trade thwarted the Crown's efforts. In each phase, according to Sampeck (2007), Izalco-centered interests exerted a gravity the world-system could not alter.

5.2 *Spanish Industries*

Early Spanish colonial cities were important political and administrative centers, but it was in matters of the economy that they exerted the strongest local force. Dedicated to the exploitation of local land and labor, Spanish conquest towns served the predatory economic interests of both the Crown and the individual colonists. Issues of power, ethnicity, gender, and identity were integrated into all aspects of the economy of San Salvador.

One well-documented activity of the Spanish and indigenous men of San Salvador was warfare. From the founding of the first San Salvador in 1525 until the defeat of Lempira's uprising in 1539 – the historical record documents several raids and battles between San Salvador and indigenous resistance forces (Barón Castro 1996, 103–104; Lardé y Larín 2000, 111). It was through war materiel, specifically made of iron, that warfare intersected with the San Salvador economy. During the Lempira revolt of 1537–1539, Governor Francisco de Montejo of Honduras appealed to the *vecinos* of San Salvador, apparently more secure against rebellion, for support and war materiel. They responded by sending a Spanish captain with 100 native auxiliaries and 1000 indigenous (probably Pipil) porters carrying gunpowder, harquebuses and balls, crossbows and bolts, swords, lances, shields, armor, and iron bars from which to manufacture points for crossbow bolts.

Blacksmithing has been identified in Structures 6F2 (Fowler 2011c) and 3D2 (Fowler 2011d). The northern smithy, at Structure 6F2, eventually closed his shop during the occupation of the town, and the structure came to be used as a trash dump. It was located near the largest residence identified at the site, Structure 6F1. A higher proportion of exotic indigenous ceramics, possibly from elsewhere in the region, and transport jars support the interpretation of this structure as the home of an *encomendero* receiving tribute from indigenous communities (Card 2007, 500, 526, 555; Gallardo Mejía 2004). The smith's shop on the main *plaza* in Structure 3D2 included charcoal manufacturing and was part of a compound also including a tavern (Card 2007, 482–499; Card and Fowler 2012; Fowler 2011d). In both cases it is notable that these "dirty" industries were located close to zones of Spanish wealth and cultural display. The blacksmith's shop and the tavern in Structure 3D2 were located across the *plaza* from the *cabildo* and likely near the public marketplace. Thus, it was located squarely in the economic, political, and social heart of San Salvador. European imported objects are rare at Ciudad Vieja, but excavations of the tavern in Structure 3D2 yielded a fragment of a Venetian crystal goblet with *vetro a fili* decoration, dating to the second quarter of the sixteenth century (Figure 9.4).

Local Spanish industries attested archaeologically, such as brick making, roof and floor tile manufacturing, blacksmithing, and charcoal making would have also depended on male indigenous and *mestizo* labor. The Dominican chronicler Antonio de Remesal (1964–66:vol. 2, bk. 9, Ch. 3, p. 202), a secondary source of the early seventeenth century who had access to the cabildo books, mentioned the following trades in the town: shoemakers, carpenters, tailors, tanners, and blacksmiths (see also Lardé y Larín 2000, 150–152). These all constitute activities that likely involved Spanish overseers and indigenous, *mestizo*, and perhaps African laborers. Bricks and tile are found across the site, though their exact production source has not been identified. A small minority of brick and tile may have been imported. Discolorations of one floor space in Structure 3D1 suggest a bichrome tile design that might have used imported tile (Fowler 2011b). One glazed *teja* roof tile and a handful of green tin-glazed *baldosa* floor tile also suggest some tile was imported. But the majority would have been locally produced.

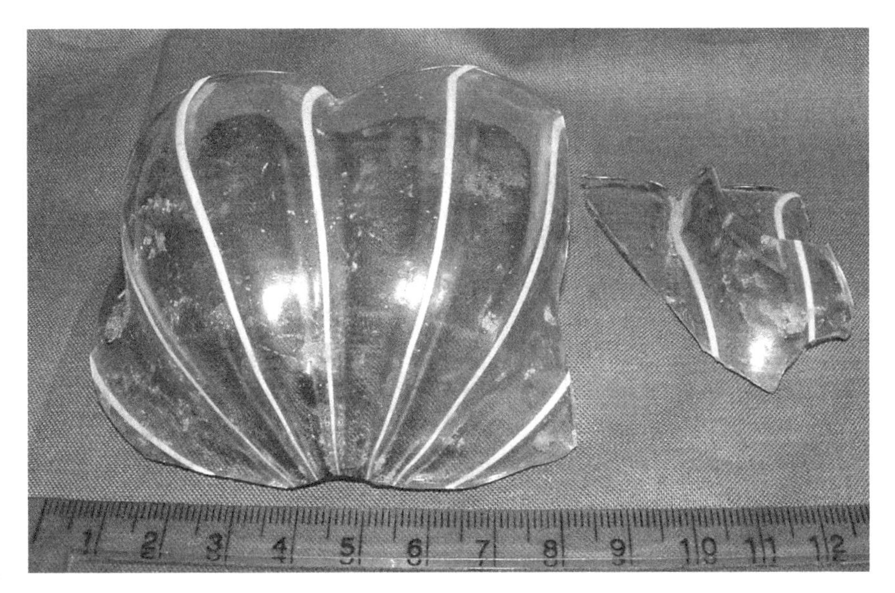

FIGURE 9.4 Fragment of Venetian crystal goblet with *vetro a fili* decoration. Structure 3D2, Room 5, Op. 03–1.127, Level 1
PHOTO BY JEB J. CARD

5.3 *Labor and Tribute*

Subsistence production in or near San Salvador probably would have included local cultivation of garden crops, tending of fruit trees, hunting, and fishing – activities of both men and women – but the majority of food consumed in the town probably was obtained from tribute rendered by conquered Pipil and Lenca towns in the surrounding region and the hinterlands. Tribute payments to the 56 Spanish encomenderos of San Salvador are enumerated in the 1532 *Relación Marroquín* (Fowler 1989, 25–26, 155–186). This extraordinary document from the Archivo General de Indias in Seville (Audiencia de Guatemala, legajo 52), published in paleographic transcription by Francis Gall in 1968, contains rich economic and demographic data on approximately ninety indigenous settlements that had been assigned in *encomienda* (a formal grant of labor and commodity tribute) to the Spanish *vecinos* of San Salvador. The document shows that maize, beans, chili peppers, turkeys, venison, salt, dried fish, honey, beeswax, firewood, and a variety of woven cotton cloth and clothing were paid in tribute by Pipil and Lenca conquered communities. Tribute was delivered to San Salvador by male porters from these communities. Failure to pay tribute was met by threat or application of violent force.

Encomienda labor was virtual slavery (Jones 1994, 99; Sherman 1979, 85–92), and labor tribute was rendered by almost all Pipil communities, taking the form of male agricultural service, construction, mining, or trade labor (in addition to burden bearing) and female domestic service such as food preparation, cleaning, and child-rearing. Many if not most Spanish households would have had numerous servants from *encomienda* communities as well as indigenous slaves obtained as war captives or through barter.

A significant but poorly known aspect of the economy of San Salvador, and early colonial Central American cities in general, consisted of slaving and slave trafficking. Slaving operations began in Guatemala and Cuscatlan during the first Spanish invasion led by Pedro de Alvarado in 1524 (Sherman 1979, 27). Alvarado himself is credited with being the largest slaver in Guatemala and one of the largest slaveholders in Central America in the early post-conquest years.

The construction and maintenance of San Salvador would not have been feasible without a large slave labor force. Wealthy *vecinos* of Santiago and San Salvador also had trained gangs of indigenous slaves working in gold and silver extraction operations in Honduras. During Alvarado's time in Honduras-Higueras there were 20 or more of these gangs each with 15–25 slaves on average (Chamberlain 1953, 112). Slaving was the most important economic activity in Nicaragua in the 1530s; slaves taken in Nicaragua were shipped to Panama and Peru (MacLeod 1973, 51–52). Spanish encomenderos of San Salvador participated in the same activity, although on a relatively smaller scale, taking slaves in the eastern Lenca territory of Chaparrastique or Popocatepet, until slaving was curtailed by indigenous population collapse in the 1540s and the implementation of the New Laws in 1548 (Barón Castro 1996, 136, 166–167; Kramer 1994, 186–190). Cerrato liberated 500 Indian slaves from about 40 *vecinos* of San Salvador in 1548 (Sherman 1979, 148). Diego de Holguín, the first *alcalde* of San Salvador in 1525, claimed that 50 slaves had been taken away from him when Alonso López de Cerrato arrived as president of the audiencia of Guatemala in 1548 to enforce the New Laws, thus liberating indigenous slaves (Sherman 1979, 71). It was probably no coincidence that just three years earlier the city began the move from its original location at La Bermuda to the more economically and demographically prosperous location in the Acelhuate valley that it occupies today.

6 Conclusions

Conquest-era San Salvador both supports and defies expectations of early Spanish colonialism. Culture shock and change appear in the creation of a brand-new Spanish town not in accordance with the existing sociopolitical order. Tremendous labor was expended to create Spanish-style architecture and the townscape, and Spanish technologies and industries are found in some of the most prominent places in San Salvador. Yet the majority of the population was indigenous, as is the bulk of the material culture assemblage, and the site may have been chosen due to deeper memory as a sacred spot. Potters brought from elsewhere experimented within the Pipil regional ceramic style, incorporating and modifying elements of European form. With time, the children of these potters created a more homogenous ceramic product, but one still Pipil in overall style, suggesting creolization rather than *mestizaje* as the conquest shifted into the early colonial period.

Traditional historiography on the cacao boom in the Izalcos region during the early colonial period tends to present a rather static picture of the dominant Spanish *encomenderos* and merchants exploiting defenseless indigenous peoples who had no choice but to submit to the new colonial world order. The archaeological data present a more subtle, more complicated tableau. We see indigenous populations not simply suffering under colonial domination, but actively constructing and structuring their own conditions of reality as they resisted exploitation and adjusted to new social, economic, and political relations. The materiality of these transformations, with its rich and varied artifact inventory and diverse architectural remains, indicates that the inhabitants of Caluco and other Pipil communities in the Izalcos region participated actively in the local and global political economy, and that they developed both passive and active strategies to resist colonial domination.

The sudden impact of San Salvador's conquest-period political economy, dominated by the indigenous *conquistadores mejicanos* and a small number of Spaniards who sought wealth rather than stability, gave way in the late sixteenth century to the likewise indigenous early colonial cacao boom in the west. The resulting global connections and evidence of consumption found in the Izalcos region demonstrates the importance of not only temporal but also social context. Overarching narratives of change and stabilization may make some broad sense at higher levels of abstraction, but the archaeological data challenge larger narratives with smaller effective scales of community, transformation, and adaptation.

Acknowledgments

We thank Corinne L. Hofman and Floris W.M. Keehnen for the invitation to present the first version of this chapter in the symposium "Material Encounters and Indigenous Transformations in the Early Colonial Americas" which they organized for the 82nd Annual Meeting of the Society for American

Archaeology, 2 April 2017, in Vancouver, Canada, and for their fine efforts in seeing to the publication of the papers of the symposium. The Caluco and the Ciudad Vieja archaeological projects were conducted by permission from the Secretaría de Cultura, and the Dirección de Patrimonio Cultural of El Salvador. Students from the Universidad de El Salvador and the Universidad Tecnológica de El Salvador participated in the Ciudad Vieja project. We express our gratitude to the following friends and colleagues for their advice and support of the research at Caluco and Ciudad Vieja: E. Wyllys Andrews, María Isaura Aráuz Quijano, Luis María Calvo, Ricardo Castellón, the late Ana Vilma de Choussy, Gabriel Cocco, Jorge Pável Elías Lequernaqué, José Heriberto Erquicia, Pedro Antonio Escalante Arce, Marlon Escamilla Rodríguez, Francisco Estrada Belli, Roberto Gallardo, Sally Graver, Conard H. Hamilton, Gustavo Herodier, Mirta Linero Baroni, Raquel López Rodríguez, Murdo MacLeod, Laura Matthew, David Messana, Juan Ramón Muñiz Álvarez, Charles E. Orser, Jr., Kathryn E. Sampeck, Evelin Sánchez, Fernando Vela Cossío, and Inez Verhagen. Funding and material support have been provided by the Academia Salvadoreña de la Historia, the Foundation for Ancient Mesoamerican Studies, Inc., the H.J. Heinz III Charitable Fund, the National Geographic Society, the Patronato Pro-Patrimonio Cultural of El Salvador, Tulane University, Vanderbilt University, Miami University of Ohio, the National Science Foundation, and the Wenner-Gren Foundation for Anthropological Research.

References

Barón Castro, Rodolfo. 1996. *Reseña histórica de la villa de San Salvador desde su fundación en 1525, hasta que recibe el título de ciudad en 1546, 2nd ed.* San Salvador: Consejo Nacional para la Cultura y el Arte.

Bayle, Constantino. 1952. *Los cabildos seculares en la América española.* Madrid: Sapientia.

Bergman, John F. 1969. "The Distribution of Cacao Cultivation in Pre-Columbian America 1." *Annals of the Association of American Geographers* 59 (1): 85–96.

Bourdieu, Pierre. 1977. *Outline of a Theory of Practice,* translated by Richard Nice. Cambridge: Cambridge University Press.

Brewer Carías, Alan R. 2008. *El modelo urbano de la ciudad colonial y su implantación en Hispanoamérica.* Bogota: Universidad Externado de Colombia.

Card, Jeb J. 2007 "The Ceramics of Colonial Ciudad Vieja, El Salvador: Culture Contact and Social Change in Mesoamerica." PhD diss., Tulane University.

Card, Jeb J. 2011. "Excavaciones y arquitectura de la Estructura 6F4." In *Ciudad Vieja: Excavaciones, arquitectura, y paisaje cultural de la primera villa de San Salvador,* edited by William R. Fowler, 116–125. San Salvador: Secretaría de Cultura de la Presidencia/ Editorial Universitaria.

Card, Jeb J. 2013. "Italianate Pipil Potters: Mesoamerican Transformation of Renaissance Material Culture in Early Spanish Colonial San Salvador." In *The Archaeology of Hybrid Material Culture*, edited by Jeb J. Card, 100–130. Carbondale: Southern Illinois University Press.

Card, Jeb J. and William R. Fowler. 2012. "Conquistador Closing Time: Wealth, Identity, and an Early Sixteenth-Century Tavern at Ciudad Vieja, El Salvador." Paper presented at the *77th Annual Meeting of the Society for American Archaeology, Memphis, April 18–22*.

Chamberlain, Robert S. 1953. *The Conquest and Colonization of Honduras, 1502–1550*. Washington, D.C.: Carnegie Institution of Washington, Publication 598.

Crumley, Carole L. 1979. "Three Locational Models: An Epistemological Assessment for Anthropology and Archaeology." *Advances in Archaeological Method and Theory* 2: 141–173.

Domínguez Compañy, Francisco. 1981. *Estudios sobre las instituciones locales hispanoamericanas*. Caracas: Academia Nacional de la Historia.

Elliott, J.H. 1963. *Imperial Spain, 1469–1716*. Harmondsworth, Middlesex: Pelican Books.

Escalante Arce, Pedro Antonio. 1992. *Codice Sonsonate*, 2 vols. San Salvador: Consejo Nacional para la Cultura y el Arte.

Escalante Arce, Pedro Antonio. 2014. *Crónicas de Cuzcatlán-Nequepio y del Mar del Sur, 2nd ed.* El Salvador: Editorial Delgado, Antiguo Cuscatlán.

Fowler, William R. 1987. "Cacao, Indigo, and Coffee: Cash Crops in the History of El Salvador." *Research in Economic Anthropology* 8: 139–167.

Fowler, William R. 1989. *The Cultural Evolution of Ancient Nahua Civilizations: The Pipil-Nicarao of Central America*. Norman: University of Oklahoma Press.

Fowler, William R. 1991. "The Political Economy of Indian Survival in 16th-Century Izalco, El Salvador." In *The Spanish Borderlands in Pan-American Perspective, Columbian Consequences, Vol. 3*, edited by David Hurst Thomas, 187–204. Washington, D.C.: Smithsonian Institution Press.

Fowler, William R. 1993. "'The Living Pay for the Dead': Trade, Exploitation, and Social Change in Early Colonial Izalco, El Salvador." In *Ethnohistory and Archaeology: Approaches to Postcontact Change in the Americas*, edited by J. Daniel Rogers and Samuel M. Wilson, 181–199. New York: Plenum Press.

Fowler, William R. 1994. "La región de Izalco y la villa de la Santísima Trinidad de Sonsonate." In *Dominación española: Desde la Conquista hasta 1700*, edited by Ernesto Chinchilla Aguilar, 601–610. *Historia general de Guatemala, 6 Volumes*, edited by Jorge Luján Muñoz. Guatemala City: Asociación de Amigos del País, Fundación para la Cultura y el Desarrollo.

Fowler, William R. 1995. *Caluco: Historia y arqueología de un pueblo pipil en el siglo XVI*. San Salvador: Patronato Pro-Patrimonio Cultural.

Fowler, William R. 2006. "Cacao Production, Tribute, and Wealth in Sixteenth-Century Izalcos, El Salvador." In *Chocolate in Mesoamerica: A Cultural History of Cacao*, edited by Cameron L. McNeil, 307–321. Gainesville: University Press of Florida.

Fowler, William R. 2008. "Spanish American Urbanism in the Conquest Period: Ciudad Vieja, El Salvador." In *El urbanismo en Mesoamérica/Urbanism in Mesoamerica, Vol. 2*, edited by Alba Guadalupe Mastache, Robert H. Cobean, Ángel García Cook, and Kenneth G. Hirth, 651–680. Mexico City: Instituto Nacional de Antropología e Historia and University Park: Pennsylvania State University.

Fowler, William R. 2009. "Historical Archaeology in Yucatan and Central America." In *International Handbook of Historical Archaeology*, edited by Teresita Majewski and David Gaimster, 429–447. New York: Springer.

Fowler, William R. 2011a. "Excavaciones de la Estructura 3D1: Una posible tienda." In *Ciudad Vieja: Excavaciones, arquitectura, y paisaje cultural de la primera villa de San Salvador*, edited by William R. Fowler, 167–181. San Salvador: Secretaría de Cultura de la Presidencia/Editorial Universitaria.

Fowler, William R. 2011b. "Excavaciones de la Estructura 3D2." In *Ciudad Vieja: Excavaciones, arquitectura, y paisaje cultural de la primera villa de San Salvador*, edited by William R. Fowler, 194–204. San Salvador: Secretaría de Cultura de la Presidencia/Editorial Universitaria.

Fowler, William R. 2011c. "Excavaciones y arquitectura de la Estructura 4E1: La casa del cabildo." In *Ciudad Vieja: Excavaciones, arquitectura, y paisaje cultural de la primera villa de San Salvador*, edited by William R. Fowler, 79–84. San Salvador: Secretaría de Cultura de la Presidencia/Editorial Universitaria.

Fowler, William R. 2011d. "Excavaciones y arquitectura de la Estructura 6F2." In *Ciudad Vieja: Excavaciones, arquitectura, y paisaje cultural de la primera villa de San Salvador*, edited by William R. Fowler, 130–142. San Salvador: Secretaría de Cultura de la Presidencia/Editorial Universitaria.

Fowler, William R. 2011e. "Introducción." In *Ciudad Vieja: Excavaciones, arquitectura y paisaje cultural de la primera villa de San Salvador*, edited by William R. Fowler, 23–57. San Salvador: Editorial Universitaria/Secretaría de Cultura de la Presidencia.

Fowler, William R. 2014a. "Central America: Historical Archaeology of Early Colonial Urbanism." In *Encyclopedia of Global Archaeology*, edited by Clare Smith, 1197–1208. New York: Springer.

Fowler, William R. 2014b. "Ciudad Vieja, El Salvador, la primera villa de El Salvador: Investigaciones de paisaje y traza (1996–2013)." In *Arqueología de los primeros asentamientos urbanos españoles en la América Central y Meridional* (Actas del I Seminario Internacional, Red Iberoamericana de Investigación del Urbanismo

Colonial), edited by Fernando Vela Cossío and Alejandro García Hermida, 109–144. Madrid: Escuela Técnica Superior de Arquitectura de Madrid, Mairea Libros.

Fowler, William R. 2015. "Ciudad Vieja, Estructura 4G1: Una posible capilla de indios en la Villa de San Salvador." In *V Congreso Centroamericano de Arqueología en El Salvador*, edited by José Heriberto Erquicia Cruz and Shione Shibata, 125–137. San Salvador: Museo Nacional de Antropología "Dr. David J. Guzmán," Secretaría de Cultura de la Presidencia.

Fowler, William R. and Howard H. Earnest. 1985. "Settlement Patterns and Prehistory of the Paraíso Basin of El Salvador." *Journal of Field Archaeology* 12: 19–32.

Fowler, William R. and Eugenia Zavaleta Lemus. 2016. "Habitus, campo y capital en las primeras fundaciones urbanas hispanoamericanas: El caso de Ciudad Vieja de San Salvador." In *Primeros asentamientos españoles y portugueses en la América central y meridional, siglos XVI y XVII*, edited by Luis María Calvo and Gabriel Cocco, 21–34. Santa Fe, Argentina: Universidad Nacional del Litoral.

Fowler, William R. and Raquel López Rodríguez. 2016. "Explorando un espacio sagrado: Arquitectura y estratigrafía de Estructura 4G1, Ciudad Vieja, El Salvador." In *Primeros asentamientos urbanos en Iberoamérica (SS. XVI y XVII): Investigación y gestión* (Actas del III Seminario Internacional de la Red Iberoamericana de Investigaciones del Urbanismo Colonial), edited by Cristina Vargas Pacheco, 53–74. Piura: Universidad de Piura.

Fowler, William R., Francisco Estrada-Belli, Jennifer R. Bales, Matthew D. Reynolds, and Kenneth L. Kvamme. 2007. "Landscape Archaeology and Remote Sensing of a Spanish-Conquest Town: Ciudad Vieja, El Salvador." In *Remote Sensing in Archaeology*, edited by James Wiseman and Farouk El-Baz, 395–422. New York: Springer.

Gallardo Mejía, Francisco Roberto. 2004. "Spanish Identity at a Sixteenth-Century Colonial House: Structure 6F1 at Ciudad Vieja, El Salvador." Unpublished Master thesis, Anthropology Department, University of Colorado.

Gallardo Mejía, Francisco Roberto. 2011. "Excavaciones y arquitectura de la Estructura 6F1." In *Ciudad Vieja: Excavaciones, arquitectura, y paisaje cultural de la primera villa de San Salvador*, edited by William R. Fowler, pp. 85–115. San Salvador: Secretaría de Cultura de la Presidencia/Editorial Universitaria.

Gramsci, Antonio. 1971. *Selections from the Prison Notebooks of Antonio Gramsci*, edited and translated by Quintin Hoare and Geoffrey Nowell Smith. New York: International Publishers.

Hamilton, Conard C. 2009. "Intrasite Variation among Household Assemblages at Ciudad Vieja, El Salvador." PhD diss., Tulane University.

Hamilton, Conard C. 2011a. "Estructura 1D1: El puesto de vigilancia sur." In *Ciudad Vieja: Excavaciones, arquitectura y paisaje cultural de la primera villa de San Salvador*, edited by William R. Fowler, 150–154. San Salvador: Secretaría de Cultura de la Presidencia/Editorial Universitaria.

Hamilton, Conard C. 2011b. "Estructura 2F1: Buscando la ocupación indígena en Ciudad Vieja." In *Ciudad Vieja: Excavaciones, arquitectura y paisaje cultural de la primera villa de San Salvador*, edited by William R. Fowler, 155–166. San Salvador: Secretaría de Cultura de la Presidencia/Editorial Universitaria.

Hardoy, Jorge E. 1978. "European Urban Forms in the Fifteenth to Seventeenth Centuries and Their Utilization in Latin America." In *Urbanization in the Americas from Its Beginnings to the Present*, edited by Richard P. Schaedel, Jorge E. Hardoy, and Nora Scott Kinzer, 215–248. The Hague: Mouton.

Jones, Oakah L. 1994. *Guatemala in the Spanish Colonial Period*. Norman: University of Oklahoma Press.

Kagan, Richard L. 2000. *Urban Images of the Hispanic World, 1493–1793*. New Haven: Yale University Press.

Kinsbruner, Jay. 2005. *The Colonial Spanish-American City: Urban Life in the Age of Atlantic Capitalism*. Austin: University of Texas Press.

Kramer, Wendy. 1994. *Encomienda Politics in Early Colonial Guatemala, 1524–1544: Dividing the Spoils*. Boulder: Westview Press.

Lardé y Larín, Jorge. 2000. *El Salvador: Descubrimiento, conquista y colonización, 2nd ed.* San Salvador: Consejo Nacional para la Cultura y el Arte.

Lister, Florence C., and Robert H. Lister. 1987. *Andalusian Ceramics in Spain and New Spain: A Cultural Register from the Third Century B.C. to 1700*. Tucson: University of Arizona Press.

Lockhart, James. 1992. *The Nahuas after the Conquest: A Social and Cultural History of the Indians of Central Mexico, Sixteenth through Eighteenth Centuries*. Stanford: Stanford Univeristy Press.

Low, Setha M. 1995. "Indigenous Architecture and the Spanish American Plaza in Mesoamerica and the Caribbean." *American Anthropologist* 97: 748–762.

MacLeod, Murdo J. 1973. *Spanish Central America: A Socioeconomic History, 1520–1720*. Berkeley: University of California Press.

Marquardt, William H. 1989. "Agency, Structure, and Power: Operationalizing a Dialectical Anthropological Archaeology." Paper presented at *Critical Approaches in Archaeology: Material Life, Meaning, and Power*, Wenner-Gren Foundation for Anthropological Research, Symposium 108. Cascais, Portugal.

Marquardt, William H. 1992. "Dialectical Archaeology." In *Archaeological Method and Theory, Vol. 4*, edited by Michael B. Schiffer, 101–140. Tucson: University of Arizona Press.

Matthew, Laura E. 2012. *Memories of Conquest: Becoming Mexicano in Colonial Guatemala. First Peoples New Directions in Indigenous Studies*. Chapel Hill: University of North Carolina Press.

Matthews, Christopher N. 2010. *Archaeology and American Capitalism*. Gainesville: University Press of Florida.

McAndrew, John. 1965. *The Open-Air Churches of Sixteenth-Century Mexico: Atrios, Posas, Open Chapels, and Other Studies*. Cambridge: Harvard University Press.

Millon, René. 1955. *When Money Grows on Trees: A Study of Cacao in Ancient Mesoamerica*. PhD diss., Department of Anthropology, Columbia University.

Morse, Richard M. 1962. "Latin American Cities: Aspects of Function and Structure." *Comparative Studies in Society and History* 4: 473–493.

Orser, Charles E. 1996. *A Historical Archaeology of the Modern World*. New York: Plenum Press.

Orser, Charles E. 2004. "The Archaeologies of Recent History: Historical, Post-Medieval, and Modern-World." In *A Companion to Archaeology*, edited by John Bintliff, 272–290. Malden: Blackwell.

Orser, Charles E. 2014. *A Primer on Modern-World Archaeology*. New York: Eliot Werner Publications, Clinton Corners.

Palm, Erwin Walter. 1951. "Los orígenes del urbanismo imperial en América." In *Contribuciones a la historia municipal de América* edited by Rafael Altamira y Crevea, Manuel Carrera Stampa, Francisco Domínguez y Compañy, Agustín Millares Cario, and Erwin Walter Palm, 239–268. Mexico City: Instituto Panamericano de Geografía e Historia.

Pauketat, Timothy R. 2001a. "A New Tradition in Archaeology." In *The Archaeology of Traditions; Agency and History before Columbus*, edited by Timothy R. Pauketat, 1–16. Gainesville: University Press of Florida.

Pauketat, Timothy R. 2001b. "Practice and History in Archaeology: An Emerging Paradigm." *Anthropological Theory* 1: 73–98.

Pauketat, Timothy R. 2009. "Wars, Rumors of Wars, and the Production of Violence." In *Warfare in Cultural Context: Practice, Agency, and the Archaeology of Violence*, edited by Axel E. Nielsen and William H. Walker, 244–261. Tucson: University of Arizona Press.

Remesal, Antonio de. 1964–66. *Historia general de las Indias Occidentales y particular de la gobernación de Chiapa y Guatemala, 2 vols*. Madrid: Ediciones Atlas.

Sampeck, Kathryn E. 2007. "An Archaeology of Conquest and Colonialism: A Comprehensive Regional Survey of Late Postclassic and Colonial Landscapes of the Río Ceniza Valley, Department of Sonsonate, El Salvador." PhD diss., Department of Anthropology, Tulane University.

Sampeck, Kathryn E. 2010. "Late Postclassic to Colonial Transformations of the Landscape of the Izalcos Region in Western El Salvador." *Ancient Mesoamerica* 21 (2): 261–282.

Sampeck, Kathryn E. 2015. "Chronology and Use of Guatemalan Maiolica: Ceramics as *Reducción* in the Izalcos Region of El Salvador." *Historical Archaeology* 49 (2): 18–49.

Sherman, William L. 1979. *Forced Native Labor in Sixteenth-Century Central America*. Lincoln: University of Nebraska Press.

Smith, Michael E. 2008. *Aztec City-State Capitals*. Gainesville: University Press of Florida.

Verhagen, Inez Leontine. 1997. "Caluco, El Salvador: The Archaeology of a Colonial Indian Town in Comparative Perspective." Unpublished PhD diss., Department of Anthropology, Vanderbilt University.

Wagner, Logan, Hal Box, and Susan Kline Morehead. 2013. *Ancient Origins of the Mexican Plaza: From Primordial Sea to Public Space*. Austin: University of Texas Press.

CHAPTER 10

Hybrid Cultures: the Visibility of the European Invasion of Caribbean Honduras in the Sixteenth Century

Russell N. Sheptak and Rosemary A. Joyce

1 Introduction

The Americas have been an especially important setting for the development of new understandings of the historical processes that followed colonization by Europeans, who acted as agents to introduce large populations of African origin, resulting in a colonial situation of great complexity. Originally conceived of as 'culture contact,' these discussions rapidly gained in sophistication (Lightfoot 1995; Lightfoot et al. 1998). Critiques of the idea of contact, in which two somewhat homogeneous entities collided, with the stronger exercising some sort of hegemony over the weaker, were accompanied by the development of detailed investigations of specific historical engagements (Silliman 2005, 2010). These blurred the lines between what could be considered original or novel, 'authentic' or hybrid. Models for the emergence of new populations with newly formed identities have been most completely developed under the framework of ethnogenesis (Palka 2005; Voss 2008; Weik 2004). Weik (2004, 36) defined ethnogenesis as, 'the formation of new or different sociocultural groups from the interactions, intermixtures, and antagonisms among people who took part in global processes of colonialism and slavery'.

Our research explores the colonial situation of a region centered on the city of San Pedro Sula, part of the Honduran province of the Captaincy General of Guatemala. Founded in AD 1536 as a Spanish *villa* (incorporated town), San Pedro flourished as the center for transmission of products of gold mines toward ports, until gold smelting was moved inland in the early 1580s to the colonial capital city, Comayagua. From that point on, the Spanish citizenry of San Pedro Sula steadily declined.

We argue that in fact, the transformation of Honduran indigenous life preceded the formal incorporation of the province of the río Ulúa into the administrative district of San Pedro. For more than a decade before the founding of the city, indigenous towns in northern Honduras had experienced impacts of disease, raiding to capture labor for mines elsewhere in Central America,

 | DOI:10.1163/9789004273689_011

and military conflicts between Spanish factions that took place in and around indigenous settlements. There was also a well-developed indigenous military resistance, and it is this social movement that produced the material traces that we argue can be seen as hybrid material culture, the earliest and most transformative visible impacts of Spanish colonial engagement.

2 Spanish *Entradas* in Northern Honduras

Sheptak (2013, 68–70) summarizes the history of Spanish presence in Honduras in the early sixteenth century. This begins in 1502 with the second voyage of Columbus, who made landfall on the mainland near what today is Trujillo (Figure 10.1). In the Bay Islands off the coast, Columbus intercepted a canoe and pressed its passengers into guiding him, before leaving them to continue their voyage (Edwards 1978). Sheptak (2013, 68) argues that unauthorized Spanish ships were likely setting in along the coast between 1502 and the first

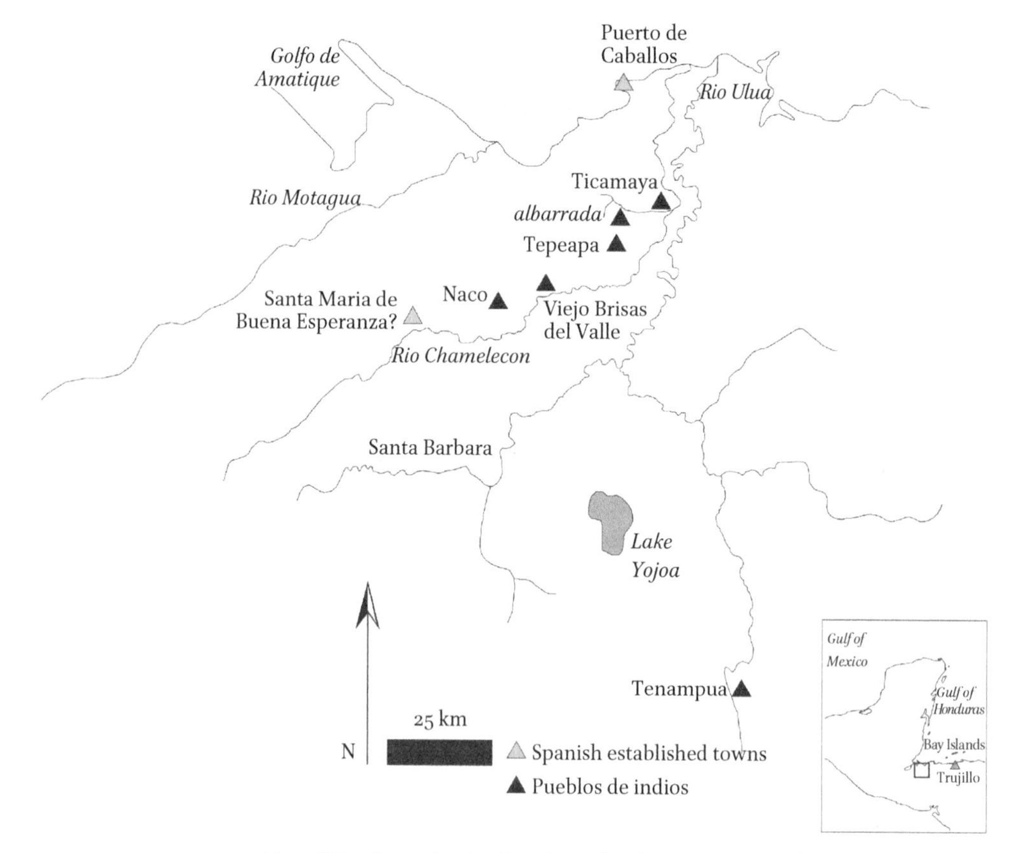

FIGURE 10.1 Map of Honduras showing locations of main sites mentioned in text

official Spanish expedition to Honduras, which arrived in 1523. This was the year Cristobal d'Olid and a group of 300 Spaniards, sent by Cortés from Mexico to "conquer and pacify" Honduras in his name, were shipwrecked and established their main settlement near Trujillo (Chamberlain 1953). Olid precipitated the first in a series of conflicts between would-be Spanish overlords by claiming the new colony for himself. Cortés sent a second officer, Francisco de las Casas, who was captured by Olid, and then finally followed himself, arriving in 1524.

Olid had in the meantime established part of his forces at the indigenous town of Naco. Cortés dispatched some of his troops, including Bernal Diaz del Castillo, to this inland town, maintaining his own heading along the Caribbean coast, founding a new Spanish town, La Navidad de Nuestra Señora, near the modern location of Puerto Cortes (Cortes 1989, 1990). Another detachment of troops were sent inland from this spot, to the indigenous town, Choloma. Cortés himself took ship to Trujillo, spending about six months regaining control over the Spanish outposts in the area. Near Trujillo, Cortés describes meeting with the leaders of indigenous towns named Papayeca and Chapagua (Cortes 1989).

When Cortés departed again for Mexico in 1525, a period of instability was begun as a series of governors of the colony were named in quick succession. By the early 1530s, the former treasurer of the colony, Andrés de Cereceda, was acting as governor. By 1533 Cereceda had relocated colonists from Trujillo to near Naco, pursuing reported gold deposits, founding a new Spanish town, Santa Maria de Buena Esperanza. In 1535, under pressure from unhappy Spanish colonists, Cereceda appealed to then-governor of Guatemala Pedro Alvarado for assistance. Alvarado received royal approval to conquer and pacify Honduras in 1532 (Chamberlain 1953). In December 1535 he arrived in Honduras, initiating military campaigns that culminated in 1536 with his attack on the indigenous resistance under a local leader named Çocamba, whose main settlement was at Ticamaya (Figure 10.1). Alvarado officially founded the city of San Pedro not far from Ticamaya, and was recognized by the remaining Spanish colonists as governor of Honduras.

3 Excavation of Sites of Spanish *Entradas*

Of the many indigenous places in the region referred to in early Spanish accounts, three have been the focus of archaeological research providing data covering the period of Spanish efforts to gain control of the territory: Naco, Ticamaya, and the Rio Claro site, identified by the excavator as the possible location of Papayeca.

3.1 *Naco*

Excavations at Naco in the 1930s encountered almost no apparent evidence of Spanish colonial presence (Strong et al. 1938, 32, Plate 4m). Two sherds of majolica ceramic were reported in these excavations, of unidentified type and date. More extensive excavations at Naco in the 1970s recovered no material remains attributable to early colonial Spanish presence at the site. Testing in what appears to be the same location that yielded the earlier majolica sample produced an eighteenth-century deposit with a single identified El Morro Ware sherd (Wonderley 1981, 23). The lack of European material in sixteenth-century contexts from Naco is notable, as Spanish archival records suggest over a year of presence of troops headed by Olid, followed by residence for some months of troops accompanying Cortés, and a subsequent period of engagement culminating in the establishment of a Spanish town not far away that drew on Naco for labor and supplies for at least three years before Alvarado's campaigns were completed.

3.2 *Papayeca*

The situation is similar at the Rio Claro site (Healy 1978). Located slightly inland from the coast in the Aguan River valley, near the site of the Spanish colonial city of Trujillo, the Rio Claro site was a tightly nucleated series of earthen platforms faced in stone arranged around two *plazas*, surrounded by a ditch measuring 1.8 to 2.5 meters in preserved depth, with three principal entries indicated by walkways (Figure 10.2). Radiocarbon samples from the site produced dates after 1000 AD (Healy 1978, Table 1). Most samples produced calibrated dates falling between AD 1100 and 1350, and were associated with ceramics diagnostic of the Early Cocal phase (Dennett 2007). The latest sample, reported as uncalibrated 450 +/- 65 BP, would encompass the period of contact when calibrated (AD 1494 +/- 77)[1]. This sample came from an excavation in the tallest, centrally located platform in the site (Healy 1978, 20). The context of the carbon sample, and the overlying context, both included Late Cocal ceramics, the only such sherds associated with radiocarbon dates at the site, where they are otherwise found as surface materials (Dennett 2007). The charcoal dated came from a hearth defined in the second excavation level. Below this, "several" hard clay floors were noted, and a second, deeply buried concentration of burned material yielded a carbon sample dating to 905 +/- 65 BP (calibrated

1 Calibrated using the Cologne Radiocarbon and Palaeoclimate Research Package Online CalPal (http://www.calpal-online.de) quickcal 2007 v1.5, with CalCurve CalPal_2007_HULU. Based on the initial 14C-age of 450 ± 65 BP, results indicate a Calendric Age of 456 ± 77 calBP, with the two-sigma (68%) range of 378–533 calBP, or a range of AD 1417–1572.

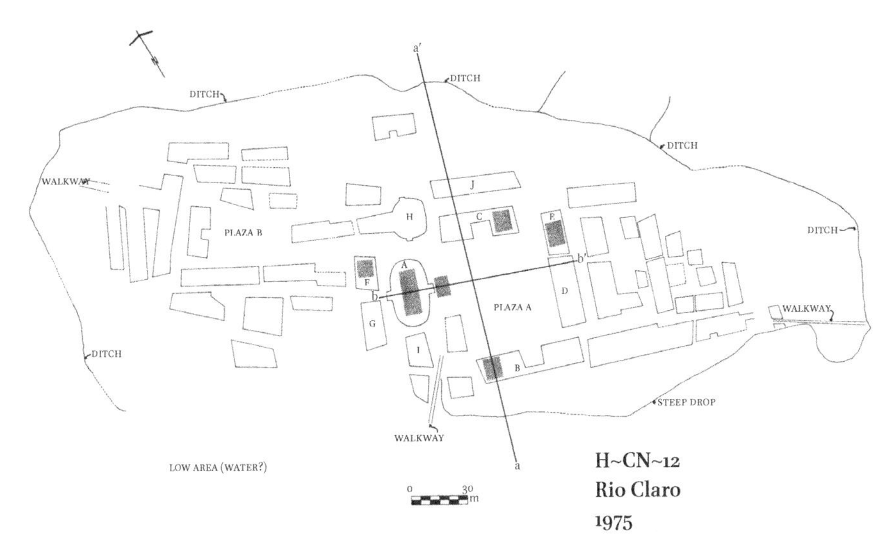

FIGURE 10.2 Plan of the Rio Claro site showing ditch surrounding compact groups of mounds (after Healy 1978, Figure 3)

as AD 1120 +/- 69)[2]. Based on its unique size, dense site plan, and late continuation of occupation, the excavator suggested that this site was possibly Papayeca, one of the places Cortés mentioned receiving visitors while he was at Trujillo (Healy 1978, 26–27).

3.3 *Ticamaya*

Ticamaya, the third site known to have been occupied during the period when Spanish troops entered Caribbean Honduras, produced the same pattern. Ticamaya is located at the ancient confluence of a former course of the Ulua River (today occupied by the Chamelecon River) and an abandoned channel of the Choloma River. Unlike the other two sites, the remains of Ticamaya are deeply buried by sediments from these two rivers. The original detection of the site was based on use of archival documents, verified by recovery of sixteenth-century ceramics in canal backdirt.

Research at Ticamaya began with systematic auguring and magnetometer survey of part of the area to determine distribution of buried deposits, and excavation of wider units where auguring and magnetometer anomalies

2 Calibrated using the Cologne Radiocarbon and Palaeoclimate Research Package Online CalPal (http://www.calpal-online.de) quickcal 2007 v1.5, with CalCurve CalPal_2007_HULU. Based on the initial 14C-age of 905 ± 65 BP, results indicate a Calendric Age of 830 ± 69 calBP, with the two-sigma (68%) range of 760–899 calBP, or a range of AD 1051–1190.

produced evidence of buried materials in situ in five dispersed areas, within an area of continuous buried remains extending 140 by 215 meters (Blaisdell-Sloan 2006). Based on AMS dating of carbon samples, stratigraphic and horizontal relationships among different excavation areas, we can identify three of the five excavated areas as including features dating to the early sixteenth century (Blaisdell-Sloan 2006, 151–155).

One AMS sample came from a burned oven in Ticamaya Operation 2C. This sample, with a reported age of 390 +/- 49 BP, yielded two probability peaks in calibration. One ranged from AD 1436–1530, the second AD 1538–1635. While Blaisdell-Sloan (2006, 155) argued for the later date span, the probability of the earlier is statistically more likely. Subsequently, Blaisdell-Sloan joined us in a re-analysis of these excavated deposits that concluded that the samples were best assigned to the earlier part of the possible range (Sheptak, Blaisdell-Sloan and Joyce 2011a).

The oven, possibly a kiln, was a one-meter diameter pit dug 50 cm. deep, lined with burned clay (Blaisdell-Sloan 2006, 131–132, 152, 169, 228–229, 249, 254). Within the outer pit of the burned oven was a second clay structure about 50 cm. in diameter that might have supported pots during firing. Surfaces in the adjacent excavation units yielded fragments of construction material typical of house construction (Blaisdell-Sloan 2006, 130, 182, 254). On these surfaces, a wide variety of plant remains were recovered, including coyol palm seeds, *Carex* sp. and *Paspalum* sp. (used for mats and bedding), *Mamillaria* sp. and lumps of tuber tissue consistent with manioc or sweet potato.

The fill inside the oven also yielded typical domestic remains, including ceramics from the late fifteenth to early sixteenth century. Also in this fill, carbonized maize seeds and tuber fragments, and bones from turtle, peccary, and white-tailed deer, reflect the same kind of domestic assemblage. Notable here was a high frequency of small obsidian dart points, which we will return to below.

A second dated carbon sample, from Ticamaya Operation 3B, has a reported age of 347 +/- 37 BP, calibrated to AD 1460–1638. The probability distribution for this sample (Blaisdell-Sloan 2006, Figure 5.1) is slightly bimodal, with one probability peak between AD 1480 and 1530 and the other between AD 1550 and 1610. Based on associated ceramics and stratigraphic relations to other excavation units, the earlier time span is more likely.

The source of this dated carbon sample was a house, Structure 3A, represented by a single posthole, burned mud wasp nests, and a hearth still containing a broken vessel (Blaisdell-Sloan 2006, 134–136, 249, 254–255). No wall rubble was found, suggesting the wasp's nests were attached to perishable walls made of pole and thatch, or that the structure was a roofed shelter for the hearth, with open walls. This house appears to have been completely burned. On the adjacent surface outside the building, more large pieces of pottery were encountered. The entire area was covered with a thin level of soil mixed with large amounts of carbon before another surface formed. The conclusion reached by Blaisdell-Sloan (2006, 152) was that this house suffered a major fire, quite possibly as part of the military campaign that took place at Ticamaya in 1535.

Surfaces around the traces of Structure 3A produced ceramics diagnostic of the early sixteenth century, including both utilitarian types and painted serving ware originating in Naco. Plant remains recovered from inside the building, near the hearth, included tissue from tubers, probably manioc, and Helianthus and Artemisia seeds (Blaisdell-Sloan 2006, 254–255). Bones from turtle and white-tailed deer were also recovered, reinforcing the resemblance to the area around the collapsed oven in Ticamaya Operation 2C. Distinguishing Structure 3A from Operation 2, a piece of sheet copper and fragments of deer antler were also recovered. These may represent evidence of craft working in this area. Like Operation 2, this area also yielded a number of obsidian projectile points unifacially chipped on blades.

A third excavated area, in Ticamaya Operation 1A, was stratigraphically related to the same period of occupation as the two previous excavation areas. Here, a new building, Structure 1A, succeeded earlier buildings with AMS dates in the fifteenth century. Structure 1A may have had unique, non-domestic use (Blaisdell-Sloan 2006, 122–124, 228, 248). It was constructed with very large (30 cm diameter) posts, placed in postholes unusual for being lined with plaster, perhaps implying that the posts themselves had been plastered. The immediate predecessor of Structure 1A, Structure 1B, had buried deposits in each corner, including tobacco seeds, ocelot and coyote teeth, and five ceramic censers (Blaisdell-Sloan 2006, 125). These features suggest this sequence of buildings could have been dedicated to ritual practices. Very few ceramics and no obsidian projectile points were found in this area. Some possible deer bone was recovered, but none of the smaller species used for food noted in other areas.

3.4 *Summary of Excavated Evidence from Contact-Period Sites*

Despite multiple reports of Spanish presence at Naco, and interactions with Ticamaya and Papayeca that included Spanish presentation of gifts, indigenous raiding and capture of Spanish prisoners, and a battle between Spanish troops and indigenous defenders, there is remarkably little in the excavated registers of these sites that directly testifies to their witnessing of these events. Only two majolica sherds at Naco have been attributed to the period, and these may actually come from a much later reoccupation of the site. However, a review of the contemporary archival documents may help us to reconsider what might constitute evidence of engagement with the Spanish in the material remains of sites engaged in the kinds of conflicts that colonization here involved.

4 Fighting and Fortifications

At Ticamaya, Blaisdell-Sloan (2006, 236–238) noted that small dart points made on obsidian blades reached their highest frequencies in the deposits assigned to the early sixteenth century. Of the total of 34 dart points, 21 (61%) came from these contexts. This form of dart point is interpreted as intended for use in battle against human opponents, a model supported by edge-wear analysis and depositional contexts for such objects related to the violent conquest of ninth-century Aguateca, Guatemala (Aoyama 2005, 204).

Subsequently, we argued that the high frequency of dart points was evidence of a newly militarized way of life ushered in at Ticamaya when the first Spanish expeditions in the early 1520s began to affect the Ulua River area (Sheptak et al. 2011b). The contexts with high proportions of these points also showed extensive burning of residential features (Blaisdell-Sloan 2006, 134, 154). These were particularly concentrated in Ticamaya Operation 3, while other areas sampled in Operation 1 and Operation 2 showed neither the extensive burning nor the presence of dart points.

Archival documents indicate that Ticamaya was surrounded by a palisade and ditches (Sheptak 2004). No archaeological evidence of a system of defensive features in the buried site remains was encountered by Blaisdell-Sloan (2006). Unlike the groups of sherds, lithics, and burned house construction material that she recovered through auguring, a ditch dug into the soil would not have provided a clear signature. Magnetometer survey might have provided evidence of anomalies, but the original magnetometer survey was limited to a 1900 square meter area due to malfunction of the equipment. This completed magnetometer survey was carried out in the area closest to the river bank, which archival sources indicate was left open for access from the river.

In 2008, Blaisdell-Sloan returned to the site with the intention of renewing magnetometer survey and excavations. Unfortunately, substantial construction had taken place, which limited additional areas where the method could be carried out. She added an additional 250 square meters to the surveyed area, but nothing suggestive of a ditch and palisade was recovered. While we cannot confirm the presence of a ditch and palisade described in archival documents, the compact nature of the area of buried remains is consistent with a densely nucleated site.

While Cortés received envoys from Papayeca, neither he nor his troops described visiting the town. Based on the plan of the Rio Claro site, it appears that it too was fortified (Healy 1978, 17, 27). A ditch reaching depths of 1.8 to 2.5 m is preserved around most of the compact cluster of buildings. In the one area where it is not found, evidence suggests there was a body of water, corresponding to the description of Ticamaya with its main entry from the river. The measured area of the Rio Claro site, 450 x 190 m, is slightly larger than the known extent of Ticamaya, 140 by 215 m.

At 8.55 hectares, the Rio Claro site, while larger than 3.01 hectare defined area of Ticamaya, is still much smaller than Naco, which is given as occupying 90 hectares (Henderson et al. 1979, 172). This difference alone suggests that Naco, unlike Ticamaya and the Rio Claro site, was not situated for defense. In addition, the Spanish word used in sixteenth-century documents describing Ticamaya and its allied towns, *albarrada*, occurs in Bernal Diaz del Castillo's account of cities in Mexico (1980), but is not used by him in his description of Naco.

Comparing the documented histories of relations between each of these towns and the early Spanish expeditions, Naco is distinguished from Papayeca and Ticamaya by a history of welcoming visitors. In contrast, the leaders of Papayeca, although initially willing to exchange gifts with Cortés, ended up leading military resistance against the Spanish colony, and were captured by Cortés (Cortes 1989).

Ticamaya has the most abundantly detailed record of military conflict with the Spanish forces (Sheptak 2004). In 1536, Pedro Alvarado described the town as the seat of a "*señor*" (lord) whose name we transcribe as Çocamba (AGI Patronato 20 N. 4, R. 6). Alvarado noted that "by visiting" (*por visitación*) he knew that Ticamaya had around eighty men. He identified it as having "some small towns subject to them" (*unos pueblos pequeños a ellos sujeto*) with fifteen, eight, and six houses, respectively.

Letters from Andres de Cereceda, governor of the Spanish colony in 1535 (AGI Guatemala 39 R. 2 N. 4), and Diego Garcia de Celis, treasurer, in 1534 (AGI Guatemala 49 N. 9) provide more details of hostilities between Ticamaya and its allies and the Spanish colonists. These letters attribute military campaigns resulting in the death or capture of the residents of a settlement that Cortés had established in 1525 near the location of modern Puerto Cortes (Puerto Caballos) to the leader of Ticamaya.

Cereceda described the sacking of Puerto Caballos sometime before 1533, saying that Çocamba had killed ten men of the settlement and captured and kept a Spanish woman prisoner (AGI Guatemala 39 R. 2 N. 4). He gave this as an excuse for his own aggression against outlying settlements that were subordinate to Ticamaya during a march from Trujillo to the Naco valley in 1533, writing that

> ... on the Rio Balahama, where our path went, we found a palisade (*albarrada*) of those that I have written about to your majesty, that the indians of that region and of the Rio Ulua make for their stronghold.
>
> AGI Guatemala 39 R. 2 N. 4 1/1

Cereceda nonetheless decided to camp out within the vicinity of this palisade, and then reports that:

> ... a little after midnight ... certain indians came from down river in canoes with a great shouting and throwing arrows (*tirando flechas*) that fell on our camp that injured some of the indians in servitude that we had with us and horses.
>
> AGI Guatemala 39 R. 2 N. 4

With this attack as his pretext, Cereceda says "I went over the palisade with fifty men," putting the indigenous troops to flight, and taking two leaders (*principales*) prisoner. He then sent these two prisoners to "the *cacique* Çocamba of the Ulua River at his *albarrada* which was two leagues from there" (AGI Guatemala 39 R. 2 N. 4).

The treasurer De Celis provides his own account of the same incident:

> ... we went to block a town subject to the greatest *cacique* that they have in all that governance as they say, who the indians call the great merchant Çocamba, and so we took fifty prisoners two leagues from his house (*casa*) ... they informed the governor that this Çocamba is very fortified with with heavy palisades (*albarradas*) of thick wood and that they had made a large quantity of pits covered with bark from them ... many people say that his *albarrada* is very fortified, of seven or eight rows of very rough wood with their towers and openings, and that it would be a very difficult thing to enter because there is no entry except over the river on which this is located on the barranca of the river.
>
> AGI Guatemala 49 N. 9

Cereceda noted that previous to the arrival of Alvarado, another Spanish troop from Guatemala had gone to try to punish the deaths of the people of Puerto Caballos, to "break the palisaded fortress of Çocamba and others of the *caciques* of that river in which they were making forts" (*a romper el albarrada de*

Çocamba y otras de caciques de aquel rio en que se hacia fuertes; AGI Guatemala 39 R. 2 N. 4). This passage gives a sense that the building of fortifications was an ongoing process along the Ulua River.

It appears that Ticamaya headed an organized resistance involving multiple fortified towns that had been underway for a decade when Alvarado undertook the campaign that defeated Ticamaya in 1536 and resulted in the capture of its leader, Çocamba. The events of this final battle were described in another letter by Cereceda, written in 1536. Although the document has been worn along the right edge, enough remained for us to produce a transcription. It describes the indigenous warriors attempting to flee on the river, but being caught by Alvarado using a very large canoe with artillery in the prow to attack Ticamaya from the river, preventing "the entry or exit from the *albarrada* to the river" (AGI Guatemala 39 R. 2 N. 6). Witnessing the deaths of his people under fire, the leader of Ticamaya surrendered.

5 Fortifications as Hybrid Tactics

Reading the archival record, it is clear that early sixteenth-century towns in the Ulua valley were fortified places. The Rio Claro site, identified as historic Papayeca, shows us what the remains of such a fortified place would have looked like if all the features were still visible on the surface, instead of buried by river sediment as is the case with Ticamaya. Yet in the Ulua and the Aguan valleys, there is no previous history of fortified sites. Nor are fortifications a practice in other parts of Honduras during the early sixteenth century: Naco appears to have been composed of a dispersed group of structures along the Naco River. Another, even larger fifteenth- to sixteenth-century site, Viejo Brisas del Valle, located along the Chamelecon River between the Naco and Ulua Valleys, has no sign of fortifications or even close spacing of buildings that might suggest there had once been perishable defensive walls (Neff et al. 1990).

One much earlier Honduran site from far inland is noted for its defensive walls. This site, Tenampua, occupies a mesa overlooking the Comayagua valley (Dixon 1989, 264–266). Between AD 900–1000, Tenampua grew to contain more than 400 buildings distributed in clusters across the mesa. The approach to the site is protected by a system of stone walls (Dixon 1987). The main wall was described in an early report as 225 meters long, up to 3 meters tall, and 8 meters thick, composed of rocks 45 to 60 cm in diameter joined in a mud plaster (Popenoe 1935, 562). Nothing about the walls of Tenampua matches the archaeological features from the Rio Claro site, or the description of the Ticamaya palisade.

Farther afield, walled precincts protecting the residences of the wealthy nobility are described for the northern Yucatan lowlands (Cortes Rincon 2007, 179–180). These are said to have been added to sites originally having open plans, in response to military threats, some described as "hastily" built. They have thick stone walls which could have supported palisades but lack the ditches, towers, and slit windows for shooting described for Ticamaya or known from the Rio Claro site. A review of lowland Maya sites with fortifications (Webster 1976, 368) concluded that earthen ditch and embankment defensive walls were products of early Classic (or even earlier occupation), and that the norm for the period immediately before the Spanish entry into the area was dry-laid stone walls, not unlike the one described for Tenampua.

While not the kind of material marker of initial Spanish contact that archaeologists have traditionally expected, we argue that the enclosure of late sites like Ticamaya and Rio Claro by ditch and palisade defensive walls should be seen as an innovative product of the engagement of indigenous Honduran peoples with Spanish troops seeking to invade and control their communities. From reports of two battles, described above, in which Spanish troops attacked such *albarradas* at Ticamaya and a nearby outlying site, we can see that these fortifications were not able to indefinitely repel either a larger number of troops willing to storm the walls, or (in the case of Ticamaya) an attack by boats carrying ordinance. Yet the Spanish sources are also clear that Hondurans employing such fortifications were able to ward off attempts at colonization for a decade.

Part of a shift toward militarization also seen at Ticamaya in the increase in obsidian points appropriate for use against human targets (Sheptak et al. 2011a), the investment in fortification at this site and others it organized in resistance to colonization is presented in documentary sources as an ongoing process. This process itself was, we have suggested, a product of cultural hybridity in the first decades of the sixteenth century. The same documents that describe Ticamaya and its defeat also describe the death in the final battle there of a shipwrecked Spanish sailor turned member of the indigenous military. In his 1536 letter describing the defeat of Ticamaya, Cereceda adds that after he surrendered,

> ... the *cacique* Çocamba said that in the battle inside the *albarrada* the previous day that, hit by fire from an arquebus, there had died a Spanish Christian named Gonzalo Aroca who is the one who walked among the indians in the province of Yucatan for twenty years and more that it is he that they say destroyed the Adelantado Montejo and how that having depopulated the Christians there he came to aid those here with a fleet of fifty canoes to kill those of us that were here.
>
> AGI Guatemala 39 R. 2 N. 6

Gonzalo Aroca is more commonly identified as Gonzalo Guerrero, shipwrecked in Yucatan before 1520, who refused an offer from Cortés to rejoin the Spanish saying he was married and had children, and had pierced his ears and tattooed his body. We view Gonzalo as a cultural mediator who may well have been critical in helping shape new emergent tactics of military defense against an enemy he knew well, having been part of it. The development of ditch and palisade defensive works in Honduras in sites that continued to resist Spanish invasion long after Naco had become a support for the new colony is a sign of contact, not in the form of imported European goods (which are rare in the country until the late eighteenth century), but in the form of new ways of living to adjust to new and deadly threats.

6 Conclusions

We have argued that we need to question what material registers might indicate "Spanish" or "colonization" in terms of emergent practices in relevant times for each place. The use of chronometric methods to establish dating is critical, as there can be no reliance on novel introduced material culture to identify the chronological placement of sites if the residents do not have access to, prize, or desire these things.

We have previously argued that expectations about how material practices would change during initial engagement of indigenous societies with Europeans and Africans in Honduras have impeded understanding colonial material registers (Joyce and Sheptak 2014). For example, research in Santa Barbara, Honduras, southwest of the Ulua valley, documented a series of colonial churches in towns where residential remains, from houses to pottery, were indistinguishable from those of the period before Spanish missionization (Weeks 1997; Weeks and Black 1991; Weeks et al. 1987).

We suggest that Michel de Certeau's (1984) concept of everyday practices as "tactics" helps anticipate the improvisational and emergent properties of new ways of doing things that may be the most common material sign of European-indigenous encounters. Tactics, we note,

> ... are not extraordinary, but ordinary; they are the continuing ways that human subjects occupy social landscapes that they do not entirely control. Tactics can be conceived of as the "appropriation" of what is offered in places like the colonial settings we examine, exceeding the intentions of those who seek control, seizing the moment for one's pragmatic ends.
>
> SHEPTAK, et al. 2011b, 149–150

The construction of fortifications in anticipation of attacks by Spanish troops was tactical adaptation during a period of challenges to the existing political order. It cannot be easily identified as either a wholly indigenous or introduced trait. The obsidian assemblage from Ticamaya provides a second illustration of the emergent and hybrid nature of material evidence of these relations. The kind of obsidian dart points made from blades that reached their peak frequency at a moment when a sector of the settlement was burned have been identified in Belize as products made for the purpose of military defense against the invading Spanish there as well (Simmons 1995). A completely indigenous material and a form unknown in Europe, it is the intensified production and use of these blades that marks the early period of invasion and conflict.

We suggest that the situation that arises is productively viewed using concepts of hybridity and ethnogenesis. Ethnogenesis places an emphasis on what emerges, not what preceded. As Barbara Voss (2008) demonstrated in her study of the new 'Californio' identity shaped at the Spanish Presidio of San Francisco, what emerges cannot be separated into component parts. Our emphasis on the emergence of new forms through tactical engagement in material practices aligns us with the tradition represented by William Hanks (2010, 93–94), who sees the attempt 'to divide an indigenous inside from a Hispanicized exterior' as 'sundering the person into two parts,' possible only if each belongs to a distinct social field. In the beginning of the Honduran colony, what we see is the taking up of positions in fields that linked people coming from different spaces in practices that we can say are innovative indigenous works, reframings of Spanish renegade knowledge, and repurposing of traditional techniques of construction all at the same time.

Archival Documents

Abbreviations used:

AGCA Archivo General de Centroamerica, Guatemala City, Guatemala
AGI Archivo General de Indias, Sevilla, Spain

1534 "Cartas de oficiales reales de Honduras: Diego Garcia de Celis, Puerto de Caballos 6/20/1534" AGI Guatemala 49 N. 9

1535 "Cartas de gobernadores: Andres de Cereceda, Buena Esperanza 8/31/1535" AGI Guatemala 39 R. 2 N. 4

1536 "Cartas de gobernadores: Andres de Cereceda, Puerto de Caballos 8/14/1536" AGI Guatemala 39 R. 2 N. 6

References

Aoyama, Kazuo. 2005. "Classic Maya Warefare and Weapons; Spear, Dart, and Arrow points of Aguateca and Copan." *Ancient Mesoamerica* 16 (2): 291–304.

Blaisdell-Sloan, Kira. 2006. *An Archaeology of Place and Self: The Pueblo de Indios of Ticamaya, Honduras (1300–1800 AD)*. Ann Arbor: The University of Michigan Press.

Certeau, Michel de. 1984. "Making Do: Uses and Tactics." In *The Practice of Everyday Life*, translated by Steven F. Rendall, 29–42. Berkeley: University of California Press.

Chamberlain, Robert S. 1953. "The Conquest and Colonization of Honduras: 1502–1550." Carnegie Institution of Washington Publication 598. Washington, D.C.: Carnegie Institution of Washington.

Cortes, Hernán. 1989. *Letters from Mexico*. Translated and edited by Anthony R. Pagden New Haven: Yale University Press.

Cortes, Hernán. 1990. "Ordenanzas para las Villas de La Navidad y Trujillo en Honduras." In *Documentos Cortesianos, Volume 1, 1518–1528*, edited by José Luis Martinez, 347–351. Mexico: Universidad Nacional Autonoma de Mexico, Fondo de Cultura Economica.

Cortes Rincon, Marisol. 2007. "A Comparative Study of Fortification Developments Throughout the Maya Region and Implications for Warefare." PhD diss., University of Texas, Austin.

Dennett, Carrie L. 2007. "The Río Claro Site (AD 1000–1530), Northeast Honduras: A Ceramic Classification and Examination of External Connections." Master thesis, Trent University.

Diaz de Castillo, Bernal. 1980. *Historia verdadera de la Conquista de la Nueva España.* Mexico: Editorial Porrua.

Dixon, Boyd. 1987. "Conflict along the Southeast Mesoamerican Periphery: A Defensive Wall System at the Site of Tenampa." In *Interaction on the southeast Mesoamerican Frontier, Part 1*, edited by Eugenia J. Robinson, 142–153. Oxford: British Archaeological Reports, International Series, 327.

Dixon, Boyd. 1989. "A Preliminary Settlement Pattern Study of a Prehistoric Cultural Corridor: The Comayagua Valley, Honduras." *Journal of Field Archaeology* 16 (3): 257–271.

Edwards, Clinton R. 1978. "Precolumbian Maratime Trade in Mesoamerica." In *Mesoamerican Communication Routes and Culture Contacts*, edited by Thomas A. Lee and Carlos Navarrete, 199–210. Papers of the New World Archaeological Foundation 40. Orinda: New World Archaeological Foundation.

Hanks, William F. 2010. *Converting Words: Maya in the Age of the Cross*. Berkeley: University of California Press.

Healy, Paul H. 1978. "Excavations at Rio Claro, Northeastern Honduras: Preliminary Report." *Journal of Field Archaeology* 5 (1): 15–28.

Henderson, John, Ilene Sterns, Anthony Wonderley, and Patricia A. Urban. 1979. "Archaeological Investigations in the Valle de Naco, Northwestern Honduras: A Preliminary Report." *Journal of Field Archaeology* 6 (2): 169–192.

Joyce, Rosemary A. and Russell N. Sheptak. 2014. "History Interrupted: Doing 'Historical Archaeology' in Central America." In *The Death of Prehistory*, edited by Peter R. Schmidt and Stephen A. Mrozowski, 161–182. Oxford: Oxford University Press.

Lightfoot, Kent G. 1995. "Culture Contact Studies: Redefining the Relationship Between Prehistoric and Historical Archaeology." *American Antiquity* 60 (2): 199–217.

Lightfoot, Kent G., Antoinette Martinez, and Ann M. Schiff. 1998. "Daily Practice and Material Culture in Pluralistic Social Settings: An Archaeological Study of Culture Change and Persistence from Fort Ross, California." *American Antiquity* 63 (2): 199–222.

Neff, Hector, Patricia Urban, and Edward Schortman. 1990. "Late Prehistoric Developments in Northwestern Honduras: Preliminary Report on the 1990 Investigation at Viejo Brisas del Valle." Tegucigalpa, Honduras: Instituto Hondureño de Antropología e Historia. Report in the Archives of the Institute.

Palka, Joel W. 2005. *Unconquered Lacandon Maya: Ethnohistory and Archaeology of Indigenous Culture Change.* Gainesville: University Press of Florida.

Popenoe, Dorothy H. 1935. "The Ruins of Tenampua." Smithsonian Institution, Annual Report, 559–572. Washington, D.C.: Smithsonian Institution.

Sheptak, Russell N. 2004. "Noticias de un cacique indígena de la época colonial: Una contribución a la historia colonias de Honduras." Paper presented at the *VII Congreso Centroamericano de Historia, Tegucigalpa, Honduras, 19–23 July.*

Sheptak, Russell N. 2013. "Colonial Masca in Motion: Tactics of Persistence of a Honduran Indigenous Community." PhD diss., Faculty of Archaeology, Leiden University.

Sheptak, Russell N., Kira Blaisdell-Sloan, and Rosemary A. Joyce. 2011a. "In-Between People in Colonial Honduras: Reworking Sexualities at Ticamaya." In *The Archaeology of Colonialism: Intimate Encounters and Sexual Effects*, edited by Barbara L. Voss and Eleanor Casella, 156–172. Cambridge: Cambridge University Press.

Sheptak, Russell N., Rosemary A. Joyce, and Kira Blaisdell-Sloan. 2011b. "Pragmatic Choices, Colonial Lives: Resistance, Ambivalence, and Appropriation in Northern Honduras." In *Enduring Conquests*, edited by Matthew Liebmann and Melissa S. Murphy, 149–172. Santa Fe: School for Advanced Research.

Silliman, Stephen W. 2005. "Culture Contact or Colonialism? Challenges in the Archaeology of Native North America." *American Antiquity* 70 (1): 55–74.

Silliman, Stephen W. 2010. "Indigenous Traces in Colonial Spaces: Archaeologies of Ambiguity, Origins, and Practice." *Journal of Social Archaeology* 10 (1): 28–58.

Simmons, Scott E. 1995. "Maya Resistance, Maya Resolve: The Tools of Autonomy from Tipu, Belize." *Ancient Mesoamerica* 6 (2): 135–146.

Strong, William D., Alfred V. Kidder, and Anthony J.D. Paul. 1938. *Preliminary Report on the Smithsonian Institution-Harvard University Archaeological Expedition to Northwestern Honduras 1936*. Washington, D.C.: Smithsonian Institution.

Voss, Barbara L. 2008. *The Archaeology of Ethnogenesis: Race and Sexuality in Colonial San Francisco.* Berkeley: University of California Press.

Webster, David. 1976. "Lowland Maya Fortification." *Proceedings of the American Philosophical Society* 120 (5): 361–371.

Weeks, John M. 1997. "The Mercedarian Mission System in Santa Bárbara de Tencoa, Honduras." In *Approaches to the Historical Archaeology of Mexico, Central and South America*, edited by Janine Gasco, Greg C. Smith, and Patricia Fournier Garcia, 91–100. Los Angeles: The Institute of Archaeology, UCLA.

Weeks, John M. and Nancy J. Black. 1991. "Mercedarian Missionaries and the Transformation of Lenca Society in Western Honduras, 1550–1700." In *Columbian Consequences, Vol. 3: The Spanish Borderlands in Pan American Perspective*, edited by David H. Thomas, 245–261. Washington, D.C.: Smithsonian Institution Press.

Weeks, John M., Nancy J. Black, and J. Stuart Speaker. 1987. "From Prehistory to History in western Honduras: The Care Lenca in the colonial province of Tencoa." In *Interaction on the Southeast Mesoamerican Frontier, Part 1*, edited by James A. Robinson, 65–94. Oxford: British Archaeological Reports, International Series 327.

Weik, Terrance. 2004. "Archaeology of the African Diaspora in Latin America." *Historical Archaeology* 38 (1): 32–49.

Wonderley, Anthony W. 1981. *Late Postclassic Excavations at Naco, Honduras*. Latin American Studies Program Dissertation Series No. 86. Ithaca, NY: Latin American Studies Program, Cornell University.

CHAPTER 11

Exotics for the Lords and Gods: Lowland Maya Consumption of European Goods along a Spanish Colonial Frontier

Jaime J. Awe and Christophe Helmke

1 Introduction

In the volume *The Lowland Maya Postclassic*, Arlen Chase and Prudence Rice (1985, 5) contend that Spanish presence in the Maya lowlands "is not clearly detectable in the archaeological record until the nineteenth century." To this they add that: "This is partially a consequence of an apparent reluctance on the part of the Maya to accept European trade items or at least to deposit them in the archaeological record." This point of view echoes the previous observation by Nancy Farris (1984, 110) that "Except for some simple metal tools [...] one can find little European material impact" on Maya culture during the early colonial period. Farris (1984, 45) also argued that the Maya of the Yucatan generally had a "cultural bias against European goods" and that the few tools and trinkets that were acquired "were passed on through generations as treasured heirlooms." Farris (1984, 45) further noted that, with the exception of metal tools and gunpowder, "which came to be regarded as a requirement for any fiesta, besides its use in hunting," there were only a few items that the Maya actually desired from the Spaniards.

While we would agree that the volume and diversity of European goods were limited along the lowland Maya colonial frontier, considerable ethnohistoric and archaeological evidence that has come to light in recent years, demonstrate both increasing acquisition and integration as well as desire, if not demand, for European objects by the contact period Maya. Avendaño y Loyola (1987, 29; see also Means 1917, 131) went even further in his assessment of the Maya interest in obtaining Spanish goods, reporting that the Itza demonstrated an "insatiable desire" for these objects. Whereas all such assertions must be tempered by the relative ubiquity or scarcity of European objects in archaeological contexts, we can nevertheless identify a series of different driving factors that fueled the Maya desire for European goods. Among these was the acquisition of European goods as status symbols, for practical and mundane or quotidian purposes, as well as for their incorporation in ceremonies

DOI:10.1163/9789004273689_012

and ritual deposits to harness what Timothy Pugh (2009, 373) refers to as the "power of alterity."

Both the ethnohistoric sources and the archaeological record indicate that the Maya acquired European goods in different ways. A review of these sources suggests that these can be divided into major categories or headings. The two most common methods included: 1) gifts from the Spanish and 2) by means of trade or barter between indigenous peoples and Europeans, and down-the-line between indigenous groups (Blacker and Rosen 1962; Clendinnen 1987; Jones 1998, 503; Oland 2014, 2017, 127; Pugh 2009). Other methods of acquisition were as 3) payment for services, 4) tokens or rewards for conversion to Christianity, and as 5) "spoils of war" following violent confrontations between the two groups (Avendaño 1987; Awe and Helmke 2015, 347–348; Jones 1990, 188; López de Cogolludo 1688, bk. 11, Chap. 14, 648; Thompson 1972, 12). Below, we review the material evidence for each of these headings in turn.

2 The Ethnohistoric Evidence: Acquisition of European Goods by Way of Gifts and Trade

Despite its relative lack of detail, the ethnohistoric record actually provides a wealth of information regarding the presentation and exchange of European goods in the Maya lowlands. The first such example can be traced back to July of 1502 when, during his fourth voyage to America, Columbus encountered a group of Maya merchants near the island of *Guanaja* off the coast of Honduras (Chamberlain 1948, 9–12; Keen 1959) (Figure 11.1). Columbus' son later provided an eyewitness account of this meeting, plus an informative description of the seaworthy Maya canoes and the merchandise they were carrying (Columbus 1960, 231–232). He reported that when Columbus met the Maya merchants, "He greeted them with great kindness and presented them with some objects from Castille in exchange for some of the strange-looking things, to take with him in order to show what kind of a people he had discovered" (Hammond 1988, 221). He also noted that Columbus retained one of the elderly Maya merchants to serve him as a guide, that he renamed the old man Juan Pérez, and that on his release of the so-called Juan Pérez, Columbus bestowed upon him presents for his assistance.

The initial encounter between Columbus and Maya merchants was followed by three subsequent Spanish expeditions to the east coast of Mesoamerica from the island of Cuba, then known as Fernandina. The first of these expeditions was the 1517 voyage of Francisco Hernández de Córdoba. According to Bernal Díaz del Castillo (1956, 19), in preparation for the journey

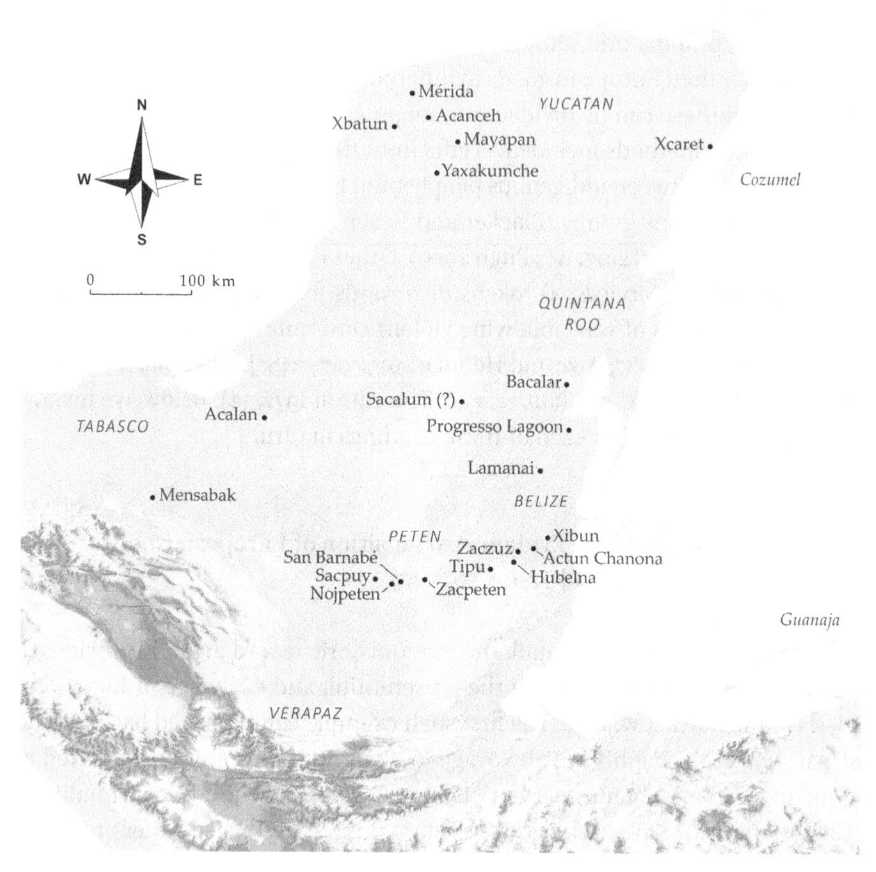

FIGURE 11.1 Map of the Maya area showing the locations of the most prominent settlements and localities mentioned in the text
MAP BY CHRISTOPHE HELMKE

to the Yucatecan coast, Córdoba provisioned the boats with various supplies, and "bought trinkets to be used for barter." This practice followed one established early on by Columbus who extensively traded in glass beads and copper bells in the Bahamian Islands and the Greater Antilles during his initial voyages (Berman and Gnivecki this volume; Bedini 1992, 27; Keegan 1992, 183–205; Keehnen 2012). In the second expedition to the Maya coast from Cuba, this time under the command of Juan de Grijalva, Díaz del Castillo (1956, 29) notes that off the coast of Tabasco they showed the indigenous peoples who approached them in canoes "small mirrors and strings of green beads that they thought were of jadeite, on which they placed great value."

The third, and most consequential Spanish expedition, was that under the command of Hernán Cortés in 1519. For this trip, the Spaniards also provisioned their vessels with the various articles that were intended for trade (Díaz del Castillo 1956, 26). This customary practice is confirmed by the dispatches sent by Cortés to the Spanish Crown. In his first letter to Charles V, which was meant, in part, to curry favor with the Emperor at the expense of Diego Velásquez, governor of Cuba, Cortés noted that their little *armada* was provisioned with "boxes of laced shirts," beads and other merchandise (Blacker and Rosen 1962, 6). He also states that "We make special mention of this so that Your Majesties may know that the *armada* fitted out by Diego Velásquez was intended as much for trading merchandise as for privateering" (Blacker and Rosen 1962, 6). On his arrival at Cozumel, Cortés gave some beads, little bells and Spanish shirts to two native men and a woman who were asked to convey an invitation to meet with their chief and community (Díaz del Castillo 1962, 58). While at Cozumel, Cortés was informed of two Spaniards, Gerónimo de Aguilar and Gonzalo Guerrero, who, following their shipwreck and subsequent stranding on the coast of Quintana Roo, had been taken captive and were living in Maya communities along the coast. In an effort to secure their freedom, Cortés provided "all kinds of beads to ... two Indian merchants of Cozumel" who were to go and barter for their ransom on behalf of Cortés (Díaz del Castillo 1956, 41; 1962, 60).

The ethnohistoric literature also mentions that following his landing along the coast of Veracruz, Cortés gave Aztec emissaries "a Florentine glass cup, much decorated and gilded, three Holland shirts, and other things" (Díaz del Castillo 1956, 59–60; also see Díaz del Castillo 1962, 94). A few years later, while en route to Honduras in 1524–1525, Cortés (Blacker and Rosen 1962, 241–243) visited Nojpeten, capital of the Itza. During this first Spanish *entrada* to the Peten, Cortés gifted to the Itza the *ajaw*, or 'king,' Kan Ek', "a shirt, and a cap of black velvet, and some little things of iron, such as scissors and knives" (Means 1917, 34). When Cortés departed, he also left his injured horse with the Itza, a statue of which was said to eventually become an object of worship (e.g., Bennett 1998, 189–190).

Fray Agustín Cano's account of Díaz de Velasco's *entrada* from highland Guatemala to the Peten in 1695 noted that while traversing Mopan territory – in what is now the southeastern Peten – an advance Spanish patrol encountered a group of Itza hunters. When the Itza drew their weapons, the Mopan guide asked that they "be peaceable and not fight, because those were merchants who sold axes and machetes" (Jones 1998, 135; Means 1917, 98). According to Means (1917, 98), the soldiers then exchanged "knives and many other little trifles" for blankets from the Indians. A few days later, Velasco's party captured a Maya by the banks of Lake Peten. When questioned, the prisoner replied: "That he had gone to look for merchants to buy axes and machetes" (Jones 1998, 138). Scholes and Roys (1948, 245) also report that Acalan merchants traded knives, axes and machetes, among other things, with the Maya of northern Yucatan.

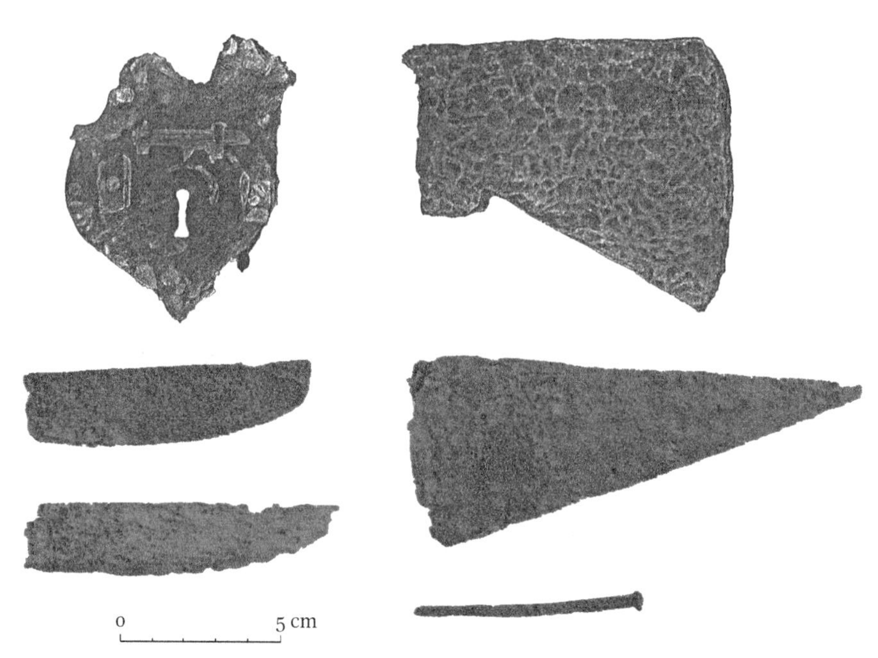

FIGURE 11.2 A selection of metal objects found at Lamanai. To the left is a European lock plate, from refuse around the later church, above two knife blades. To the right is an axehead above a hatchet and a nail
DRAWINGS BY EMIL HUSTIU AND CHRISTOPHE HELMKE BASED ON PHOTOS BY BRIAN BOYLE

During the conquest of the Itza, generals Martín de Ursúa y Arizmendi and Melchor de Mencos, and Alejandro Pacheco complained that the "metal tools and other barter items for the Itzas had still not arrived from Verapaz by March 18th" (Jones 1998, 364) (Figure 11.2). Being short on food, and eager to acquire maize for their small army, Ursúa y Arizmendi and Mencos sent a party to Sacpuy where they bartered machetes, knives, axes, silver coins and salt for maize (Jones 1998, 366).

3 The Acquisition of European Goods to Encourage Conversion

The ethnohistoric literature contains numerous references of gifts given by priests to encourage Maya conversion to Catholicism (Graham et al. 1989, 1258). For example, Villagutierre Soto-Mayor (1983, 16, also see Jones 1989, 265–266) provides a list of non-perishable European objects that were gifted to the Itza prior to their conquest. The items included axes, knives, machetes, glass beads, earrings and necklaces. Van Oss (1986, 16) reports that a common Spanish practice in highland Guatemala was the gifting of axes, blankets, hats, knives, needles, rings, scissors, and other goods to *caciques* following their acceptance of Christianity. Elizabeth Graham (2011) further notes that in addition to glass beads and needles, the Spanish friars often carried other gifts, including mirrors, religious paintings and "other paraphernalia" that were awarded to the Maya following their conversion to Catholicism (Figure 11.3b–c). Some of these "other paraphernalia" included rings, lace tags and objects made from jet and amber (Graham 2011, 234).

In Fray Andrés de Avendaño y Loyola's description of the *entrada* to Tipu and Lake Peten by friars Bartolomé de Fuensalida and Juan de Orbita in 1618, he noted that the friars presented the Itza *cacique* Kan Ek' with "the trifles that had been given them in Mérida for this purpose and also a little cacao from Tipu […] and a very good *hanger* (cutlass)" (Means 1917, 69) (Figure 11.3a). Avendaño y Loyola further notes that Kan Ek' "was the first to receive, with great pleasure, a Cross which the Padres placed in his hands, and afterwards some of his men received others" (Means 1917, 73).

In his second *entrada* to the Itza capital in Lake Peten, Avendaño y Loyola reported that:

> I gave them, as they came up to the novel sight, some necklaces and other trinkets and trifles for their wives and daughters, and for the men some knives, for the desire to possess which all came again, thus obliging me to give them presents a second time, all which I did with pleasure, one reason being the abundance of what our benefactors in their kind zeal had given me, and the other in order to draw them to our Catholic faith … They approached me to get what I had remaining in some hampers, in which I carried for the petty King an entire suit of clothes … and other things which I was carrying for the chiefs of Peten Ytza, in order the better to gain their good will, besides other things necessary for our ministry and support.
>
> MEANS 1917, 131

It is also apparent that Spanish priests took bells for the *visita* churches they had constructed in various Maya communities. Fray Diego López de Cogolludo (1688, bk. 11, Chap. 13, 645), for example, recounts that when Fuensalida's party arrived to Zaczuz, they found that the church had been burnt down and the bell thrown into the bushes. Graham (2011, 246) notes that when the Maya of Xibun (present day Sibun River in Belize) abandoned their community in 1630–1631, they took the "church bells and church ornaments with them." Jones (1998, 161) also notes that in preparation for a trip to Nojpeten in the 1690s,

FIGURE 11.3 (a) Reconstruction of the sixteenth-century *visita* church at Tipu (watercolor by Louise Belanger, after Jones et al. 1986, 43); A selection of glass beads from Christian burials associated with the church at Tipu: (b) Necklace of glass beads with jet from the burial (B139) of a juvenile (5–7 years of age); (c) Nueva Cádiz glass beads from the burial of a male (18–22 years of age) (after Smith et al. 1994, Plate IVB and Figure 8)

Fray Juan de San Buenaventura requested a list of religious paraphernalia from Fray Antonio de Silva, "the provincial of the Franciscan order in Yucatan." The list included "communion tables, chalices, chrismatories (vessels for holy oils), surplices [white vestment worn over cassock by clergy], images of saints and the pope." Fray Juan de San Buenaventura also pleaded with de Silva to send him bells for newly constructed churches in several Itza towns (Jones 1998, 161).

4 The Acquisition of European Goods as Payment for Services

Other references indicate that European goods were sometimes used as payment for services rendered to the Spanish by the Maya. Juan de Villagutierre y Soto-Mayor (1983, 101), for example, mentions that a machete was given to a "mayor" of a Maya village to help in clearing a path through the jungle in Ch'ol territory in Verapaz. This same source (Villagutierre y Soto-Mayor 1983, 312) also implies that the Maya would barter for hatchets and machetes with the Spanish. According to Jones (1989, 284) Ursúa y Arizmendi reported that iron tools were in high demand by the Itza Maya of the Peten Lakes region, and that Itza road builders assisting the Spaniards were "paid in machetes and axes" (Ursúa y Arizmendi 1697).

5 The Acquisition of European Goods by Force

The aftermath of the massacre of Spaniards at Sacalum in 1624 provides one of the best examples of Maya acquisition of European objects by force. As Thompson (1972, 12) and Jones (1989, 185–187) report, the Maya seized the weapons of some of the Spaniards while they were attending mass in Sacalum. They then slaughtered Francisco Mirones and his weaponless Spaniards in the church and took off with their arms and various other objects. When Ajk'in Pol, leader of the rebellious Maya, was captured several months later, they found in his possession the "chalices and silver from the Sacalum church, as well as a silver-plated dagger and some clothing belonging to Mirones" (Jones 1990, 188).

Similarly, Restall (1998, 74) notes that along the west coast of the Yucatan peninsula, renegade Maya raided communities and travelers and took their "knives, machetes, clothing, and whatever else" they carried.

During Avendaño y Loyola's second *entrada* to Nojpeten, he noted that in a small bay of Lake Peten

> ... a nephew of the King, whom I had rewarded with some Spanish trinkets, coveting the image of a Santo Christo, which I wore on my neck, and which I had refused to give him on two occasions when he had asked me for it, on my giving a cutlass with its blade to the petty King, his uncle, seized the hand of his uncle with excessive insolence, and snatching the blade from its sheath, turned it to my breast, and passing the blade across my throat, cut the string with one blow and took the image of Christ from me.
>
> MEANS 1917, 133.

This incident recalls another similar confrontation between Father Fuensalida and the Maya of Hubelna. In 1641, while attempting to reach Tipu from Bacalar, Fuensalida, who was accompanied by three other Franciscans and several Maya porters, found many of the towns along the way burnt and abandoned. When they arrived at Zaczuz, Fuensalida was informed that they were not welcome at Tipu. The Spanish party subsequently arranged to travel up the Yaxteel Ahau River to the newly established community at Hubelna (see Awe and Helmke 2015, 348). Following their arrival at Hubelna, Fuensalida and his party were bound, insulted, and their possessions confiscated and destroyed (López de Cogolludo 1688, bk, 11, Chap. 14, 648; Jones 1989, 53). López de Cogolludo (1688, bk. 11, Chap. 14, 647, 649) also remarked that "The one that they most mistreated and stripped bare was Lázaro [Pech], whom they knew was servant to the friars, and they took from him of a good machete that he had, lest in anger he should kill one of them" (translation by the authors).

Interestingly, Restall (1998, 74) has also noted that a Maya noble named Don Juan Xiu of Yaxakumche, a community near Oxkutzcab, petitioned the Spanish authorities in 1662 for permission to carry a musket. Whereas the petition in itself is noteworthy enough, it is possession of the firearm that is interesting and raises questions as to its origin and the means of its acquisition.

6 Archaeological Evidence for Maya Consumption of European Objects

In a 1983 publication, David Pendergast commented that "The principal defects in the evidence regarding Spanish impact on native material and nonmaterial culture in sixteenth- and seventeenth-century Belize afflict both the excavated evidence and the documentary picture" (Pendergast 1983, 113). While Pendergast's remark accurately represented the state of archaeological affairs in the late 1970s and early 1980s, a number of subsequent archaeological projects that focused on contact period sites have significantly improved our knowledge on the consumption of European goods by the contact period Maya. In spite of this development, some of the best archaeological information on Maya – Spanish interaction still derives from the almost two decades of research at the Belize sites of Lamanai and Tipu, both of which were investigated by Elizabeth Graham and David Pendergast along with several of their colleagues (Pendergast 1983, 1985, 1998; Graham 1987, 2011; Graham et al. 1989; Jones 1989, 1990, 1998). Other projects that have contributed to this growing record include the work of Marilyn Masson and Maxine Oland at Progresso Lagoon, Patricia McAnany's and her students research in the Sibun River Valley, Jaime Awe and Christophe Helmke's investigations in the Roaring Creek Valley of western Belize, E. Wyllys Andrews' projects in the Yucatan, INAH's explorations of cenotes in Quintana Roo, and Timothy Pugh's work in the Peten Province of Guatemala.

One of the primary research questions addressed by all these projects concerns the timing of the arrival of European objects to this part of the Maya world. Both the ethnohistoric documents and archaeological record suggest that communities in the Yucatan were among the first to acquire European goods, and that this process started with the expedition of Francisco Hernández de Córdoba in 1517 (Díaz del Castillo 1956). For sites in Belize where Spanish-made objects have been discovered, Pendergast and his colleagues argue that archaeological investigations "permit fairly precise fixing of the time of use of the community within the span of the sixteenth and seventeenth centuries" (Pendergast et al. 1993, 70). This is particularly "true of olive jar and majolica types, which at Lamanai and very probably also at Tipu cannot be later than the 1630s and are most likely to have reached the sites between 1544 and 1600" (Pendergast et al. 1993, 70). These dates are corroborated by John Goggin's (1960, 20–24, 1968, 101–114) analysis of olive jars and majolica which he suggests were being imported into the area from ca. 1580 to 1850 (see also James 1988).

For the Peten, the initial introduction of European objects takes place during the 1524–1525 *entrada* of Cortés into this area of the Maya lowlands (Blacker and Rosen 1962). Recall that during his brief stay at Nojpeten, Cortés gifted several objects to Kan Ek', and left his horse in the care of the *cacique* (Bennett 1998, 189–190). Subsequently, and particularly during the years preceding the 1697 conquest of the Itza Maya, an array of Spanish goods, including clothing, glass beads, machetes and a variety of other objects were gifted to the Itza to encourage their conversion to Catholicism and their capitulation to Spanish control (Jones 1998; Means 1917, 131; Pugh et al. 2012, 6; Villagutierre y Soto-Mayor 1983).

Another question that archaeologists have tried to address concerns the purpose(s) for which the Maya employed the various exotic goods acquired from the Spanish. Pendergast (1983, 113) previously noted that the ethnohistoric literature is quite vague to this end, and "are never descriptive of Maya or European goods as they were understood or utilized within the native cultural context." Indeed, the ethnohistoric sources mostly inform us about the reasons why the Spanish provided their European goods to the Maya, and we now know that these efforts were primarily to establish or promote acquiescence or good relationships. Fortunately, the recently improved archaeological record is now allowing us to fill this void, and our study of the contextual distribution of European objects at contact period sites suggests that the Maya utilized Spanish objects for three main purposes. These include a) functional/mundane reasons, b) as status markers, and c), for ritual purposes.

7 Functional Uses of European Goods by the Maya

David Pendergast (1983, 116; also Graham et al. 1989) observed that the intrusion of Spaniards into the Maya area appears to have affected pre-Hispanic trade in obsidian. Given the predominantly utilitarian function of this commodity, its absence from local markets would certainly have driven the lowland Maya to seek out and acquire axes and machetes from the Spanish. The ethnohistoric literature unquestionably reflects this increasing demand for metal tools and is rife with passages describing the constant efforts of the Maya to acquire them. We have, for example, already referred to the Maya who, when taken captive by Díaz de Velasco's party along the shores of Lake Peten, informed the Spaniards that he was looking "for merchants to buy axes and machetes" (Jones 1998, 138). Recall too that during that same expedition, Velasco's Mopan guides asked a group of Itza warriors to "be peaceable and not fight, because those were merchants who sold axes and machetes" (Jones 1998, 135). In Fray Juan de San Buenaventura's letter to the provincial Silva, he wrote: "They also say that for the past six months the Itzas who came to look for iron tools among these Cehaches told them not to run away from the Spaniards when they came" (Jones 1998, 159). Jones (1998, 205) adds that "Trade for metal tools," particularly axes and machetes, was the primary motivator "for increasing contact with the Spaniards." Avendaño y Loyola also reported that on his last day in Nojpeten, several leaders confronted Kan Ek' deriding him for his friendship with the Spanish, and questioned whether the reason for his cozy relationship with the foreigners was to acquire "axes and machetes for their cultivations" or "the goods and clothing of Castile" (Jones 1998, 209).

Metal tools have been found at several sites in Belize and in the Peten. At Tipu, for example, Graham (1998) as well as Graham et al. (1989, 1256; 1985, 207–210) note that "metal hooks, iron nails, locks, and other Spanish ironwork occur in refuse deposits and building debris." Graham (2011, 365, n. 184) also recovered metal needles in middens at Tipu. At Lamanai, Pendergast (1983, 129, Figs. 7–8) found several iron artifacts in Str. N11-18, including a possible knife handle, two knife blades, what may be a horseshoe and an axe. The contexts of discovery of these objects strongly suggest that their primary function served practical mundane purposes.

For the Lacandon area of Chiapas and Peten, Joel Palka (2005) reports that throughout the colonial period, and into the early twentieth century, the people of this region acquired a variety of European objects that were used for various purposes. Among these European goods were white earthenware ceramics, glass bottles, metal cooking pots, machetes, axes and files. Here again, the very nature of the latter four objects leave little doubt that their primary function was utilitarian.

In a recent analysis of skeletal remains from a burial site in the Mensabak area of the Lacandon forests in Chiapas, Cucina et al. (2015) concluded that the remains displayed evidence of violent deaths and wounds caused by metal weapons such as machetes and swords. Maxine Oland and Palka (2016, 480) note that the Mensabak region was an unconquered zone and that this type of violence was a result of the use of acquired metal tools in "local indigenous conflicts." The Lacandon Maya of Chiapas continued to obtain European goods, particularly metal tools and majolica well into the twentieth century (Palka 2005).

Nancy Farris (1984, 121) notes that "Steel axes and other Spanish tools [...] were introduced via a clandestine trade with the conquered areas to the north" and that these were likely "substituted for the manufactured goods formerly imported from highland Mexico." She (Farris 1984, 70) also notes that the Maya acquired steel machetes from the Spanish for everyday use, and that "despite official prohibitions" the colonial Maya were able to acquire guns and gunpowder for hunting (Farris 1984, 70).

8 European Goods Used as Status Markers

Following his excavations of Str. N11-18 at Lamanai, Pendergast concluded that the concentration of iron objects in the building "underscores the importance of the structure and its occupants, not as regards European impact on local technology but rather in terms of use of imported objects as physical manifestations of rank or status that derived from Spanish interest" (Pendergast 1983, 130) (Figure 11.4a). In addition to metal objects, 91% of all glass beads found at Lamanai were recovered "within and around Str. N11-18" which

FIGURE 11.4 (a) Structure N11-18 at Lamanai, where many European objects have been found, is the probable residence of the site's *cacique* (after Pendergast 1983, Figure 2); (b) Gilded brass book hinges decorated with figurative medallions, discovered within Str. N11-18. Based on the style of these pieces these date to no later than ca. AD 1550

PHOTOS COURTESY OF DAVID PENDERGAST

Pendergast (1983, 128) identified as the primary residence of the *cacique*. Other materials recovered from the *cacique*'s house included leaves from two brass book hinges (Pendergast 1983, 129) (Figure 11.4b). This clear concentration of European goods around Str. N11-18, *vis-a-vis* other residences at the site, provides compelling evidence that these exotics represented important status objects.

The archaeological context of European objects at Tipu differs from that at Lamanai. At the former, most Spanish imports were found in association with burials (Graham 1991, 2011; Graham et al. 1985, 1989; Pendergast 1983). For example, approximately 720 glass beads were recovered at Tipu, the majority of which were associated with nine child burials (Smith et al. 1994, Table 1). The other beads were found in association with three male and three female adults, leading Pendergast (1983, 128) to contend that the presence of these adults "among the bead-associated burials also suggests the use of beads as markers of rank or status." Excavations of a large house designated as Str. H12-7, at Tipu, also yielded "olive jar sherds, a copper ring, a glass bead, and a lock plate for a chest" (Graham 2011, 230).

At both Lamanai and Tipu, Pendergast (1983, 125), and Graham (1991, 323) recovered fragments of majolica bowls and dishes. At both sites, however, the frequency of these glazed Spanish wares was low, suggesting that the importation of European pottery was not very significant at either of these communities. The low frequency of majolica at other sites, seems to confirm this observation. This is undoubtedly partly caused by the continued and relatively expedient manufacture of ceramics from local clays as well as the difficulty of transporting European glazed wares over large distances of uneven terrain. Despite their low frequencies, however, the contextual distribution of majolica seems to be highest in buildings associated with individuals of higher status. This is certainly the case in the Sibun River Valley where Steve Morandi (2003, 151–152; 2010) found numerous fragments of Spanish majolica, a small copper star, and olive jar fragments in an elite residence at the Spanish colonial settlement of Cedar Bank.

At Progresso Lagoon in northern Belize, investigations by Oland and Masson (2005) recovered several artifacts of Spanish origin. With the exception of olive jar fragments, most of the European goods included "luxury items" such as glass beads, a glass earring, and majolica plates. Here again, all goods of European manufacture were discovered in Structure 1, a large and impressive building that Oland (2012, 188–189) associated with elite residence. The fact that this large residence "had more exotic consumption and craft production than any other house" at the site is a pattern that mirrors the distribution of exotics at Lamanai (Oland 2012, 188–189).

To the west, in Guatemala, Pugh (2009, 382) noted that "No evidence at Zacpeten documents that European goods "trickled down" to non-elites at the site; the goods appear to have been restricted to public ceremony and elite power play." In Str. T29 at San Bernabé on the Tayasal peninsula, Pugh et al. (2012, 15) discovered "10 majolica sherds, four pieces of glass, a mirror fragment, lead shot, a square nail, four pieces of unidentified iron, a copper alloy ring, and a silver coin." They (Pugh et al. 2012, 15) also concluded that Str. T29 was likely "a residence of the San Bernabé parish."

Interestingly, Pugh et al. (2012, 15–16) suggest that most of the Spanish-style objects found at Tayasal were likely produced in the Americas. This includes the coin whose weight, which does not confirm to royal decree, suggests that it was minted in Santiago de Guatemala sometime during the early colonial period. They (Pugh et al. 2012, 17) further suggest that because most of these objects were found in middens, it is possible "that the value of exotic goods changed over the contact and colonial periods as they became more common." We must consider, however, that these middens were adjacent to some of the largest platforms in the community, thus were likely associated with higher status residences.

Another example of the association of European objects with elite Maya residences comes from our own work in the Roaring Creek Valley of western Belize. At a site, which could possibly be the location of the contact period community known as Hubelna (see Awe and Helmke 2015, 345–346, Figure 1), one of our workers discovered several gray-glazed olive jar sherds. The plazuela group where the sherds were found are among the largest mounds at the site, thus likely representing the residence of elite members of this contact period community. At the Cedar Bank site along the Sibun River, McAnany and her colleagues (2004, 306) also discovered several olive jar fragments in association with a large contact period platform. Here again, McAnany et al. (2004) and Morandi (2010; also Jones 1989, 200) note that Cedar Bank may have been the location of a Spanish *visita* church that was constructed at the end of the sixteenth century.

Colonial documents provide equally compelling evidence that European objects often represented important status markers in contact period Maya society. The ethnohistoric sources, for example, mention several cases where Castilian clothing and sheathed knives were gifted to Maya leaders. This is particularly true in the case of the aforementioned Kan Ek', *cacique* of the Itza. We already noted above that Cortés gifted Kan Ek' some Spanish clothes on his brief stop at Nojpeten in 1524–1525. During Avendaño y Loyola's preparation for his trip to Nojpeten, Ursúa, then Governor of Yucatan, also gave him a "suit of Spanish clothing for Kan Ek', complete with a hat and staff of office. The intention was to dress the Itza ruler up as a typical Yucatan Maya *alcalde* [...]

and a means of co-optation. Other gifts [...] included a machete and sheath, a knife with a belt, and three yards of taffeta" plus "numerous smaller gifts, such as necklaces and knives, intended as general handouts" (Jones 1998, 188).

9 European Goods Used for Ritual Purposes

Pugh (2009) argues that although several scholars have suggested that the primary factor which drove the Maya to desire and acquire European objects was the technological "superiority" of these goods, the location of these objects in excavated Maya communities indicate that they "often employed these tools for "non-utilitarian" – particularly ceremonial – purposes" (Pugh 2009; also see Miller and Hamell 1986, 314).

At Zacpeten, which was occupied by the Kowoj Maya until about 1650, Pugh (2009; also see Oland and Palka 2016, 482, Figure 4) recovered several colonial period artifacts of Spanish origin. Among the latter was the modified jawbone of a cow, iron tools, a white clay ball and a lead musket ball, along with Maya objects that were recovered in public and elite residential architecture (Figure 11.5). Pugh (2009, 382) argues that the Maya incorporated these exotic Spanish

0 5 cm

FIGURE 11.5 Partial cow mandible found in deposits dated to ca. AD 1650 at Zacpeten
PHOTO BY PRUDENCE RICE, COURTESY OF TIMOTHY PUGH

objects in caches and offerings to consecrate "the sacred spaces at Zacpeten," and to "harness the power of the Europeans" (Oland and Palka 2014, 481). According to Pugh (2009, 111) "The Kowoj appropriated the peripheral materials and planted them in their most central ceremonial spaces as positive contagion."

Pugh (2009, 382) also comments on the interesting Kowoj practice of incorporating European fauna, in this case cow remains, in cache deposits at Zacpeten. Interestingly, a similar pattern has been found at Cozumel where the remains of cattle, horses and sheep were recovered in ritual contexts (Hamblin 1984, 142–143). Yet another example of this practice at Cozumel is reflected by the discovery of contact period ceramic vessels along with olive jars, fragments of majolica and cow bones that were ritually deposited into local cenotes. The latter were recovered by an underwater archaeology project co-directed by Luis Alberto Martos López (2008, 107).

Investigations by the authors (Awe and Helmke 2015) indicate that the ritual deposition of olive jars was a relatively common practice during and after the Spanish conquest of the Maya lowlands. For example, we previously reported on a complete olive jar that was cached and ritually killed in a cave in the Roaring Creek Valley of western Belize (Awe and Helmke 2015) (Figure 11.6a). Other complete and fragmentary remains of olive jars, also from sacred cave contexts, are known from Xcaret, Quintana Roo (Andrews and Andrews 1975, 72, figure 88), the Crocodile Cenote system in Cozumel (Martos López 2008, 107), from the Cenote Canun near the sites of Acanceh and Mayapan in Yucatan (Anthony Andrews, personal communication 2010), and from the cenote Xbatun (CONACULTA 2008; González et al. 2004; Alfredo Barrera Rubio, personal communication 2010). It is important to note that one of the Xbatun olive jars also displays a small kill hole on its side similar to that on one of the Roaring Creek specimens. In addition to the above, Oland (2014, 2017) also recovered an olive jar fragment cached beneath an altar in a shrine next to the "primary elite residence" at Progresso Lagoon in northern Belize.

Investigations in Belize provide various other examples of the deposition of exotic Spanish objects in ritual cave contexts. In the Roaring Creek Valley, for example, we (Awe and Helmke 2015) previously reported on the discovery of a European rapier sword that was cached in a small cave that is located in proximity to Olive Jar Cave (Figure 11.6b). East of Roaring Creek, Peterson (2006, 26) investigated several cave sites in the Gracy Rock area of the Sibun River drainage that contained both Spanish and British colonial remains, including olive jar fragments. One of the sites, Hickatee Cave, reported by Peterson (2006, 90) contained several European artifacts, including majolica. Another particularly interesting find is the discovery of historic-period graffiti written on flowstone

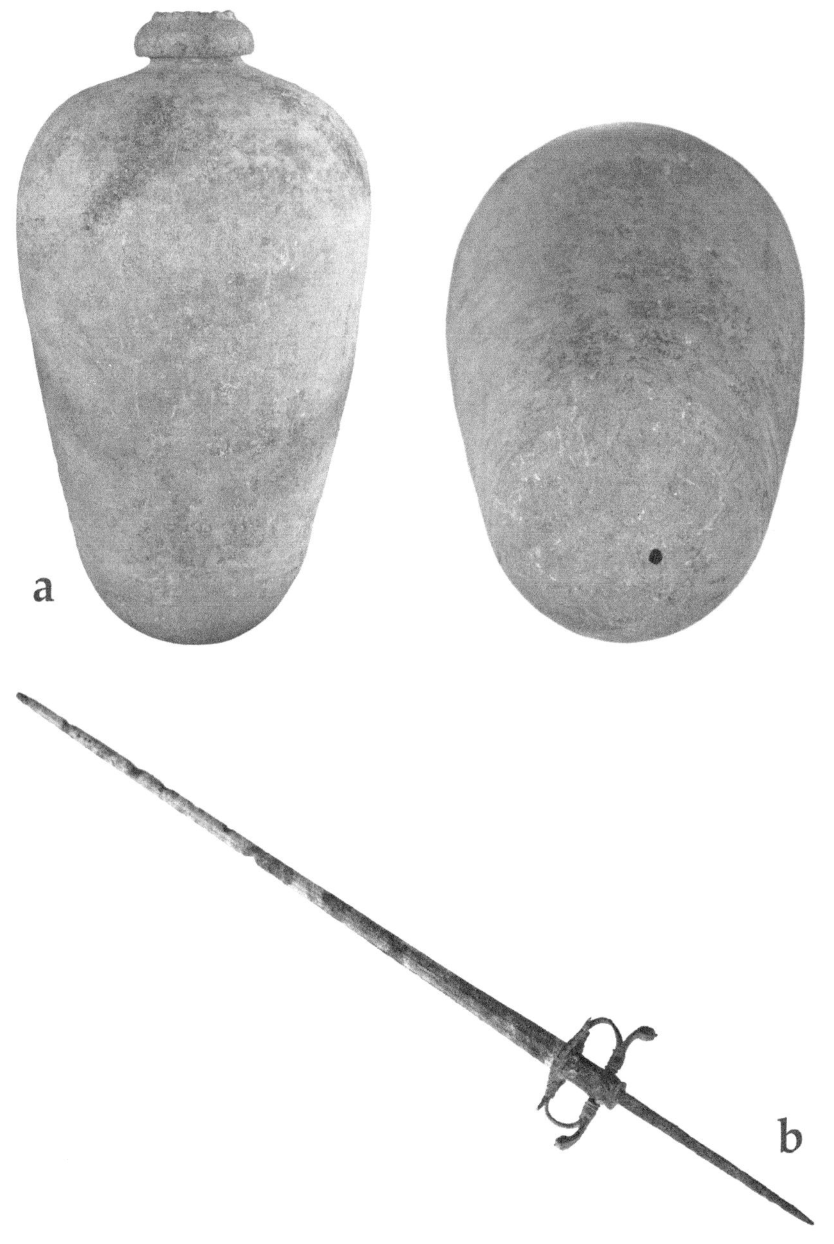

FIGURE 11.6 (a) Two views of a Spanish olive jar dated to ca. AD 1540–1630, ritually killed and deposited within a cave in the Roaring Creek Valley of western Belize. The jar is 48 cm tall (after Awe and Helmke 2015, Figure 2); (b) European rapier sword that was cached in a small cave located in the Roaring Creek Valley of western Belize. Total length is 87.8 cm
PHOTO BY JAIME AWE AND CHRISTOPHE HELMKE

within Actun Chanona. The text of the graffiti includes the Spanish Word "Dios" (Peterson 2006, 36). In all these cases, we believe that the Maya purposely deposited these European objects in sacred cave contexts as part of a long established set of cave rituals (Helmke 2009; Prufer and Brady 2005), but here replacing indigenous objects with European material culture, because the exotic nature of the foreign objects may have increased their "ritual value."

10 Conclusion

A major aim of this chapter was to demonstrate that both the ethnohistoric literature and the archaeological record contain substantial information on the acquisition of European-made objects by the contact period Maya. Based on this brief review, we would argue that both these sources of information certainly confirm this position. Furthermore, these highly complementary data indicate that although the peripheral locations of most sixteenth- and seventeenth-century Maya communities in Belize and the Peten placed them outside of, and away from, the major centers of early colonial Spanish control, the frontier Maya of Belize and the Peten were still able to acquire Spanish goods. The historic documents, in fact, suggest that the Spanish keenly gifted and traded items to prompt acquiescence and peaceable relations. At the same time, the local vantage reveals that these frontier Maya regularly sought out these exotic objects, and they actively and purposely expended efforts to acquire them. This is perhaps best illustrated by Avendaño y Loyola's comment that the Maya demonstrated an "insatiable desire" for European goods (Means 1917, 131).

The ethnohistoric record also informs us that the Maya acquired European objects by way of five main methods. These included direct gifts from the Spanish, through trade or barter, and sometimes as payment for services rendered to the Spanish. In other cases, the Spanish provided European goods to the Maya as rewards for accepting conversion to Christianity. Occasionally, however, the Maya also obtained these goods by forceful means, particularly during skirmishes as well as periodic revolts and uprisings against the Spaniards.

Our research further indicates that the impetus that fueled the Maya desire for European goods was for their use as status objects, for practical mundane purposes and for their use as special offerings in ritual contexts. Both the ethnohistoric and archaeological record indicate that European objects may have actually replaced native exotics as key indicators of rank and status. As Oland (2017, 129) has noted: "When exotic objects were used in processes of defining and identifying indigenous rank, European objects were often easily adapted to elite purposes." Indeed, Maya lives were transformed with the introduction of Europeans into their world. One aspect of Maya life that was definitely impacted, and which Pendergast (1983, 116) previously noted, was the trade in exotics that derived from other areas of Mesoamerica. In many cases, European goods replaced these native objects and possession of the foreign objects quickly became one of the standards for measuring and displaying rank and status. For this reason, the distribution of some Spanish objects was likely guarded and controlled by the Maya elite. Pugh (2009, 382), for example, has argued that "No evidence at Zacpeten documents that European goods "trickled down" to non-elites at the site; the goods appear to have been restricted to public ceremony and elite power play." The contextual distribution of European objects at the archaeological sites of Lamanai, Tipu, Cedar Bank and Progresso Lagoon in Belize, and at Zacpeten and Tayasal in the Peten supports this position and provides compelling evidence for the predominantly elite consumption of these exotic objects.

The introduction of European objects into the native value system, also made them worthy of inclusion in ritual and religious contexts. The caching of the European sword in Rapier Cave, and the numerous olive jars that have been found in caves across Belize and the Yucatan are ample testimony of the increasing inclusion of European exotica in sacred Maya contexts. This change in the caching tradition of the contact period Maya is also evident at surface sites and is clearly reflected by the placement of olive jar fragments beneath an altar at Progresso Lagoon, by the European grave goods found in the burials at Lamanai and Tipu, as well as by the various Spanish objects that were used to consecrate "the central axes of sacred spaces" at Zacpeten (Pugh 2009, 382). Beside the special value that their exotic nature imbued them with, the Maya likely incorporated European objects in their sacred contexts to harness the power they represented and to revitalize "themselves and their world through the contagion of alterity" (Pugh 2009, 383). What is also intriguing to consider in this regard, is how the idea of exoticism, or the attractiveness of these foreign material goods, was maintained throughout the period of contact and how it was fostered or altered in later times.

In spite of the apparent elite control over some European objects, and their ritualized incorporation in caches, burials and sacred locations, it is also evident that the Maya utilized several European goods for basic mundane and utilitarian purposes. This is particularly true of axes and machetes that provided the Maya with more hardy and efficient tools for preparing their kitchen gardens, fields or *milpa*. In later years, the colonial Maya were also able to acquire guns and gunpowder for hunting (Farris 1984, 70). The subsequent use of axes, machetes and guns for violent confrontations, however, unquestionably affected the lives of the lowland Maya more than we can ever measure, and they continue to impact the world of the Maya into the twenty-first century.

Acknowledgments

Many kind and heartfelt thanks to Elizabeth Graham, Grant Jones, David Pendergast, Mathew Restall, Maxine Oland and Timothy Pugh for insightful comments and productive conversations on European-Maya interactions over the years. We are particularly grateful to Floris Keehnen and Corinne Hofman for inviting us to contribute this chapter to the volume. We also extend our thanks to Elizabeth Graham, David Pendergast, Prudence Rice and Timothy Pugh for permission to include illustrative materials that have greatly enriched this chapter. All shortcomings remain the responsibility of the authors.

References

Andrews IV, E. Wyllys and Anthony P. Andrews. 1975. *A Preliminary Study of the Ruins of Xcaret, Quintana Roo, Mexico – With Notes on Other Archaeological Remains on the East Coast of the Yucatan Peninsula.* Middle American Research Institute Publication 40. New Orleans: Tulane University.

Avendaño y Loyala, Fray Andrés de. 1987. *Relation of Two Trips to Peten: Made for the Conversion of the Heathen Ytzaex and Cehaches*, edited by Frank E. Comparato; translated by Charles P. Bowditch and Guillermo Rivera. Culver City: Labyrinthos.

Awe, Jaime J. and Christophe Helmke. 2015. "The Sword and the Olive Jar: Material Evidence of Seventeenth-Century Maya – European Interaction in Central Belize." *Ethnohistory* 62 (2): 333–360.

Bedini, Silvio A. 1992. *The Christopher Columbus Encyclopedia, Volume I.* New York: Simon and Schuster.

Bennett, Deb. 1998. *Conquerors: The Roots of New World Horsemanship.* Solvang: Amigo Publications.

Blacker, Irwin R. and Harry M. Rosen. 1962. *Conquest: Dispatches of Cortez from the New World.* New York: The Universal Library, Grosset and Dunlap.

Chamberlain, Robert S. 1948. *The Conquest and Colonization of the Yucatan 1517–1550.* Washington, D.C.: Carnegie Institution of Washington.

Chase, Arlen F. and Prudence Rice. 1985. "Postclassic Temporal and Spatial Frames for the Lowland Maya: A Background." In *The Lowland Maya Postclassic*, edited by Arlen F. Chase and Prudence M. Rice, 9–22. Austin: University of Texas Press.

Clendinnen, Inga. 1987. *Ambivalent Conquest: Maya and Spaniard in the Yucatan 1517–1570*. New York: Cambridge University Press.

Columbus, Ferdinand. 1960. *The Life of the Admiral Christopher Columbus by his son Ferdinand*. Translated and annotated by Benjamin Keen. London: Folio Society.

CONACULTA (Consejo Nacional para la Cultura y las Artes). 2008. Cuevas y cenotes mayas. Temporary exhibition, October 2008 – February 2009. Mexico, D.F.: Museo del Templo Mayor.

Cucina, Andrea, Vera Tiesler, and Joel Palka. 2015. "The identity and worship of human remains in rockshelter shrines among the northern Lacandons of Mensabak." *Estudios de Cultura Maya* 45: 141–169.

Díaz del Castillo, Bernal. 1956. *The Bernal Diaz Chronicles: The True Story of the Conquest of Mexico*. Translated by Albert Idell. Garden City: Doubleday & Co., Inc.

Díaz del Castillo, Bernal. 1962. *The Conquest of New Spain*. Translated by J.M. Cohen. Middlesex: Penguin Books.

Farris, Nancy M. 1984. *Maya Society under Colonial Rule: The Collective Enterprise of Survival*. Princeton: Princeton University Press.

Goggin, John M. 1960. *The Spanish Olive Jar: An Introductory Study*. Publications in Anthropology No. 62. New Haven: Yale University.

Goggin, John M. 1968. *Spanish Majolica in the New World*. Publications in Anthropology No. 72. New Haven: Yale University.

González, Arturo González, Carmen Rojas Sandoval, and Octavio del Río Lara. 2004. *Atlas Arqueológico Subacuático para el registro, estudio y protección de los cenotes en la península de Yucatán*. Mexico: Instituto Nacional de Antropología e Historia, Subdirección de Arqueología Subacuática.

Graham, Elizabeth. 1987. "Terminal Classic to Early Historic-Period Vessel Forms from Belize." In *Maya Ceramics: Papers from the 1985 Maya Ceramic Conference, Part 1*, edited by Prudence M. Rice and Robert J. Sharer, 73–98. Oxford: British Archaeological Reports, International Series 345(i).

Graham, Elizabeth. 1991. "Archaeological Insights into Colonial Period Maya Life at Tipu, Belize." In *Columbian Consequences, Vol. 3: The Spanish Borderlands in Pan-American Perspective*, edited by David Hurst Thomas, 319–335. Washington, D.C.: Smithsonian Institution Press.

Graham, Elizabeth. 1998. "Mission Archaeology." *Annual Review of Anthropology* 27: 25–62.

Graham, Elizabeth. 2011. *Maya Christians and Their Churches in Sixteenth-Century Belize*. Gainesville: University Press of Florida.

Graham, Elizabeth A., Grant D. Jones, and Robert R. Kautz. 1985. "Archaeology and Ethnohistory on a Spanish Colonial Frontier: An Interim Report on the Macal-Tipu Project in Western Belize." In *The Lowland Maya Postclassic*, edited by Arlen F. Chase and Prudence M. Rice, 206–214. Austin: University of Texas Press.

Graham, Elizabeth, David M. Pendergast, and Grant D. Jones. 1989. "On the Fringes of Conquest: Maya-Spanish Contact in Colonial Belize." *Science* 246: 1254–1259.

Hamblin, Nancy L. 1984. *Animal Use by the Cozumel Maya.* Tucson: University of Arizona Press.

Hammond, Norman. 1988. *Ancient Maya Civilization.* Third edition. Cambridge: Cambridge University Press and Rutgers University Press.

Helmke, Christophe G.B. 2009. "Ancient Maya Cave Usage as Attested in the Glyphic Corpus of the Maya Lowlands and the Caves of the Roaring Creek Valley, Belize." PhD diss., Institute of Archaeology, University of London.

James, Stephen R., Jr. 1988. "A Reassessment of the Chronological and Typological Framework of the Spanish Olive Jar." *Historical Archaeology* 22 (1): 43–66.

Jones, Grant D. 1989. *Maya Resistance to Spanish Rule: Time and History on a Spanish Frontier.* Albuquerque: University of New Mexico Press.

Jones, Grant D. 1990. "Prophets and Idol Speculators: Forces of History in the Lowland Maya Rebellion of 1638." In *Vision and Revision in Maya Studies*, edited Flora S. Clancy and Peter D. Harrison, 179–192. Albuquerque: University of New Mexico Press.

Jones, Grant D. 1998. *The Conquest of the Last Maya Kingdom.* Stanford: Stanford University Press.

Jones, Grant D., Robert R. Kautz, and Elizabeth A. Graham. 1986. "Tipu: A Maya Town on the Spanish Colonial Frontier." *Archaeology* 39 (1): 40–47.

Keegan, William H. 1992. *The People Who Discovered Columbus: The Prehistory of the Bahamas.* Gainesville: University Press of Florida.

Keehnen, Floris W.M. 2012. "Trinkets (f)or Treasure? The role of European Material Culture in Intercultural Contacts in Hispaniola during Early Colonial Times." Master thesis, Faculty of Archaeology, Leiden University.

Keen, Benjamin. 1959. *The Life of the Admiral Christopher Columbus by his Son Ferdinand.* New Brunswick: Rutgers University Press.

López de Cogolludo, Diego. 1688. *Historia de Yucathan.* Madrid: Juan García Infanzón.

Martos López, Luis Alberto. 2008. "Underwater Archaeological Exploration of the Maya Cenotes." *Museum International* 60 (4): 100–110.

McAnany, Patricia A., Eleanor Harrison, Polly A. Peterson, Steven Morandi, Satoru Murata, Ben S. Thomas, Sandra L. López Varela, Daniel Finamore, and David G. Buck. 2004. "The Deep History of the Sibun Valley." *Research Reports in Belizean Archaeology* 1: 295–310.

Means, Philip Ainsworth. 1917. *History of the Spanish Conquest of Yucatanand of the Itzas.* Papers of the Peabody Museum of Archaeology and Ethnology Vol. 7. Cambridge: Harvard University.

Miller, Christopher L. and George R. Hamell. 1986. "A New Perspective on Indian-White Contact: Cultural Symbols and Colonial Trade." *Journal of American History* 73 (2): 311–328.

Morandi, Steven J. 2003. "Colonial-Period Occupational Debris at Cedar Bank (Operation 40)." In *Between the Gorge and the Estuary: Archaeological Investigations of the 2001 Season of the Xibun Archaeological Research Project*, edited by Patricia A. McAnany and Ben S. Thomas, 153–158. Boston: Department of Archaeology, Boston University.

Morandi, Steven J. 2010. "Xibun Maya: The Archaeology of an Early Spanish Colonial Frontier in Southeastern Yucatan." PhD diss., Boston University.

Oland, Maxine H. 2012. "Lost among the Colonial Maya: Engaging Indigenous Maya History at Progresso Lagoon, Belize." In *Decolonizing Indigenous Histories: Exploring Prehistoric/Colonial Transitions in Archaeology*, edited by Maxine Oland, Siobhan M. Hart, and Liam Frink, 178–200. Tucson: University of Arizona Press.

Oland, Maxine H. 2014. "'With the Gifts and Good Treatment That He Gave Them': Elite Maya Adoption of Spanish Material Culture at Progresso Lagoon, Belize." *International Journal of Historical Archaeology* 18: 643–667.

Oland, Maxine H. 2017. "The Olive Jar in the Shrine: Situating Spanish Objects within a 15th-to-17th Century Maya Worldview." In *Foreign Objects: Rethinking Indigenous Consumption in American Archaeology*, edited by Craig N. Cipolla, 127–142. Tucson: University of Arizona Press.

Oland, Maxine and Joel W. Palka. 2016. "The Perduring Maya: New Archaeology on Early Colonial Transitions." *Antiquity* 90: 472–486.

Oland, Maxine H. and Marilyn A. Masson. 2005. "Late Postclassic-Colonial Period Maya Settlement on the West Shore of Progresso Lagoon." *Research Reports in Belizean Archaeology* 2: 223–230

Palka, Joel. 2005. *Unconquered Lacandon Maya: Ethnohistory and Archaeology of Indigenous Culture Change*. Gainesville: University Press of Florida.

Pendergast, David M. 1983. "Worlds in Collision: The Maya/Spanish Encounter in Sixteenth and Seventeenth Century Belize." *Proceedings of the British Academy* 81: 105–143.

Pendergast, David M. 1985. "Lamanai, Belize: An Updated View." In *The Lowland Maya Postclassic*, edited by Arlen F. Chase and Prudence M. Rice, 91–103. Austin: University of Texas Press.

Pendergast, David M. 1998. "Intercessions with the Gods: Caches and Their Significance at Altun Ha and Lamanai, Belize." In *The Sowing and the Dawning: Termination, Dedication and Transformation in the Archaeological and Ethnographic Record of Mesoamerica*, edited by Shirley Boteler Mock, 55–64. Albuquerque: University of New Mexico Press.

Pendergast, David M., Grant D. Jones, and Elizabeth Graham. 1993. "Locating Maya Lowlands Spanish Colonial Towns: A Case Study from Belize." *Latin American Antiquity* 4 (1): 59–73.

Peterson, Polly. 2006. "Ancient MayaRitual Cave Use in the Sibun Valley, Belize." PhD diss., Boston University.

Prufer, Keith M. and James E. Brady (eds). 2005. *Stone Houses and Earth Lords: Maya Religion in the Cave Context.* Boulder: University Press of Colorado.

Pugh, Timothy W. 2009. "Contagion and Alterity: Kowoj Maya Appropriations of European Objects." *American Anthropologist* 111: 373–386.

Pugh, Timothy W., José Romulo Sanchez, and Yuko Shiratori. 2012. "Contact and Missionization at Tayasal, Peten, Guatemala." *Journal of Field Archaeology* 37 (1): 3–19.

Restall, Matthew. 1998. *Maya Conquistador.* Boston: Beacon Press.

Scholes, France V. and Ralph L. Roys. 1948. *The MayaChontal Indians of Acalan-Tixchel: A Contribution to the History and Ethnography of the YucatanPeninsula.* Carnegie Institution of Washington, Publication 560. Washington, D.C.: Carnegie Institution of Washington.

Smith, Marvin, Elizabeth Graham and David M. Pendergast. 1994. "European Beads from Spanish-Colonial Lamanai and Tipu, Belize." *Journal of the Society of Bead Researchers* 6: 21–47.

Thompson, J. Eric S. 1972. *The Maya of Belize: Historical Chapters since Columbus.* Belize: Benex Press.

Ursúa y Arizmendi, Martín de. 1697. *Testimonia de Martín de Ursúa y Arizmendi.* Seville: Archivo General de Indias, Guatemala 151.

Van Oss, Anna. 1986. *Catholic Colonialism: A Parish History of Guatemala, 1524–1821.* Cambridge: Cambridge University Press.

Villagutierre y Soto-Mayor, Juan de. 1983. *History of the Conquest of the Province of the Itza.* Translated by Robert D. Wood; edited by Frank E. Camparato. Culver City: Labyrinthos.

CHAPTER 12

Resignification as Fourth Narrative: Power and the Colonial Religious Experience in Tula, Hidalgo

Shannon Dugan Iverson

1 Introduction

Scholars have typically described colonial religious change in Mesoamerica in one of three major narrative frames:

(1) the "spiritual warfare narrative": a top-down imposition of Christianity;
(2) the "core-veneer narrative": a largely failed colonial project in which indigenous subjects retained many of their essential religious traits; or
(3) the syncretism narrative: a passive, relatively equal blending of two originally coherent belief systems.

These debates are, at their core, ideas about the way that power operates in early colonial situations, and each constitutes a narrative of power that is enabled, strengthened, challenged, and refined by empirical data.

However, as I worked through the data I collected from two early Franciscan sites in Tula, Hidalgo in central Mexico (Figure 12.1), I found that existing narratives of religious change were inadequate to interpret the full extent of the transformations and continuities that I was observing. These data pointed toward a complex but unequal exchange: indigenous subjects clearly did not have full autonomy in early colonial Christian contexts, yet their diverse preexisting religious ontologies shaped the New World Church to a remarkable degree. This finding, though shared with many other researchers with similar topics (Graham 2011; Tavárez 2011; Wernke 2007), did not fit well with established narratives of colonial religious power. This was not an "ideal-type" problem: that is, the inherent mismatch between real-world data and inherently inadequate "ideal-type" models. Rather, there seemed to be a gap where a fourth narrative should be. Even so, the old "commonsense" narratives of religious change seemed to stubbornly persist despite ample data and careful refutations of existing models. Finding an interpretation of colonial power that articulated honestly with my data became my most challenging task.

To contextualize the Tula case, I explain existing narratives of religious change in the region. I then contrast two forms of material culture from Tula – buildings and ceramics – that, at least superficially, appear to tell opposite

DOI:10.1163/9789004273689_013

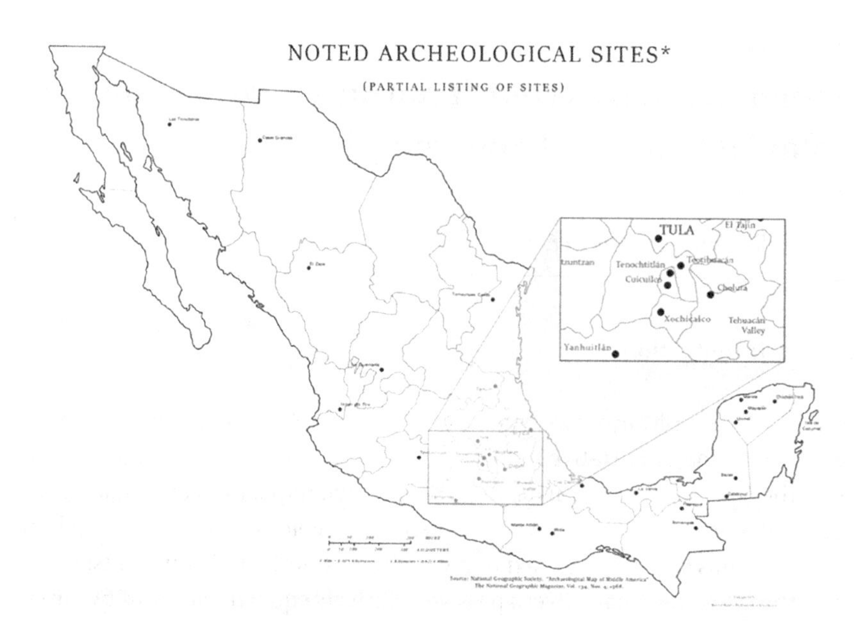

FIGURE 12.1 The location of Tula, Hidalgo in central Mexico. Adapted from "Noted Archaeological Sites," Board of Regents, University of Texas 1975
IMAGE COURTESY OF UNIVERSITY OF TEXAS LIBRARIES, UNIVERSITY OF TEXAS AT AUSTIN

stories about the evangelization program in that city. Closer analysis revealed that indigenous preferences played a significant role in shaping each material category. Nonetheless, that influence took place within the context of "social and cultural conversion of entire ethnic groups as a part of colonial domination" (Hansk 2010, 5): indigenous contributions to Christianity were not necessarily intentional, peacefully negotiated, or the product of overt or covert resistance. In the concluding section, I propose a "fourth narrative" based on Judith Butler's (1990) concept of resignification that provides a balanced account of these two known phenomena (indigenous contributions to Catholicism and the domination of the Church), and in so doing offers a more powerful explanation of the material patterns at Tula.

2 Religion in Early Colonial Mexico

For over 1500 years, Mesoamerican and Spanish religious traditions developed independently of one another. The former's roots developed out of their Teotihuacano and Toltec heritage, pan-regional interactions, and tensions between the rural and state-oriented aspects of Aztec religious systems (Berdan 2014, 33–36; Brumfiel 2001; López Austin and López Luján 2000). The Spanish traditions were the product of tensions and interactions between idiosyncratic rural paganism, Moorish influences, Sephartic Jewish traditions, the Crusades, and the evolving relationship between the Church and the Spanish Empire (Christian 1989). These two complex, heterogeneous traditions did not come into regular contact until the Spanish conqueror Hernán Cortés arrived on the shores of Veracruz and began his campaign to overthrow the Aztec Empire.

The Spanish Empire used Christianity to justify its own expansion, and Cortés' campaign traveled with a Franciscan mendicant who haphazardly fulfilled the Crown's religious obligations to convert the indigenous subjects of the New World to the "true faith" (Díaz 1963). In 1521, once the military conquest was complete, Cortés requested that the Crown send a group of mendicants to begin the "spiritual conquest" of the new territory. The band of Franciscans, known as "The Twelve" after the apostles of Christ, arrived in 1524 and quickly set about demolishing the outward remnants of the Aztec state religion: the temples, priesthood, sacred books, and monthly public ceremonies (Ricard 1966). In their place, they established new Christian places of worship, performed mass baptisms, and changed the focus of worship to the Christian God and saints (Ricard 1966; Schwaller 2011).

But friars' Utopian ambitions (Gómez-Herrero 2001) to remold the very foundations of indigenous worship met many challenges. Friars faced outright resistance. They also faced the "resistance of culture", that is, the resistance asserted by the force of preexisting social and geographical structures (see Wernke 2007 for an analogous case in Peru). Epidemic disease decimated the indigenous population, which resulted in major population shifts (Hanks 2010, 32), just as the plagues had in the Old World (Christian 1989). The two societies faced one another with fundamental misunderstandings: concepts such as sin were alien to the indigenous peoples, while the Spanish could not understand practices such as sacrifice (Cervantes 1997; Gibson 1964). Linguistic translation was a theological minefield (Hanks 2010; Ricard 1966). The friars also faced very low ratios of monastics to indigenous populations (Hanks 2010, 41) which made friars' individual ideas and preferences more salient and bred variation (Graham 2011, 286). Then, too, there were philosophical differences between the orders themselves, between the orders and the secular priests, and squabbles between the Crown and the orders (Ricard 1966).

Most importantly, the friars faced the proactive engagement of indigenous subjects in Christianity, which fundamentally changed the nature of ritual and the nature of Christianity itself (Burkhart 1989, 1998; Cervantes 1997; Gibson 1964; Graham 2011; Lockhart 1992; Pardo 2006). As scholars have learned more about pre-Columbian religions, languages, and lifeways it has become increasingly clear that indigenous ways of being informed the foundations of colonial Mexican societies (Burkhart 1989, 1998; Gibson 1964; Lockhart 1992).

The violence and domination of the Church, the active contributions of indigenous subjects to Christianity, the reinterpretations and innovations within novel religious contexts: these were all realities of the early colonial evangelization program that existed simultaneously. We need a way out of established narratives that overstate the monolithic power of the Church (Schwaller 2011), as well as narratives that set up a false dominance/resistance binary. And we need new narratives that can honestly account for the complexities of the material record, which point to indigenous influence that was not the result of intentional resistance, peaceful cultural exchange, or an accident of circumstance. Rather, heterogeneous indigenous religious traditions informed and transformed Catholicism. I argue that this transformation was the direct result – intentional or not, coerced or not – of active indigenous participation in the early Church.

3 Narratives of Colonial Religious Change

Narrative, Frederic Jameson (2013, xiii) tells us, is "the central function or instance of the human mind"; stories are what our brains are built to do (Bruner 2004). Narrative allows us to make sense of mundane and extraordinary events, and through repetition, allows societies to collectively make sense of the past (Connerton 1989). This process takes place at levels ranging from the individual – the autobiographical story (Bruner 2004) – to the structural: the "national narrative," for example.

Studies from multiple disciplines reveal that our narrative repertoires are limited. Hayden White (2009) has argued that traditional historical narratives are constructed according to a limited set of archetypes that structure historical events into recognizable storylines ("progress," for example, or "the hero's journey"). Naming these limited narratives and providing empirical evidence that counteracts them has been a productive strategy for scholarly and popular texts. For example, Restall's (2004) "Seven Myths of the Spanish Conquest" identified and dismantled narratives of the early colonial past that continue to persist in the popular imagination. Similarly, Enrique Rodríguez-Alegría identified a problematic notion within recent popular media, "the narrative of quick replacement," that wrongly assumed that "superior" European tools and technologies quickly replaced indigenous technologies. In fact, evidence shows that the production of indigenous obsidian tools increased in the colonial era (Rodríguez-Alegría 2008). Material culture is an effective tool for challenging inaccurate, "commonsensical," and partial narratives about the past. However, if narrative's efficacy is constructed through repetition and by building order, it is not enough to test old narratives and reveal their gaps and inaccuracies. Ultimately, we have to construct new narratives.

Existing narratives of colonial religious conversion practices are based primarily on documents compiled by sixteenth- and seventeenth-century mendicants (e.g., de Alarcón 1987; de Mendieta 1870; Durán 1971; Sahagún 1950) as well as men who had participated in the conquest (Díaz 1963). Louise Burkhart (1989, 3) has succinctly summarized the tensions on the mendicants' side of the colonial process as "an odd mix of medieval theology, which insisted that all human souls were equal, Renaissance humanism, which suggested that something of worth might be found in another way of life, and Catholic intolerance, which justified – or excused – the study of pagan things on the grounds of facilitating their eradication." Colonial documents authored by indigenous peoples reveal some of the anguish suffered by those groups as they were forced to abandon their gods and rites (e.g., Klor de Alva 1980), as well as their participation in religious brotherhoods in the later colonial period (Lockhart 1992). Scholars have also examined indigenous religious agency through creative readings of Spanish-authored documents: court records from idolatry trials, for example (Tavárez 2011).

Based on Spanish and indigenous colonial documents, scholars have produced analyses, interpretations, and syntheses of the religious experience in colonial Mexico, out of which three persistent themes emerge (Graham 2011, 289–281 provides a more extensive list).

The invisible war narrative. Robert Ricard's (1966) scholarship was the foundation for this model, which interpreted the early colonial religious experience as a war that Spanish friars won unequivocally: "at least in the field of religion, therefore, a complete rupture occurred" (1966, 286). This perspective was rooted in the ways that many friars understood themselves: as soldiers combating the forces of the devil and idolatry (Cervantes 1997, Ephesians 6). Ricard's interpretation, therefore, depended upon adopting the yardstick that the friars themselves used to measure the "successes" of the spiritual conquest. David Tavárez (2011) summarized the problems with this theoretical stance: "take [the friars] at their word ... would be to adopt several troublesome assumptions: that the stakes in this war were evident and transparent to both sides, that native idolaters sought to present a united front against Christianity, and that this united front depended on an antipodal version of Christianity implanted by the Devil in the natives' less discerning minds." Ricard's scholarly legacy set up and naturalized a simplistic domination/resistance binary that even his opponents (as in the "core/veneer" model below) implicitly adopted, though they arrived at conclusions opposite to his.

Superficial conversion, or the "core-veneer" narrative. Against Ricard's inadequate model, scholars proposed a radical alternative: the "core-veneer" model, or the notion that indigenous peoples acceded to Christianity out of coercion, but in fact maintained their own pre-Columbian beliefs. Charles Gibson was an early proponent of this idea (Gibson 1964, 98–135); Jorge Klor de Alva (1982) later emphasized this model as the most common indigenous response to Christian conversion. However, active engagement with Christianity is a near impossibility within this model, despite the fact that indigenous children grew up in the Christian faith very shortly after the conquest and claimed to be Christians (Burkhart 1998, 362; Graham 2011, 290–291). As William Hanks (Hanks 2010, 8) has noted for the Maya area: "Maya engagement with Christianity was anything but superficial or short-lived, even if it was partial, contradictory, and put to uses never envisioned by the friars."

Some scholars have modified the core-veneer model to emphasize the "resistance of culture," or the idea that colonized peoples reinterpret new events and ideas in light of their own history and social logics (Burkhart 1989; Sahlins 2009; Wernke 2007). Indeed, while continuing indigenous traditions within Christianity cannot be seen as "survivals" (i.e., an authentically indigenous "core") many aspects of modern Mexican Catholicism would be unthinkable without the influence of pre-Columbian ontologies.

Syncretism. Mounting evidence of indigenous "idolatry", resistance, and specifically indigenous contributions to Mexican Catholicism, manifested in present-day Mexican Catholicism in material culture related to ritual, elements of the physical landscape, religious art, and myriad other traditions (Burkhart 1989; Pardo 2006; Tavárez 2011, 270), led some scholars to propose a model of syncretism (Andrews and Hassig 1984), or the blending of religious traditions. Inga Clendinnen (1990, 109) called syncretism "that familiar mix-and-match model" because it typically focuses on the outcome of the combination of two originally static cosmologies and bodies of material culture (see also Graham 2011, 72; Hanks 2010, 94). Much recent research from archaeology and ethnohistory has shown that Spanish and indigenous religious and material traditions were not internally coherent before the Spanish military conquest, that each tradition continued to evolve both internally and as a result of contact; and that material exchanges were not politically neutral: they took place within highly unequal power structures.

4 Case Study

The data that I discuss in this chapter come from my excavations and analyses in Tula, Hidalgo (Figure 12.1). Tula is most famous as the capital of the Toltec civilization, which flourished there between AD 900–1150. The city also had important Late Aztec-era (AD 1350–1521) and colonial-era (AD 1521–1810) occupations, and it remains an important industrial center today. In 2013, I directed a project at two different archaeological sites in Tula: an Open Chapel constructed in AD 1530, and a monastery constructed in AD 1550 that continues to serve as Tula's cathedral and as the headquarters of the Diocese of Tula (Figure 12.2).

Because of the standing architecture and intact deposits from the earliest stages of the Christian evangelization program, Tula offers a unique opportunity to study the material culture of colonial conversion efforts in close diachronic comparison in order to understand how the evolving authority of the Church impacted material culture. In addition, it affords the opportunity to compare documentary-based accounts of sixteenth-century conversion processes against material evidence of those processes. I have chosen to highlight two bodies of material culture from my project, buildings and ceramics, because taken individually and superficially they appear to tell very different

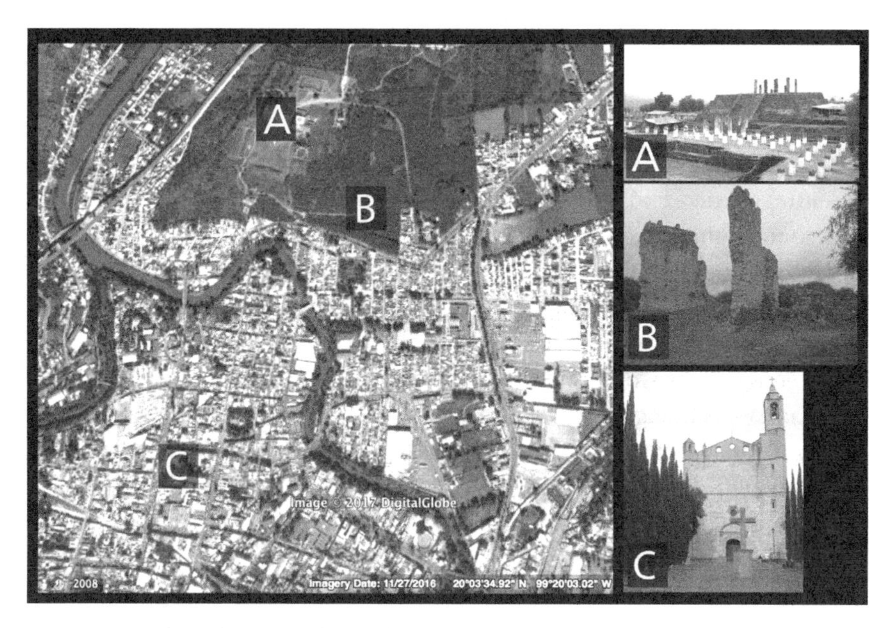

FIGURE 12.2 Relative locations of (A) Tula Grande, the Toltec ceremonial center; (B) the Open Chapel in 2008, before Carol Vazquez's restorations; (C) The Cathedral of San José
MAP ADAPTED FROM DIGITALGLOBE DATA, COPYRIGHT 2017

stories about the colonial evangelization program. Archaeologists and other scholars also tend to use these categories of material culture in very different ways: the built environment is frequently studied as a site of ideology and resistance (Hutson 2002; Solari 2013), while ceramics help us to understand ethnic and class negotiations, lived experience, and trade dynamics (Deagan 1987; Rodríguez-Alegría 2005).

5 Architecture

In the Late Aztec period (AD 1350–1519), prior to the arrival of Spanish mendicants, Aztec royals ruled over a majority-Otomí population in Tula. We have very little information regarding religious activities there, but we can establish that Aztec rituals took place at the by-then-ancient Toltec ceremonial center of Tula Grande (Acosta 1956; Iverson 2017). Within Tula Grande, they made offerings and practiced rituals (such as a New Fire ceremony) that are consistent with ritual activities at other major centers, such as Templo Mayor in the Aztec capital city of Tenochtitlan (Acosta 1956; Elson and Smith 2001; López Austin and López Luján 2009). They also added an altar to the Pyramid C, one of Tula Grande's major temples (Acosta 1956, 109). The Toltec ceremonial center is, to date, the only known locus of Aztec-era ceremonial activities in Tula (Iverson 2015, 2017).

The Franciscan mendicant Friar Alonso Rengel and indigenous builders constructed the Open Chapel, the first Christian building in Tula, around AD 1530 (Ballesteros García 2003), only nine years after the Spanish military conquest of central Mexico, and only six years after the first Franciscan friars arrived in Mexico to begin a mass evangelization program in earnest. Open chapels, a form of religious architecture unique to the New World, are widely thought to be a compromise between Spanish and indigenous forms of worship (Edgerton 2001). The three walls of the Open Chapel in Tula opened to a large patio, where congregants would gather for religious festivals and to hear mass (see Figure 12.2) – echoing practices in the pre-Columbian era. The increased theatricality, ceremonialism, and pageantry of religious rites in this period is also widely thought to be an indigenous contribution to Mexican Catholicism that open chapels facilitated (Burkhart 1998; Clendinnen 1990; Córdova Tello 1992; Edgerton 2001). Another important indication of the accommodative nature of evangelism in this period at Tula is that the Chapel was built within sight of, but did not destroy or displace, the pre-Columbian Toltec ceremonial center.

My colleague Carol Vázquez has worked to excavate, consolidate, and restore the Open Chapel using an approach known as arqueotectura, or the archaeology of architecture (Vázquez Cibrián 2013). Her excavations revealed that the chapel was constructed in three phases, in part revealing the exigencies of the period. My excavations at the Open Chapel in 2013 complemented Vázquez's work by seeking to understand the longer constructive sequence at the chapel: two of my excavation units at the Open Chapel site were situated to understand the Chapel's articulation with earlier Aztec- and Tollan-phase occupations. In both units, we observed that the Open Chapel was built on top of pre-Hispanic architectural features. The sacristy or living quarters[1] of the Open Chapel was built directly on top of a pre-Hispanic wall that was likely an Aztec construction (see Figure 12.3). We also discovered that the foundations of the northern wall the Open Chapel rested very close to was a Tollan-phase structure that featured adobe floors and a pillar (see Figure 12.3).

The Toltec building that we encountered in Operation 1 (Figure 12.3) at the Open Chapel was not a temple. Its wide adobe floors and circular pillar echo, instead, Toltec constructions that are popularly known as "palaces," such as the Palacio Quemado in Tula Grande, more likely served as elite meeting houses (Healan 2012, 101). Likewise, we did not find evidence to show that the walls and floors that we encountered in Operation 3 (Figure 12.3) served an institutional religious function. It was most likely a residence or civic building.

The early use of the Open Chapel in Tula corresponds with the broader colonial religious history of the region, in which mendicants were focused on gaining footholds in the urban areas of New Spain, learning the local languages, and eradicating the vestiges of institutional pre-Columbian religions: destroying the "pagan" temples was a major priority. Some scholars consider open chapel constructions to have served primarily as expedient preludes to the larger, more permanent, and more formal religious architecture that began to be built in the second and third decade after the conquest (Kubler 2012).

Indeed, in 1550, Friar Motolinía (the regional provincial, or regional religious authority) issued an order that a new monastery be built in Tula (Ballestros García 2003, 128). This building, the Cathedral of San José (Figure 12.2), is a large, enclosed structure that today serves as the regional headquarters of the Diocese of Tula. The fortress-like appearance of the Cathedral was not coincidental, but rather a deliberate symbolic reference to spiritual warfare based on the writings of Saint Paul (Ballesteros García 2003, 131; Ephesians 6:10–20). The Cathedral of San José likely slightly displaced the local population at Tula, as it shifted the town center approximately one kilometer to the south (Figure 12.2), much further away from the Toltec center of Tula Grande.

1 The small rooms that abutted Open Chapels could be used as sacristies, living quarters for itinerant priests, or both (Solari 2013, 15).

FIGURE 12.3 (Clockwise from top left) Operations 1 and 3 at the Open Chapel and Operations 6 and 7 at the Cathedral of San José. In the foreground of Operation 1, several courses of adobe, stucco, and stone foundations – likely Toltec-era – are visible. The wall in the background of the picture is the 9.4 m-tall northern wall of the Open Chapel. At Operation 3, the wall of the sacristy is to the east (right), and can be seen as having been constructed directly over two intersecting walls of pre-Hispanic (likely Aztec) origin. In Operations 6 and 7, the Toltec "box" technique for platform construction is visible.

Was the abandonment of the Open Chapel and the construction of the Cathedral a straightforward symbol of the successes of the spiritual war?

Our excavations revealed a more complicated answer. First, we found that the Cathedral was constructed on what appeared (based on ceramics and construction techniques) to have been a large Toltec-era platform. This platform may have supported temples or other superstructures in the past, though evidence from my excavations was inconclusive. Second, we found the foundations of a previously unknown structure in the atrium of the Cathedral that probably served as an open chapel or for some other outdoor religious purpose during the colonial era. This indicated that the stylistic and ritual compromises that resulted from indigenous religious prerogatives (that is, emphasis on outdoor worship) had not disappeared once the Church consolidated its power in later decades. Instead, those changes were built into the fabric of New World Christianity from the earliest years, and continued to influence architecture even when the Church was more established and had less need to acquiesce to indigenous influence.

Questions regarding the relationships between indigenous and colonial buildings are not trivial. It is often taken as a given, particularly in colonial studies, that the act of placing important Christian monuments directly on top of the ruins of the religious monuments of conquered indigenous cultures is a transparent act of ideological warfare (Low 1995, 749 presents a concise summary of these assumptions). But buildings and landscapes are inherently multivalent; their presence and use holds different meanings for actors in different social positions (e.g. Hutson 2002, 58–60) and the meanings of the built environment change over time (Meskell 2003, 50). Even if the destruction of "idolatrous" buildings and their replacement with Christian monuments was part and parcel of religious imperialism, multiple indigenous and European understandings of the same spaces meant that the semantic significance of colonial religious buildings was never stable. As Elizabeth Graham has noted for the Maya area, "...it is unlikely that places believed by the Maya to have accumulated power would have lost their force. New spirits or supernaturals were likely to have become associated with traditional places of power..." (Graham 2011, 288).

In the Tula case notions of "ideological warfare" become even muddier because of the layered occupational sequence. On one hand, both the Open Chapel and the Cathedral were built on top of early Toltec remains that had been modified in the Aztec era, and the Open Chapel incorporated Toltec-era building materials into its fabric (Vázquez Cibrián 2013, 178). And, as many scholars have pointed out in similar cases, both buildings likely took advantage of existing religious connotations of the existing built environment and the surrounding landscape. On the other hand, however, the Open Chapel was built on top of architecture that was civic rather than religious, and neither building resulted in the destruction of the only known locus of Aztec-era ceremonial practice, the ancient center at Tula Grande, as we might expect from an evangelization program that was attempting to eradicate idolatry.

The complexities of the Tula architectural case make it hard to argue for a straightforward narrative of increasing Church authority (the top-down, "spiritual warfare" narrative of Church power). Yet both buildings represent indisputable European interventions in the Mesoamerican landscape that make it difficult to argue that indigenous subjects in Tula simply retained their core spiritual beliefs while participating in Church rites.

6 Ceramics

The architectural evidence at the Open Chapel and the Cathedral of San José document the nearly immediate intervention of Spanish friars in the built landscapes of Mesoamerica. In contrast, the ceramic assemblages at both sites are remarkable because, superficially, they register almost complete continuity with indigenous pre-Columbian ceramic traditions.

The collections from my sites at Tula are made up of three major ceramic groups: the Toltec ceramic family, the Aztec-tradition family, and colonial ceramics comprised of European and Asian imports and their New World iterations. Toltec wares have a thick, slightly coarse, light-tan colored paste; this tradition has "little or no continuity" with the thin, hard orange pastes of the later Aztec tradition (Healan 2012, 94). The Aztec tradition is further subdivided into three major wares based on surface decoration: Plain Orange, Black-on-Orange (the ware that is still most useful for chronometric dating), and Red Wares (Minc 2017; Parsons 1966).

All three Aztec wares continued to be manufactured and used well into the colonial period, with some significant changes. First, indigenous potters developed Aztec IV, a popular Black-on-Orange decorated ware, which sometimes featured naturalistic motifs (such as birds and leaves). Second, Red Wares became much more popular, while Black-on-Orange styles began to wane. Third, indigenous potters added translucent, shiny lead glazes to thin orange vessels. However, though lead glaze was a European introduction, potters in indigenous workshops creatively innovated by using this finishing technique on existing indigenous-tradition wares (Hernández Sánchez 2011, 111–112); and thus cannot be said to be properly "European tradition." Imported colonial ceramics, such as majolica and porcelain, used materials and techniques (such tin glazes) that were not employed in the Americas until the advent of Spanish colonialism. Majolicas were soon manufactured in the New World as well – chiefly in workshops in Puebla and Mexico City (Deagan 1987).

Many scholars have posited that Spanish colonists would have preferred to maintain a separation between their own material culture and that of indigenous subjects (Rodríguez-Alegría 2005 provides a summary of this assumption). Based on empirical evidence that there was no such separation, other scholars have proposed that friars would have rejected expensive imports and "elite" ceramics, such as majolica, and instead adopted indigenous ceramics because of their vows of poverty (Charlton and Fournier 1993). Rarely, however, are the changes and continuities in ceramic patterns in religious spaces attributed to indigenous preferences.

Ceramics were not neutral objects from a religious perspective: so-called mundane objects, including ceramics, also tied into a broader concern with the sacred. Everyday serving vessels were used as offerings in burials, the New Fire Ceremony, and termination rituals (Elson and Smith 2001; Iverson 2017). Durán wrote of pre-Columbian offerings of "little bowls" to the Aztec priests (Durán 1971). Friars knew of the sacred meanings of everyday objects in celebratory contexts, but were able to convince themselves that Nahua religious customs that immediately seeped into Christianity were actually markers of true faith and enthusiastic conversion: a part of what Louise Burkhart (Burkhart 1998, 368) has called the friars' "ontological sleight-of-hand." For example, Friar Diego Durán wrote "it must be noted that the offerings of strings of ears of corn and flowers on the Day of Our Lady in September and during the festivities in that month are a survival of the [pagan] custom. But I believe they have been turned into an offering to His Divine Majesty" (Durán 1971, 228). It is important to note that the vessels themselves, the foods that they were used to serve, and the ceremonial contexts of which they formed a part had residual meanings that carried over from pre-Columbian celebrations: they were always multivalent. These meanings were not "survivals" – that is, the meaning of material culture was radically altered in the Christian context. On the other hand, traces of earlier sacred connotations of quotidian objects made the friars' vision of a "tabula rasa" in the New World completely untenable.

My own ceramics collections revealed some surprising findings. First, I found that it was nearly impossible to distinguish pre-Columbian and colonial contexts on the basis of ceramics alone; I had to rely on a combination of ceramic types, architectural context, and other materials, such as faunal remains (Iverson, 2015, 236). We also found a complete absence of prized imported European ceramics at both sites – this was particularly unexpected because we did find imports at non-religious sites in Tula. That is, the absence was a choice, not a question of access. I found that the entire assemblage of over 48,000 sherds from both sites contained only 23 sherds of New World majolica types, several of which would not have entered the archaeological record until the late colonial or early republican period – that is, the eighteenth or nineteenth centuries (Seifert 1977).

Then, I narrowed my analysis to include only early colonial contexts at both sites, excluding earlier and later periods. To my surprise, early colonial contexts contained no examples of majolica or European imports whatsoever. Instead, I found that the overwhelming majority of ceramics were of indigenous tradition. The sole contribution of European-tradition ceramic production was glazing technology, which appealed to indigenous potters and very gradually became more popular. However, most colonial ceramic change evident at my sites occurred within the indigenous tradition rather than as the result of imports or imitation. For example, Aztec Black-on-Orange ceramics acquired new motifs and gradually waned in popularity, while Aztec Red Wares increased in popularity (Figure 12.4). Finally, we also found several examples of Aztec-tradition censers – explicitly religious artifacts – in colonial contexts at the Open Chapel (Figure 12.5). In a plural society that included Spaniards, castas, Africans, and a majority of Otomí and Nahua-speaking peoples (Ballesteros García 2003), these patterns are significant: they mean that ceramics used in religious contexts were evolving according to indigenous ideas about the proper vessels for religious feasting and ceremony, despite access to European imports and majolica.

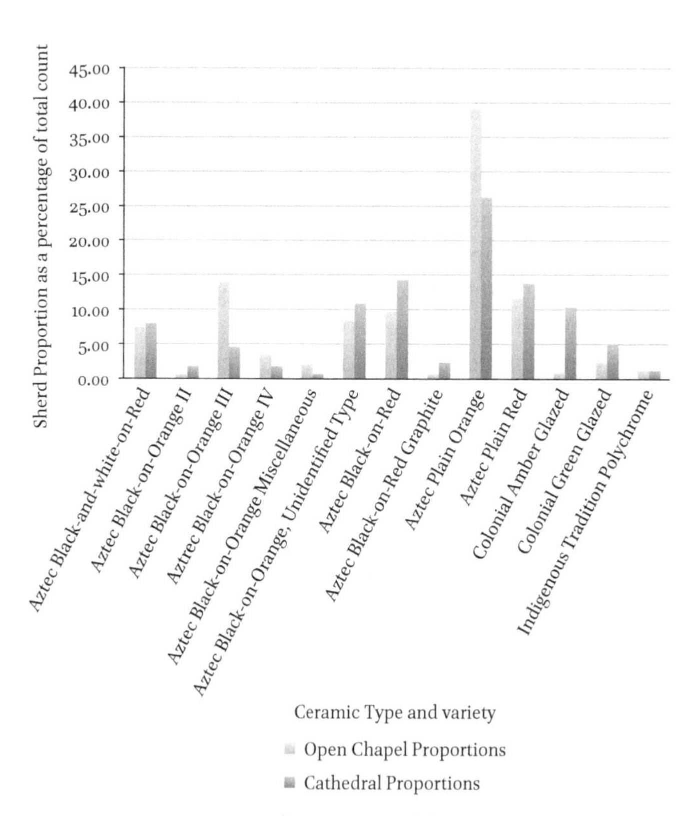

FIGURE 12.4 A comparison of proportions of diagnostic Aztec-tradition and colonial ceramic serving ware sherds encountered in colonial-era contexts at the Open Chapel (n=438) and the Cathedral of San José (n=176) in Tula, Hidalgo. Note that Black-on-Orange types III and IV decrease over time, while four Red Ware types increase in popularity at the later Cathedral location.

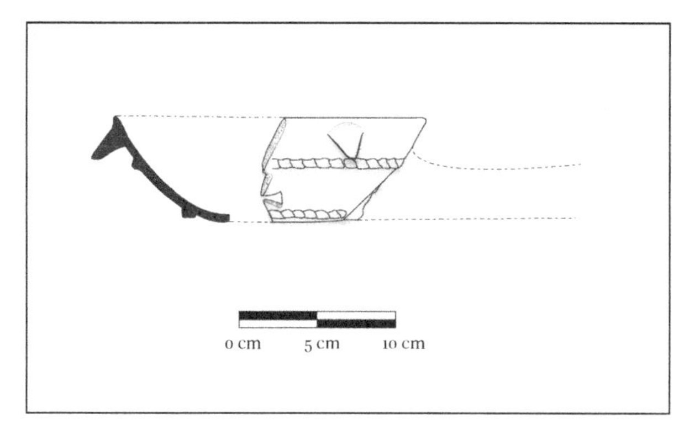

FIGURE 12.5 Aztec-tradition Texcoco Molded-Filleted style *sahumador* (censer) from a colonial-era context at the Open Chapel.
DIGITAL ILLUSTRATION BY SHANNON DUGAN IVERSON BASED ON A DRAWING BY DANIEL CORREA

How do the ceramics findings compare to the architectural evidence at Tula? Superficially, the ceramics seem to tell the opposite story: ceramics were by far the most conservative material category, conforming to indigenous tastes over the course of several centuries even as indigenous households outside of the religious centers adopted European wares. The ceramic patterns certainly do not point to top-down European influence. Nor do they point to a lack of genuine engagement in the Church: the evidence suggests indigenous subjects celebrated and feasted at both sites. A passive blending of cultural elements is also an inappropriate explanation for these patterns: indigenous subjects innovated within their own ceramic traditions, by and large rejecting the European and Asian imports that they adopted in other contexts. However, it is crucial to remember that these vessels were being used in the service of Christian celebrations and feasts, which was a radically altered context regardless of how much indigenous subjects influenced those ceremonies.

7 Conclusion: a Fourth Narrative

Because of the overt and subtle violence and coercion of the evangelization program, the bulk of secondary historical literature characterizes conversion as a top-down process (the "spiritual warfare" narrative) that could be either resisted or passively accepted by indigenous peoples (the "core-veneer" narrative), but only rarely as a process in which they actively participated. The "syncretism" model of religious change does acknowledge the participation of indigenous agents, but it fails to account for the uneven distribution of power inherent to the colonial Church and the broader colonial enterprise.

Still, in spite of and because of the violence of the Spanish evangelization campaigns, indigenous peoples became Christian – and even in a situation of severe constraint, they made the Church their own. Their authentic, diverse engagements with that religion formed the single greatest challenge to the friars' Utopian ambitions, and created a fundamentally new Christianity in the process.

I have found Judith Butler's concept of resignification to be the most powerful framework for that pattern. Briefly, Butler's work allows us to see that concepts such as 'woman' are in fact fictions without a single material example in reality; they are constituted only through performance. Real human agents appropriate, negotiate, and reject these categories according to their particular interests and social positions, but it is never possible to assume the ideal form in reality. Change happens precisely because the real-world iterations of the categories are never perfect and always becoming; in the iterative process, they destabilize the meanings of the categories themselves.

So it was, I argue, for ideal concepts in the colonial world. "Church" as ideal category could not help but take on new significance when its material form was built with indigenous labor, using stones quarried from indigenous temples, on landscapes that had preexisting supernatural connotations. On the other hand, indigenous-tradition ceramics used in religious celebrations and feasts acquired new significance within the Christian sphere. Neither the colonizers nor the colonized ever intended colonial New World Christianity to be something distinct from the European forms of Christianity that preceded it. (Moreover, a single "European form" of Christianity never existed.) Even so, indigenous peoples exerted a powerful influence, intentional or not, on colonial structures as they negotiated the material reality of Christianity. In the process, Christianity itself changed, and its meanings destabilized.

This semantic and discursive destabilization is not infinite, however; Butler has emphasized that past semantic contexts continue to exist alongside new meanings as residues and traces (Olson and Worsham 2000, 737–737). Nor do resignifications necessarily achieve an "unmooring" of established regimes: even intentional resignifications, such as the deliberate reappropriation of the term "queer," can reinscribe particular configurations of power, or take on new and unintended meanings (Olson and Worsham 2000, 737). While indigenous contributions to Christianity fundamentally changed and challenged the Church, they did not radically alter colonial configurations of power. But they did produce multiple, conflicting ontologies that were never fully resolved, and new iterations of Christianity that privileged indigenous prerogatives.

A "resignification" narrative of colonial religious change does not deny or overstate the powerful force exerted by colonial institutions, including the Church. But, unlike syncretic and "core-veneer" models, it has the capacity to explain the participation and influence of indigenous actors: rather than assuming covert resistance or a simple exchange of ideas and material culture, it presumes that indigenous subjects' very participation in the Church altered its meanings and steadily shifted its values toward their own prerogatives. In doing so, the model acknowledges that despite the unequal distribution of power in the early colonial period, Spanish colonial institutions were deeply dependent on preexisting indigenous structures, and deeply indebted to the active contributions of indigenous agents.

Acknowledgments

This research was funded by a National Science Foundation grant (NSF DDIG #1156359), the University of Texas, and a Peyton and Douglas Wright Memorial Grant. Many thanks to INAH, the Diocese of Tula, Enrique Rodríguez-Alegría, Robert Cobean, Luis Gamboa, Heath Anderson, Rocío Cara Labrador, and my excavation and analysis crew: Abraham Leura, Pedro Rodríguez, Clara Margarita Serrano, Ana Suárez, Margarita Hernández, and Maria Elena Suárez.

References

Acosta, Jorge R. 1956. "Resumen de Los Informes de Las Exploraciones Arqueológicas En Tula, Hgo., Durante Las VI, VII Y VIII Temporadas, 1946–1950." *Anales Del Instituto Nacional de Antropología E Historia* 6 (8): 37–115.

Alarcón, Hernando Ruiz de. 1987. *Treatise on the Heathen Superstitions That Today Live among the Indians Native to This New Spain, 1629. Vol. 164.* Norman: University of Oklahoma Press.

Andrews, J. Richard and Ross Hassig. 1984. *Treatise on the Heathen Superstitions That Today Live among the Indians Native to This New Spain, 1629 by Hernando Ruiz de Alarcón.* Norman: University of Oklahoma Press.

Ballesteros García, Victor. 2003. "San José de Tula: Enclave Franciscano En La Cuidad de Quetzalcoatl." In *Tula: Más Allá de La Zona Arqueológica*, edited by Laura Elena Sotelo Santos, 127–135. Pachuca: UAEH.

Berdan, Frances F. 2014. *Aztec Archaeology and Ethnohistory.* New York: Cambridge University Press.

Brumfiel, Elizabeth M. 2001. "Aztec Hearts and Minds: Religion and the State in the Aztec Empire." In *Empires: Perspectives from Archaeology and History*, edited by Kathleen D. Morrison, Susan E. Alcock, Carla M. Sinopoli, and Terence N D'Altroy, 283–310. New York: Cambridge University Press.

Bruner, Jerome. 2004. "Life as Narrative." *Social Research* 71 (3): 691–710.

Burkhart, Louise M. 1989. *The Slippery Earth: Nahua-Christian Moral Dialogue in Sixteenth-Century Mexico*. Tucson: University of Arizona Press.

Burkhart, Louise M. 1998. "Pious Performances: Christian Pageantry and Native Identity in Early Colonial Mexico." In *Native Traditions in the Postconquest World. A Symposium at Dubarton Oaks, 2nd through 4th October 1992*, edited by Elizabeth Hill Boone and Tom Cummins, 361–381. Washington, D.C.: Dumbarton Oaks Research Library and Collection.

Butler, Judith. 1990. "Gender Trouble, Feminist Theory, and Psychoanalytic Discourse." In *Gender Trouble: Feminism and the subversion of identity*, edited by Butler Judith, 324–341. London and New York: Routledge.

Cervantes, Fernando. 1997. *The Devil in the New World: The Impact of Diabolism in New Spain.* New Haven: Yale University Press.

Charlton, Thomas H., and Patricia Fournier. 1993. "Urban and Rural Dimensions of the Contact Period." In *Ethnohistory and Archaeology. Approaches to Postcontact Change in the Americas*, edited by J. Daniel Rogers and Samuel M. Wilson, 201–220. New York: Springer.

Christian, William A. 1989. *Local Religion in Sixteenth-Century Spain.* Princeton: Princeton University Press.

Clendinnen, Inga. 1990. "Ways to the Sacred: Reconstructing 'Religion' in Sixteenth Century Mexico." *History and Anthropology* 5 (1): 105–141.

Connerton, Paul. 1989. *How Societies Remember*. Cambridge University Press.

Córdova Tello, Mario. 1992. *El Convento de San Miguel de Huejotzingo, Puebla*. Arqueología Histórica. Mexico City: Instituto Nacional de Antropología e Historia.

Deagan, Kathleen. 1987. *Artifacts of the Spanish Colonies of Florida and the Caribbean, 1500–1800: Volume 1-Ceramics, Glassware and Beads.* Washington, D.C.: Smithsonian Institution.

Díaz, Bernal. 1963. *The Conquest of New Spain.* Translated by John M. Cohen. London: Penguin Books.

Durán, Fray Diego. 1971. *Book of the Gods and Rites and the Ancient Calendar*, edited by Fernando Horcasitas and Doris Heyden. Norman: University of Oklahoma Press.

Edgerton, Samuel Y. 2001. *Theaters of Conversion: Religious Architecture and Indian Artisans in Colonial Mexico*. Albuquerque: University of New Mexico Press.

Elson, Christina M. and Michael E. Smith. 2001. "Archaeological Deposits from the Aztec New Fire Ceremony." *Ancient Mesoamerica* 12 (2): 157–174.

Gibson, Charles. 1964. *The Aztecs Under Spanish Rule: A History of the Indians of the Valley of Mexico, 1519–1810.* Stanford: Stanford University Press.

Gómez-Herrero, Fernando. 2001. *Good Places and Non-Places in Colonial Mexico: The Figure of Vasco de Quiroga (1470–1565).* Lanham: University Press of America.

Graham, Elizabeth. 2011. *Maya Christians and Their Churches in Sixteenth-Century Belize.* Gainesville: University Press of Florida.

Hanks, William F. 2010. *Converting Words: Maya in the Age of the Cross.* Berkeley: University of California Press.

Healan, Dan M. 2012. "The Archaeology of Tula, Hidalgo, Mexico." *Journal of Archaeological Research* 20 (1): 53–115.

Hernández Sánchez, Gilda. 2011. *Ceramics and the Spanish Conquest: Response and Continuity of Indigenous Pottery Technology in Central Mexico.* Boston: Brill.

Hutson, Scott R. 2002. "Built Space and Bad Subjects Domination and Resistance at Monte Albán, Oaxaca, Mexico." *Journal of Social Archaeology* 2 (1): 53–80.

Iverson, Shannon Dugan. 2015. "The Material Culture of Religion and Ritual: An Investigation of Social Change in the Aztec-to-Colonial Transition at Tula, Hidalgo." PhD diss., University of Texas at Austin.

Iverson, Shannon Dugan. 2017. "The Enduring Toltecs: History and Truth during the Aztec- to-Colonial Transition at Tula, Hidalgo." *Journal of Archaeological Method and Theory* 24 (1) 90–116.

Jameson, Fredric. 2013. *The Political Unconscious: Narrative as a Socially Symbolic Act.* London: Routledge.

Klor de Alva, J. Jorge. 1980. "The Aztec-Spanish Dialogues of 1524." *Alcheringa* 4 (2): 52–193.

Klor de Alva, J. Jorge. 1982. "Spiritual Conflict and Accommodation in New Spain: Toward a Typology of Aztec Responses to Christianity." In *The Inca and Aztec States 1400 (1800): Anthropology and History (Studies in Anthropology)*, edited by George Collier, Renato I. Rosaldo, and John D. Wirth, 345–366. Cambridge: Academic Press.

Kubler, George. 2012. *Arquitectura Mexicana Del Siglo XVI.* San Diego: Fondo de Cultura Economica USA.

Lockhart, James. 1992. *The Nahuas after the Conquest: A Social and Cultural History of the Indians of Central Mexico, Sixteenth through Eighteenth Centuries.* Stanford: Stanford University Press.

López Austin, Alfredo, and Leonardo López Luján. 2000. "The Myth and Reality of Zuyuá: The Feathered Serpent and Mesoamerican Transformations from the Classic to the Postclassic." In *Mesoamerica's Classic Heritage: From Teotihuacan to the Aztecs*, edited by David Carrasco, Lindsay Jones, and Scott Sessions, 21–84. Boulder: University Press of Colorado.

López Austin, Alfredo, and Leonardo López Luján. 2009. *Monte Sagrado-Templo Mayor.* Mexico: INAH.

Low, Setha M. 1995. "Indigenous Architecture and the Spanish American Plaza in Mesoamerica and the Caribbean." *American Anthropologist* 97 (4): 748–762.

Mendieta, Gerónimo de. 1870. *Historia Eclesiástica Indiana. Vol. 3.* Antigua librería [Printed for F. Diaz de Leon and S. White].

Meskell, Lynn. 2003. "Memory's Materiality: Ancestral Presence, Commemorative Practice and Disjunctive Locales." In *Archaeologies of Memory,* edited by Ruth M. Van Dyke and Susan E. Alcock. Wiley Online Library.

Minc, Leah D. 2017. "Pottery and the Potter's Craft in the Aztec Heartland." In *The Oxford Handbook of the Aztecs,* edited by Deborah L. Nichols and Enrique Rodríguez-Alegría, 355–374. New York: Oxford University Press.

Olson, Gary A., and Lynn Worsham. 2000. "Changing the Subject: Judith Butler's Politics of Radical Resignification." *JAC* 20 (4): 727–765.

Pardo, Osvaldo F. 2006. *The Origins of Mexican Catholicism: Nahua Rituals and Christian Sacraments in Sixteenth-Century Mexico.* Ann Arbor: University of Michigan Press.

Parsons, Jeffrey Robinson. 1966. "The Aztec Ceramic Sequence in the Teotihuacan Valley, Mexico." Unpublished PhD diss., University of Michigan.

Restall, Matthew. 2004. *Seven Myths of the Spanish Conquest.* Oxford University Press.

Ricard, Robert. 1966. *The Spiritual Conquest of México.* Translated by Leslie Byrd Simpson. Berkeley: University of California Press.

Rodríguez-Alegría, Enrique. 2005. "Eating like an Indian." *Current Anthropology* 46 (4): 551–573.

Rodríguez-Alegría, Enrique. 2008. "Narratives of Conquest, Colonialism, and Cutting-Edge Technology." *American Anthropologist* 110 (1): 33–43.

Sahagún, Bernardino de. 1950. *General History of the Things of New Spain: Florentine Codex.* Translated by Arthur J.O. Anderson and Charles E. Dibble. Santa Fe: School of American Research.

Sahlins, Marshall D. 2009. *Historical Metaphors and Mythical Realities: Structure in the Early History of the Sandwich Islands Kingdom.* Ann Arbor: University of Michigan Press.

Schwaller, John Frederick. 2011. *The History of the Catholic Church in Latin America: From Conquest to Revolution and beyond.* New York: New York University Press.

Seifert, Donna J. 1977. "Archaeological Majolicas of the Rural Teotihuacán Valley, Mexico." PhD diss., University of Iowa.

Solari, Amara. 2013. *Maya Ideologies of the Sacred: The Transfiguration of Space in Colonial Yucatan.* Austin: University of Texas Press.

Tavárez, David. 2011. *The Invisible War: Indigenous Devotions, Discipline, and Dissent in Colonial Mexico.* Stanford: Stanford University Press.

Vázquez Cibrián, Carol. 2013. "La Edificación de Una Historia: Arqueoarquitectónico de La Capilla Abierta de Tula." PhD diss., Escuela Nacional de Antropología e historia, Mexico.

Wernke, Steven A. 2007. "Analogy or Erasure? Dialectics of Religious Transformation in the Early Doctrinas of the Colca Valley, Peru." *International Journal of Historical Archaeology* 11 (2): 152–182.

White, Hayden. 2009. *The Content of the Form: Narrative Discourse and Historical Representation.* Baltimore: Johns Hopkins University Press.

CHAPTER 13

Indigenous Pottery Technology of Central Mexico during Early Colonial Times

Gilda Hernández Sánchez

The Spanish colonization dramatically interrupted the autonomous development of ancient Mesoamerican culture. Nevertheless, indigenous societies learned to live with the conquest. It was not only a time of crisis, but also an extraordinary creative period. The complex interaction between the indigenous and European worlds gave way to new social systems, technologies and artistic expressions. In this process, material culture played a central role. Things provoked rather than just reflected people's particular responses and adaptations to the changing circumstances. After the Spanish conquest, for example, the encounter of Mesoamerican potters with European ceramics profoundly impacted the native pottery technology. Potters faced foreign ceramics and decided to adopt, reinterpret or reject them. This work presents insights into that process of transformation by focusing on the interaction of indigenous potters with the Spanish pottery in central Mexico during the early colonial period (AD 1521–1650). In that region, on the eve of the conquest, potters made a wide variety of objects, with many techniques and in many styles, in which dexterity, creativity and aesthetics played important roles. The Spaniards introduced new wares and new technologies to produce them. Emblematic were the potter's wheel, the glazing and the majolica ware. They were typical of the Spanish pottery technology at that time and implied a different understanding of ceramics. As will be shown, potters interacted with these novelties in different and complex ways. The pre-Hispanic ceramic technology persisted after the conquest, but the various dimensions of ceramic-making were differently impacted by such particular encounters. Clay recipes, method of forming and firing technology were maintained without change. In contrast, surface finishing and decoration evidenced great creativity.

At the time of the Spanish arrival, the central Mexican ceramic industry was flourishing. Late pre-Hispanic (AD 900–1521) archaeological remains suggest that potters from the various regions of central Mexico used similar technology for manufacturing ceramics. In general terms, they worked in family workshops and had comparable methods for forming and firing, and created vessels using similar stylistic and formal canons. Although there was a lot of local

DOI:10.1163/9789004273689_014

variation in decorative patterns, the majority showed the distinctive trends of that time in Mesoamerica, such as the use of painting to decorate vessels or the addition of three supports to bowls and plates. Here, we will explore why and how potters adopted, modified or rejected three attributes of Spanish ceramics not known in Mesoamerica before: potter's wheel, glazing and majolica ware.

The analysis is based on the integration of previous research on ceramics from central Mexico, as well as on the consulting of several archaeological collections of early colonial ceramics from many contexts in that region, in particular from the Valley of Mexico, where most research on early colonial contexts have been conducted. These collections are deposited in the *Departamento de Colecciones Comparativas* at INAH in Mexico City. These materials consist of selected samples of diagnostic ceramics found in colonial locations throughout the city. The size and variety of the samples did not respond to any statistical principle and do not provide clues on the total amount and variety of the ceramics excavated in every location. Nevertheless, those collections represent a wide and varied sample of the pottery made and used in the Valley of Mexico during the colonial period. In addition, ethnoarchaeological research was conducted in several pottery towns of central Mexico, in which part of the methods of manufacture and organization of production are still intimately attached to the Mesoamerican world (see Hernández Sánchez 2012).

The focus of study is the early colonial period, following James Lockhart's period division for central Mexico (1992, 427), from 1521 to 1640–50, during which – despite the conquest – little changed in the Nahua communities of central Mexico. At that time, the Spanish political institution *cabildo* was introduced. It concerned the political organization of Spanish settlements and indigenous communities, the monasteries and the labor draft called *repartimiento*. According to Lockhart (1992, 427), these institutions encouraged many Spanish elements to pervade indigenous communities but with limitations, thus little changed in the indigenous framework. In that context native potters encountered the Spanish potter's wheel, the glazing technique and the majolica ware.

1 Encounters with the Potter's Wheel

There are different techniques for applying pressure to plastic clay to form vessels. Potters may use only one method, such as casting a vessel in a mold or throwing it on a potter's wheel, or they may combine various methods, for example, making part of the vessels by molding and the rest by applying coils of plastic clay. In some forming techniques, potters perform several operations at different stages of the plastic range of the clay (Rye 1981, 21). That is, a section of a vessel may be dried before the rest is made in order to avoid deformation, or the lower walls of vessels made by coiling require the use of softer clay. Often, when the water content of the clay drops below the minimum for the plastic range, handles or other additions are applied. Thus, this part of the process of ceramic manufacture involves not only motor habits mastered by frequent repetition, but also detailed knowledge of the sequence of execution. Both types of expertise are learned by potters and transmitted across generations. As Olivier Gosselain explains based on ethnographic research (2000), these types of expertise are very resistant to change because they are internalized, not visible, and similar vessels can be shaped using other forming methods. The techniques used to form ancient ceramics can be identified from vessels remains alone, as they often leave clear marks on the finished products. However, these marks are often covered by later processes of surface finishing like painting or glazing.

In central Mexico, late pre-Hispanic potters made vessels using a combination of molding and coiling techniques. Molding means to press the clay into or over a mold, and coiling is the use of rolls or "coils" to build up a vessel around a circumference in order to increase the height (Rye 1981, 76, 81). In the Valley of Mexico, for example, molds to form the body and the neck of vessels were horizontal (see Charlton et al. 1992, 106–107; Hernández et al. 1999, 77), that is, they served to create horizontal sections of the vessel. It seems that in that region potters used similar techniques for vessel forming. Differences in this process were related to vessel shape (*ollas* were made out of more sections than bowls), rather than to vessel use (cooking pots were made in similar way to serving pots) (Figure 13.1). These techniques, in particular the use of molds, promoted standardization in shape and size. The fact that the methods for forming vessels were, in general terms, similar and stable suggests that this kind of knowledge was transmitted without disruption across generations.

We still do not know exactly when the Spanish ceramic technology arrived to Mexico, as this industry is scarcely mentioned in early colonial documentation, and ceramic remains do not offer fine chronological details. It seems that after the conquest, the Spanish colonizers wanted to maintain their European eating habits, and for them this meant eating from the same vessels used at home, such as glazed wares and white tin-enameled majolica wares. Jerónimo de Mendieta (1980, 404) mentions in his *Historia Eclesiástica Indiana*, written between 1571 and 1596, that a master potter from Spain settled in the colony. We can infer that he, or other Spanish potters, started a workshop for Spanish-style ceramics, such as majolica ware, and introduced the potter's wheel because these vessels were formed with the wheel.

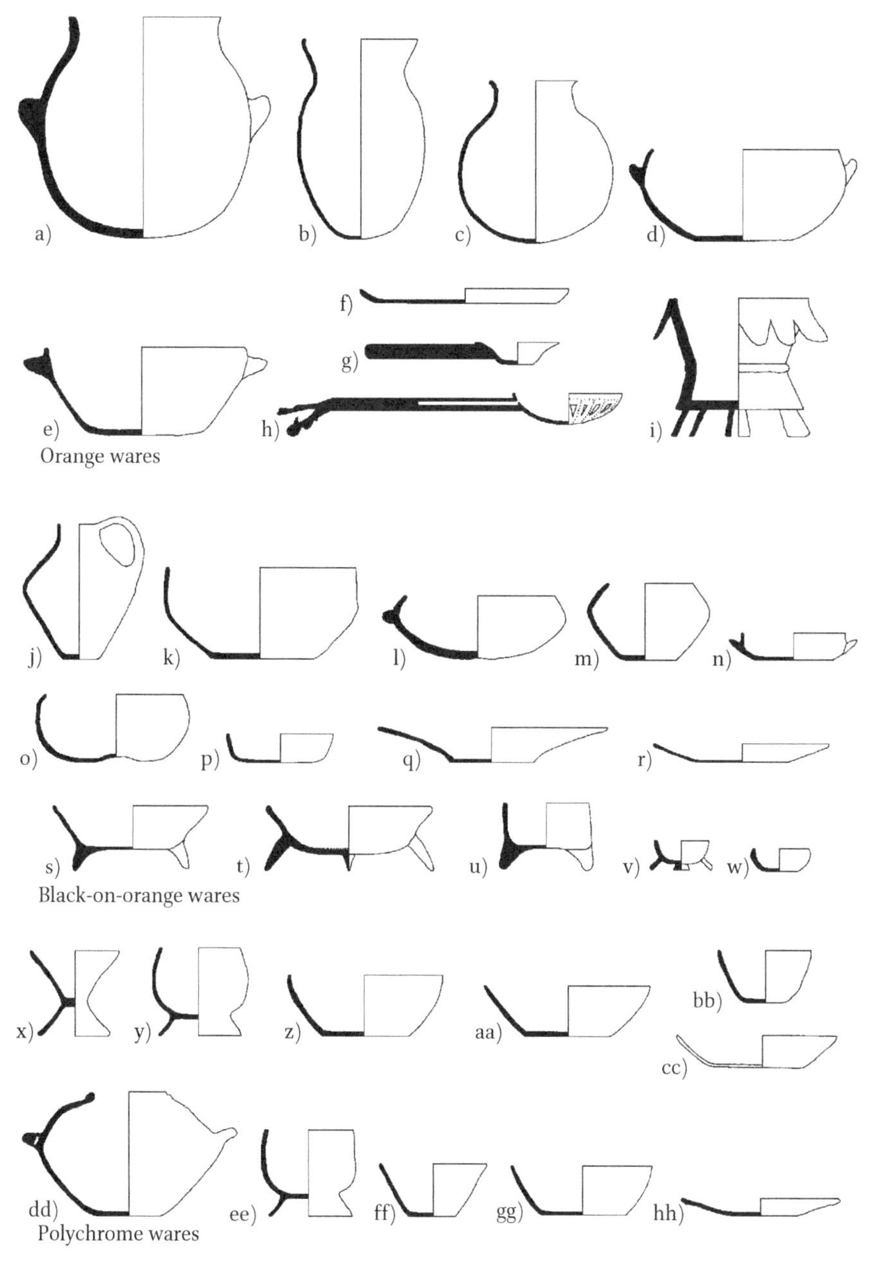

FIGURE 13.1 Late pre-Hispanic vessel shapes in the Valley of Mexico: (a-c) *ollas*; (d) basin with upright walls; (e) basin with flaring walls; (f) *comal*; (g-i) censers; (j) pitcher; (k) basin with upright walls; (l-m) hemispherical bowls; (n) bowl with upright walls; (o) hemispherical bowl; (p) bowl with upright walls; (q) bowl with flaring walls; (r) dish; (s) tripod bowl with flaring walls; (t) *molcajete*; (u) bowl with upright walls; (v-w) miniatures; (x-y) goblets; (z-cc) bowls with upright bowls; (dd) bowl with composite silhouette; (ee) goblet; (ff-gg) bowls with upright walls; (hh) plate. Not scaled

The first Spanish-style workshops for majolica ware were established in Mexico around the 1530s, considering the morphology and style of the earliest majolicas produced in Mesoamerica (Lister and Lister 1978, 22). Likely at the same time the potter's wheel was introduced to form vessels, as this implement was characteristic of those kinds of workshops. Forming vessels with the centrifugal force of the wheel was the common method for pottery manufacturing at that moment in Spain (Sánchez Cortegana 1994), and it had a long tradition in the old world. It was present in southern Levant as early as the beginning of the 4th millennium BC (Roux 2003, 2). Mexican majolica vessels, as well as other vessels made with the wheel, show the typical attributes associated with throwing: spiral rhythmic grooves and ridges on the interior of the base, compression ridges on the interior of the walls, or straight, parallel grit dragmarks on the base (Rye 1981, 75).

After the conquest, indigenous potters in central Mexico continued using the same methods for forming vessels. Remains of indigenous-style ceramics, from contexts identified as early colonial, show that vessels were made with horizontal molds (Figure 13.2). In the case of ollas, juncture marks show that they were made using two or three horizontal molds as in pre-Hispanic times. Also, as in early times, bowls were made with one horizontal mold. Even ceramics of indigenous style but with new morphological or decorative traits characteristic of the early colonial period were made with horizontal molds. For example, in colonial times there were many innovations in the manufacture of indigenous red wares; many new shapes and decorations appeared. Nevertheless, manufacturing marks visible on the vessels show that they continued to be made with molds. This means potters in indigenous-style workshops did not adopt the potter's wheel. We can propose several reasons for this. First, the new method of manufacturing did not represent a technical improvement, as some modern researchers believe (e.g., Foster 1960, 101; Katz 1977, 124–125). Some kinds of vessels, such as small bowls and pitchers, could be made faster with the wheel, however, bigger forms such as large *cazuelas* or *ollas* were difficult to make by that method, as potters consulted during ethnographic research in central Mexico clarified (Hernández Sánchez 2012, 170–172). Second, the connection between particular clay recipes, vessel shapes and method of manufacture was the result of a vast pottery experience accumulated through generations. Potters could not simply replace the ancient forming method in favor of the wheel without also incorporating important changes in the clay "recipes" and morphology of the vessels, as present-day potters further explained (Hernández Sánchez 2012, 170–172). Third, this was one of the most difficult parts of the process of manufacturing of vessels. It involved motor habits mastered and internalized by frequent repetition and required knowledge. Today potters acquire the knowledge and bodily skills required for forming

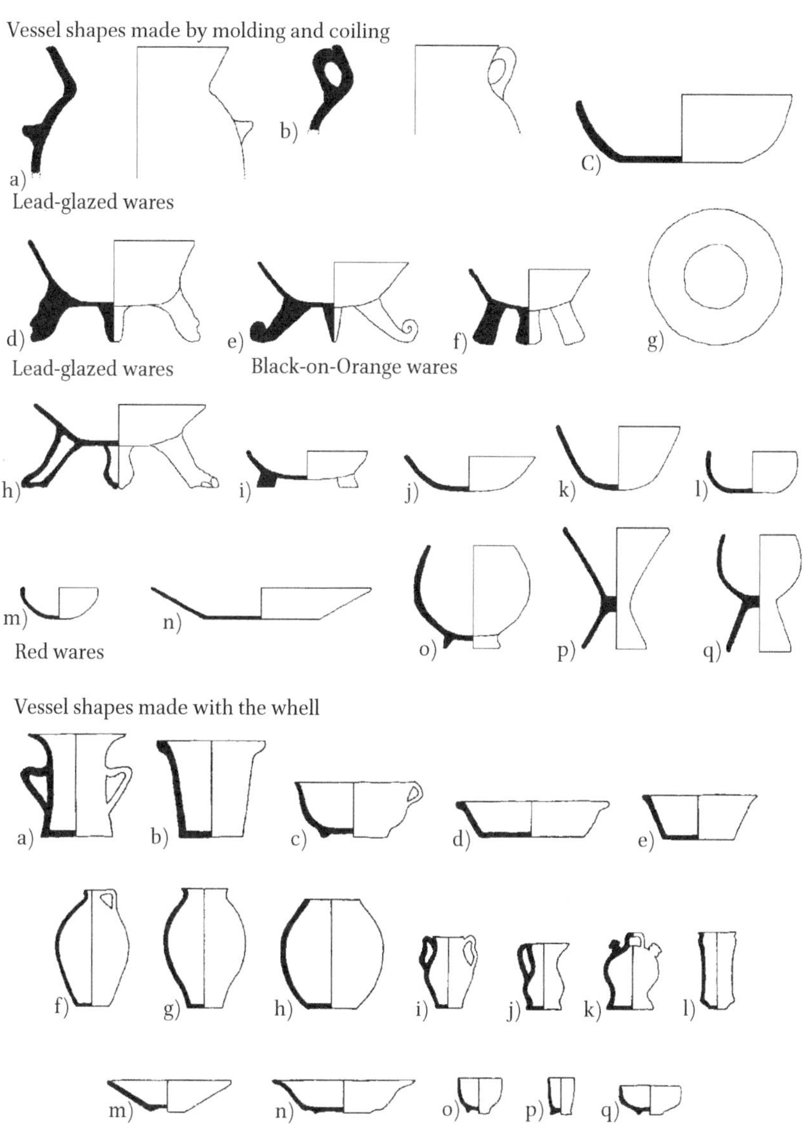

FIGURE 13.2 Shapes of early colonial serving vessels from the Valley of Mexico: vessel shapes made with molding and coiling: (a-b) *ollas* (based on Charlton et al. 2007, Figure 68); (c) hemispherical bowl (based on Charlton et al. 2007, Figure 69); (d) tripod *molcajete* (based on Charlton et al. 2007, Figure 66); (e-f) tripod bowls (based on Charlton et al. 2007, Figure 15); (g) plate; (h-i) tripod bowls (based on Charlton et al. 2007, Figure 23; Rodríguez Alegría 2002, Figure A.1); (j-m) bowls; (n) plate; (o-q) goblets (based on Charlton et al. 2007, Figure 26; Rodríguez Alegría 2002, Figure A.1). Vessel shapes made with the potter's wheel: (a-b) *bacín*; (c) *bacinilla*; (d-e) *lebrillo*; (f) *cántaro*; (g) *tinaja*; (h) *orza*; (i) *jarra*; (j) *pitchel*; (k) *hidroceramo* (*botija*); (l) *albarelo*; (m-n) *plato*; (o) *taza*; (p) *pocillo*; (q) *escudilla* (based on Deagan 1987, Figure 4.1)

vessels within the family. Potters learn from parents or uncles, often as children, as the workshop is usually the core of family life (Hernández Sánchez 2012, 172). The method of forming is therefore intimately related to their understanding of ceramic-making and to family knowledge transmitted for generations, and because of this, potters do not easily change it. In early colonial times, it probably was the same. For all these reasons, the method of forming was very conservative.

In the early colonial Valley of Mexico, some glazed vessels were made with the wheel while others were made by mold. This can be recognized in the vessels themselves; in particular in common and simple pots for cooking and serving, as their surface still has some visible marks of the process of forming (see Hernández Sánchez 2012, 112). Vessels made with a potter's wheel were majolica ware or were lead-glazed ware and had specific shapes. These were mainly plates with ring base, cups, pitchers, *lebrillos* (basins), *bacines* (basins with high walls), *botijas* (amphorae for olive oil), *albarelos* (high drug vases) and candleholders (see Figure 13.2). All of them had Spanish antecedent and names taken from the Spanish vessel repertoire (Lister and Lister 1987; Sánchez 1998). A few of them were clearly related to Spanish uses, such as the olive jars, which were lead-glazed in the interior to avoid filtration (Goggin 1960). The rest could be associated to Spanish uses, such as the *albarelos*, which were medicine containers, or the *lebrillos*, which were chamber pots, but these uses were not exclusive Spanish customs. In contrast, vessels with shapes of indigenous origin, such as *ollas*, *cazuelas* and bowls, were made with molds. This may suggest that these two groups of vessels were made in different workshops. It seems that after the conquest indigenous-style workshops continued using the same methods of manufacture for producing the known repertory of vessels. At the same time, Spanish-style workshops for manufacturing Spanish-style vessels used the wheel.

Majolica wares were all made by wheel. Thus, as a rule, vessels with typical Spanish-style shapes were made by wheel. A possible exception is a serving ware present during the sixteenth century in Mexico City but made, according to chemical analysis of the clay, in Michoacán, west of central Mexico (Fournier et al. 2007). These vessels were decorated with white slip covered with lead glaze, which has the appearance of majolica, although its glaze is not blended with tin. Archaeologists call them Indígena Ware (Lister and Lister 1978, 19). Observation suggests that these vessels were made with a mold as the typical marks of a wheel were not detectable. However, as slip and glaze have covered large parts of the vessel's surface, it is still necessary to conduct microscopic analysis to confirm this hypothesis. The vessels show Spanish-style shapes, such as plates and small bowls with Spanish-style proportions, but are decorated with motifs both of Spanish and indigenous origin.

2 Encounters with the Glazing Technique

Once a vessel is formed, but also during the process of forming, a potter generally finishes its surface by rubbing a tool against the leather hard clay or by applying a slip (fluid suspension of clay in water of different color than the vessel clay) to modify its texture and light reflecting qualities (Rye 1981, 89). Surface finishing requires particular motor habits, but also experience with clay properties. Both kinds of knowledge are learned and transmitted across generations. However, as this part of the process of ceramic-making is visible on the vessels, it may be influenced by other potters and by users. The methods used to finish ancient ceramics can be identified by observing vessel remains, although normally earlier stages of this process are covered by later stages, leaving only the later visible. Potters may give variable attention to the finishing of the surface of a vessel. They may only smooth the surface, that is, rub the vessel until it acquires a regular texture but a matte appearance. They may burnish the surface, that is, rub the surface regularly but the tool is used directionally so that a pattern may be produced, and the appearance is a combination of matte and luster. On the other hand, they may polish the vessel, producing a regular surface with uniform luster (Rye 1981, 89–90).

Late pre-Hispanic potters of the Valley of Mexico gave a similar surface finishing that was simple and without extra decoration to common vessels, such as those for cooking, storage and transportation. The surface of this kind of objects shows the natural color of the fired clay, which is generally orange-brown (Blanton and Parsons 1971, 304). The surface is relatively well smoothed but hastily burnished, that is, strikes left by the burnishing tool can be seen (Blanton and Parsons 1971, 304). Potters also made serving vessels with painted decoration. These objects show more variety in surface finishing. The most frequent were orange bowls and plates with black painted decoration; archaeologists today, name them today Black-on-Orange vessels (Whalen and Parsons 1982, 441) (Figure 13.3). These objects maintained the natural orange-brown color of the fired clay. Potters also made serving vessels with red decoration; archaeologists today name them Red Ware (Cervantes et al. 2007, 279; Tolstoy 1958; Whalen and Parsons 1982, 446) (Figure 13.3). They were, as a rule, better finished than vessels with black decoration. Their orange-brown surface was well smoothed and relatively well burnished. Some of these vessels were additionally decorated with black or white paint or with incisions. Potters also made other serving vessels with more complex decoration and higher quality. These objects were painted with red, orange and black designs in the same style as the famous late pre-Hispanic polychrome "codex style" ceramics and were finely polished to reach a glossy finish (Whalen and Parsons 1982, 441, 446).

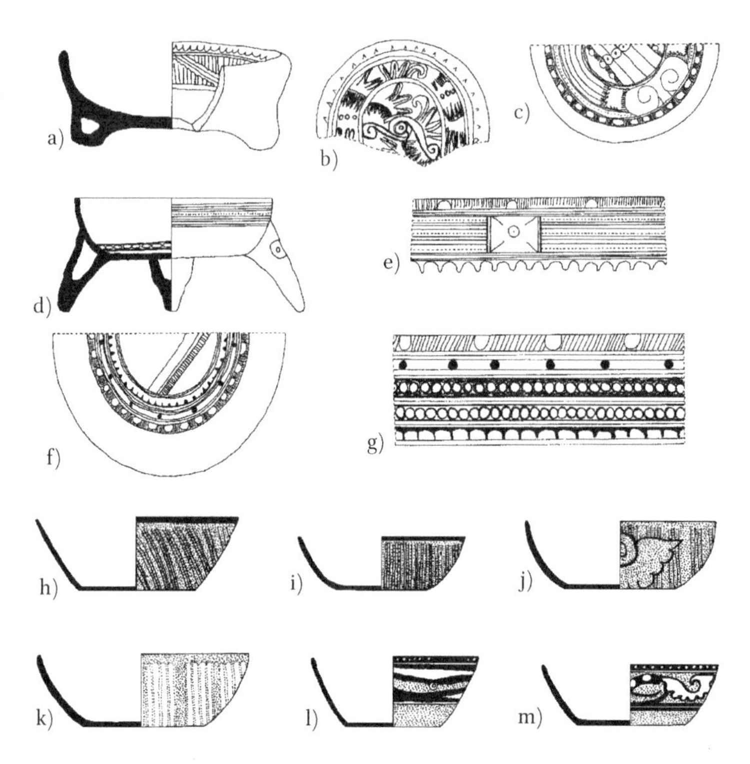

FIGURE 13.3 Patterns of decoration on late pre-Hispanic ceramics from the Valley of Mexico: (a) early Aztec Black-on-Orange tripod bowl (based on Minc et al. 1994, Figure 6.2a); (b) early Aztec plate (based on Cervantes et al. 2007, Figure 14); (c) early Aztec plate (based on Cervantes et al. 2007, Figure 23); (d) late Aztec tripod bowl (based on Cervantes and Fournier 1995, Figure 2); (e) decoration pattern on late Aztec plate (based on Cervantes and Fournier 1995, Figure 8); (f) late Aztec plate (based on Cervantes and Fournier 1995, Figure 5); (g) decoration pattern on late Aztec bowl (based on Cervantes and Fournier 1995, Figure 9); (h-j) late Aztec Black-on-Red bowls (based on Cervantes et al. 2007, Figure 61); (k) late Aztec White-on-Red bowl (based on Cervantes et al. 2007, Figure 39); and (l-m) late Aztec White-and-Black-on-Red bowls (based on Cervantes et al. 2007, Figure 43). The early Aztec period corresponds to ca. AD 900–1350 (Brumfiel 2005, Sanders et al. 1979); the late Aztec period to ca. AD 1350–1521 (Cervantes et al. 2007, 280; Charlton 2000; Hare and Smith 1996)

On the eve of the conquest, potters in central Mexico finished their vessels in similar manner. They shared not only the methods of rubbing the surface, but also the high attention given to the surface of some serving objects. That is, in every region of central Mexico there were repertories of fined polished vessels with polychrome painted decoration (see Hernández Sánchez 2012, 60–62). Surfaces were not decorated with vitreous glaze in pre-Hispanic Mesoamerica. Plumbate ware, a widespread trade pottery made in the Soconusco region on the coast of Guatemala during the tenth and eleventh centuries, has some similarity to glazed ceramics. These ceramics have a ferruginous slip, which after firing has an iridescent gray-orange color (Shepard 1948). Although this slip is vitrified in some places due to the particular mineralogical composition and firing, it is not a high-fired vitreous glaze (Rice 1987, 20). This technology, however, did not continue in Mesoamerica after the eleventh century.

We still do not exactly know when the Spanish ceramic technology arrived to Mexico, but it seems that there was some very early Spanish interest to produce glazed vessels. In a document sent by Alonso Figueroa, Chantre of Oaxaca, to Charles V in 1529 he states: "*Con trabajo e ingenio alcancé el vidriado que no tenían, un plato en que comer sino venía de Castilla*"[1] (cited in López Cervantes 1976, 15). However, as he says, wares were still imported from Spain. A few later sources, the *Florentine Codex* (Sahagún 1961, X, 839), apparently prepared as early as 1547 and completed in 1569 (D'Olwer and Cline 1973, 193), the *Historia Eclesiástica Indiana* (Mendieta 1980 [1571–1596], 404) and a letter of Viceroy Lorenzo Suárez de Peralta sent dated in 1583 (Cervantes 1939, I, 18); show that by 1570s–1580s the production of glazed wares was already established in the colony. In that letter sent to the *alcalde mayor* of Michoacan, Suárez de Peralta mentions:

> ... por cuanto por parte de los naturales de la ciudad de Patzcuaro, que son oficiales de hacer platos y escudillas de loza vidriada y otras piezas de barro, me ha sido dada relación que la justicia de dicha ciudad, proveyó veedores de este oficio para que viesen y visitasen la obra que se hazía, para que siendo tal se pudiese vender y no lo siendo se los quitase y no se vendiese. Y agora estos indios olleros que no son ni han sido ni pueden ser oficiales de dicho oficio ni lo saber hacer dichos platos mal hechos y de donde se sigue fraude y engaño ... y me pidieron les mandase dar y diese mandamiento para los que son tales oficiales usen el dicho oficio y no los olleros ....[2]
>
> Archivo General de la Nación, Ramo Indios, Vol. II expedient 718; LÓPEZ CERVANTES 1976, 15

1 "With work and talent I was able to make glaze, as they did not have a plate to eat if it did not come from Castile" (my translation).

2 "...concerning the inhabitants of the city of Patzcuaro, which are officials in the trade of making glaze ceramic plates and bowls and other objects of clay, it was informed to me that the justice of the mentioned city, provided observers of this trade in order to observe and visit the works made, and in the case they were right they could be sold and if not they could be taken and not sold. And now these indigenous pot makers which are not and were not and cannot be officials in this profession and cannot make those wrong made plates and where fraud and tricks are followed ... and they asked me to give an order for those who are officials could practice this profession and not the pot makers..." (my translation).

This shows that, by that time, not only the manufacture of glazed ware was well established, but also that indigenous potters were using this technique. The glaze technique consisted of the application of a mixture of lead oxide, silicate and clay to the surface of a fired vessel (Charlton et al. 2007, 485–486). After that, the vessel was fired again at a higher temperature, and the glaze material melted and fused to the surface, obtaining a physical structure similar to glass (Rye 1981, 44). The result was a glossy vessel with brownish or greenish glaze.

After the conquest, the surface finishing and decoration of indigenous-style vessels manifested notorious changes. We can recognize two major trends. On one side, serving vessels evidence great impulse of creativity and innovation. Potters modified parts of the surface finishing and decoration that existed before the conquest, and experimented with new techniques. In particular, this was the case for the red wares in which new styles and motifs of decoration are observed (see Charlton et al. 1995, 143; Fairbanks 1966). At the same time, the manufacture of other decorated serving wares decreased and was a bit simplified, such as the orange vessels with black decoration (see Charlton et al. 2005, 2007, 440; Garraty 2006, 368) and the polychrome vessels (Lind 1994, 81). On the other side, a second trend was that the surface finishing of cooking vessels was simplified while a new technique, the lead glaze, became quite popular.

The application of lead glaze for decorating vessels was an early Spanish introduction. It seems that this technique was readily accepted by native potters and was established in indigenous workshops by the second part of the sixteenth century. Sahagún (1961, 839), in his description of indigenous potters and the pottery craft, mentioned that they made a variety of pre-Hispanic wares but also glazed vessels. This reference could date the establishment of this technique among indigenous potters between 1547 and 1569, when the production of the *Florentine Codex* is estimated (D'Olwer and Cline 1973, 193), although it could occur earlier. Unfortunately, the morphology and archaeological context of indigenous-style glazed vessels do not provide more chronological detail. For example, in several places in the Valley of Mexico typical pre-Hispanic vessels with lead glaze, specifically *molcajetes* (bowls with striated interior bottom for grinding chili sauces), have been found. These objects are dated for the early colonial period (AD 1521–1620), according to their morphology and context (Charlton et al. 2007, 486); however, we are not yet able to date them with more precision. *Molcajetes* were clearly for indigenous users. Their function as grinders for chili sauces was pre-Hispanic as well as their shape and decoration. In addition, other kinds of indigenous-style vessels were also glazed, like *ollas* and *cazuelas*.

Lead glaze as surface finishing required different types of effort and energy than the typical pre-Hispanic surface treatments. These vessels did not require finishing the surface with detailed burnishing, as the glaze covered most of the previous process of finishing. However, these vessels required two firings and higher temperatures, and therefore a larger amount of fuel. Even with the higher efforts this involved, indigenous-style workshops implemented this technique, as is evidenced by the presence of glazed vessels with indigenous-style forming methods and morphology. Although glaze notably altered the appearance of these vessels, it was relatively easy to implement without modifying other parts of the process of manufacture (with exception of firing). The fact that this technique was not only implemented in serving wares that are normally those in which potters' influences are reflected, but also in cooking wares, shows that lead glaze had an important impact on the indigenous pottery. This was not exceptional, however; Mesoamerican potters had always been open to new forms of surface finishing and decoration as the variety of pre-Hispanic vessel repertoires in different regions and epochs show.

3 Encounters with Majolica Wares

The shape of ceramic objects is the result of several interconnected variables: function, physical properties of materials, forming method and aesthetic preferences of the potters. Therefore, the form of the vessel may change even if its function remains the same. Likewise, shifts in function may not be evident in its shape. The techniques to make particular vessel shapes can be learned and transmitted through generations. Nevertheless, potters can easily modify their size, proportions and silhouette by the influence of users, relatives, neighbors or fellow potters. In addition, potters normally produce particular assemblages of vessels. That is, they make a specific variety of vessels for cooking, as well as a variety of vessels for serving food and drink, and also several objects for ritual purposes, such as censers. Although late pre-Hispanic central Mexican potters made a wide variety of vessel forms, the shape repertoire was, in broad terms, similar. The most common vessels designed for cooking, storage and transportation were high-necked and short-necked *ollas* of various sizes, as well as basins with upright walls or flaring walls (see Blanton and Parsons 1971, 299; Cervantes et al. 2007, 283–284; Whalen and Parsons 1982, 438–441, 450). There were also simple hemispherical bowls, *comales* and *molcajetes*. Common vessels for serving were bowls, with or without supports, and plates.

In late pre-Hispanic central Mexico, vessels for serving were often decorated. Overall, decorative techniques for ceramics take into account the possibilities offered by the various properties of the clay during the process of drying and the effects of firing. Knowledge related to these aspects can be learned and transmitted across generations. However, as this part of the process of ceramic-making is highly visible, potters may easily make changes influenced by users, relatives, neighbors or colleagues (see Gosselain 2000, 191). Designs and patterns executed in a particular decoration technique may even be more easily modified as this does not require extra technical knowledge but only new ideas resulting from inspiration, imitation or reinterpretation.

On the eve of the Spanish conquest, most of the decoration on ceramics in central Mexico was made with painting (see Cervantes et al. 2007; Noguera 1954; Whalen and Parsons 1982). In addition, there was often a relationship between vessel form and decoration. Common vessels designed for serving food and drink, such as the Black-on-Orange ware, were painted with patterns of black lines and curvilinear motifs (Cervantes et al. 2007, 280; Charlton 2000; Hare and Smith 1996), and were bowls with tripod supports and plates (Blanton and Parsons 1971, 294; Whalen and Parsons 1982, 441, 450) (see Figure 13.3). In contrast, the red wares were ornamented with large sections of thick and well-polished red paint, and often black designs; less frequently they also included white painting and incisions (see Blanton and Parsons 1971, 309; Cervantes and Fournier 1995, 100; Cervantes et al. 2007, 300; Whalen and Parsons 1982, 446, 450) (see Figure 13.3). These red vessels were mostly bowls and had no appendages at all. Both black-on-orange and red wares were common serving vessels. In general, the painted designs were simple, schematic and hastily done, which suggests that they did not play a special role in the communication of meaning relevant to the contexts where these objects were to be used. In contrast, potters also manufactured some objects of superior quality with complex pictographic decoration (Vega 1975, 25). The painted motifs were part of the symbolic corpus of central Mexican pictographic writing, and were associated with important meanings in the context of Mesoamerican ceremonialism, such as piety, preciousness or nobility (see Hernández Sánchez 2005). In contrast to black-on-orange and red wares, these fine polychrome vessels had a large formal inventory (see Hernández Sánchez 2005, Table 8.2).

The first Spanish colonizers wanted to maintain their eating habits, and for them this implied eating from the same vessels used at home, such as glazed wares and white tin-enameled majolica wares. In the beginning, Spanish ships brought loads of ceramics to the Americas. For example, in the Dominican Republic typical fifteenth-century Andalusian service wares, which still evidence Arabic stylistic traits, have been found (Deagan and Cruxent 2002, 139). After the conquest, Spanish ceramics also arrive to Mesoamerica (see Fournier 1996, 452; Lister and Lister 1978; Sánchez 1996), but probably not in large quantities and not common vessels as transoceanic transportation was costly and reserved for other basic items such as weapons, wine and oil (Sánchez 1996, 128). In addition, after the establishment of the Manila Galleon Trade in 1573, a few Chinese porcelains arrived in Mexico City (Charlton et al. 2005, 62; Lister and Lister 1978, 10).

Majolica wares were very popular at that time in Spain (see Pleguezuelo 1999; Sánchez 1994). These vessels, after a first fire, were covered with a mixture of tin, lead and silicate, and decorative motifs painted with metallic oxides; following a second firing, the vessel acquired a white milky glaze and decorative patterns in various colors (Lister and Lister 1982, vii). The first Spanish-style workshops for majolica ware were established in Mexico City. According to the Listers (1978, 22), this occurred around 1530s, considering the morphology and style of the vessels made. Afterwards the production was moved in the 1580s to the city of Puebla, where in the seventeenth century innovations occurred with majolica wares and new shapes and colorful decorations appeared (see Lister and Lister 1984, 87). The production became so significant that Mexican majolica wares were exported to other Spanish colonies in the Americas (e.g., Duarte and Fernández 1980; Goggin 1968, 223). However, both the indigenous and the Spanish traditions of ceramics were apparently produced in different workshops. The excavation of a colonial workshop from the end of the sixteenth century and beginning of the seventeenth century in Mexico City, for example, revealed that only majolica wares were produced there (Gámez 2003, 236). Furthermore, the guild regulations of the seventeenth century for majolica potters of Puebla suggest that they produced only common and fine-grade glazed wares. That is, the fifth statute states:

> Que haya de tener separación los tres géneros de loza fina, común y amarilla, que se entiende ollas y cazuelas, y otros vasos, jarros colorados, no pueden hacer loza fina, ni común, menos que habiéndose examinado para ello de forma que cada uno ha de labrar, sólo el género de que se examinarse, y no otro ninguno, si no es que se comprende todo en su examen.[3]
>
> NOVELO 2007, 101

The statute suggests that there were workshops specialized in Spanish-style ceramics. They produced majolica ware, but also used the potter's wheel and

3 "It should be made a separation between the three grades of ware, fine, common and yellow, which is understood as the jars, *cazuelas* and other vases, red pitchers, they cannot made fine or common wares, at least they are examined for this in that way everyone can only produce the grade of ware for which he has been examined, and not other grade at least is included in his exam" (my translation).

manufactured vessel shapes that were not made before in Mesoamerica, such as particular forms of cups and plates. They used new decorations, as well. Following the Spanish tradition (see Lister and Lister 1987), majolica ware made in Mexico was decorated with bands of curvilinear and geometric motifs painted in several colors, mainly blue, yellow and orange. Compositions were similar to those of the majolica vessels produced at that time in workshops around Seville (see Charlton et al. 2007; Lister and Lister 1978, 1982, 1987). Potters that specialized in indigenous-style red ware vessels incorporated a few decorative elements of these new ceramics (Figure 13.4). However, indigenous and Spanish decorative traditions were maintained separately until the end of the early colonial period. Red wares continued to be made in the late colonial period, but their decoration was increasingly different than that of earlier times. Some specialists consider Indígena Ware (vessels with white matte slip covered by lead glaze) as indigenous imitations of majolica ware (Charlton et al. 2007,

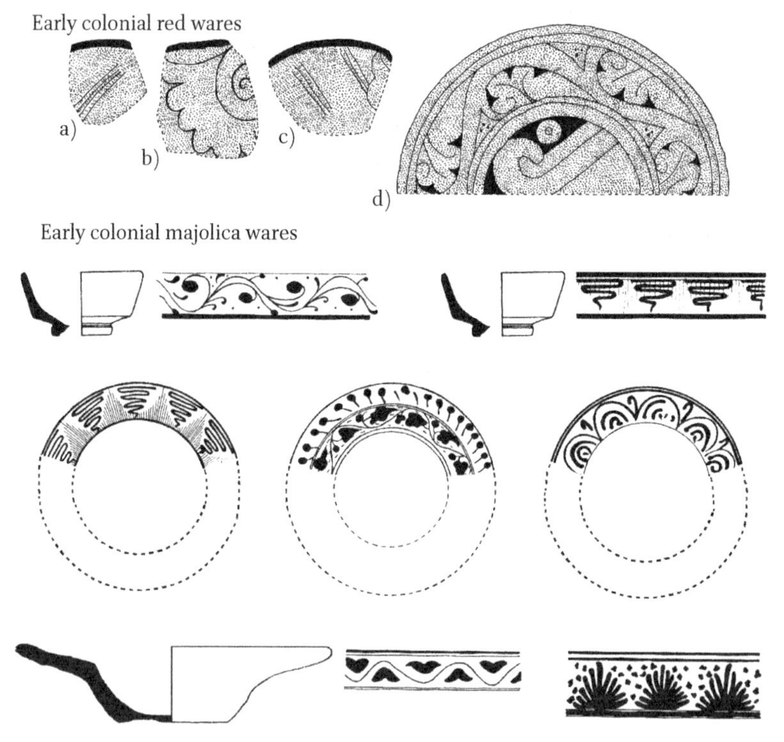

FIGURE 13.4 Patterns of decoration on early colonial wares from the Valley of Mexico. Red wares: (a-b) bowls with flared walls (based on Charlton et al. 1995, Figure 6); (c) bowl with upright walls (based on Charlton et al. 1995, Figure 5); (d) plate with interior decoration (based on Charlton et al. 1995, Figure 8). Majolica ware all based on Lister and Lister 1987, Figure 85

470–471; Lister and Lister 1978, 21). However, we do not yet have enough data to confirm that those vessels were made in indigenous-style workshops.

It seems that during the first generations after the conquest there were not many points of conflict between indigenous and Spanish potters or between indigenous potters and colonial authorities. This is suggested by the scarcity of administrative documents and other written sources related to this topic. The creation of pottery guilds and regulations in the seventeenth century shows, however, that by the late colonial period Spanish-style workshops were competing with other pottery workshops. Craft guilds, like the guild of majolica potters in Puebla, were established in particular sectors of the city. They examined their members and established a hierarchy according to their knowledge and experience, and created many rules for their work and products (Carrera Estampa 1954). This form of organization gave way to workshops not based on family relations but rather on occupation relations. This was a clear contrast to indigenous-style workshops based on the family. Likely, this resulted in the two kinds of workshops developing different forms of personal relations and knowledge transmission.

4 Indigenous Pottery Technology in Central Mexico after the Conquest

It seems that Spanish-style workshops in early colonial central Mexico were maintained separately from indigenous-style workshops. This is recognized in the use of different methods of manufacture, different shapes and different decorations in vessels from both traditions. Despite that separation, indigenous potters were well aware of the newcomer ceramic technology, and selectively incorporated and readapted various elements. The potter's wheel was not really implemented in indigenous-style workshops. This method did not represent an improvement to the known technology. Therefore, there was no reason to modify the most stable part of the production sequence, which was deeply rooted in potter families for generations and was closely associated with their own conceptualizations about pottery (see Hernández Sánchez 2008). The glazing technique was indeed a novelty that attracted the attention of indigenous potters and was widely implemented early in the colonial period. This decoration was showy and relatively easy to create. The challenging part was that vessels needed to be fired twice, and the second fire required a temperature hot enough to reach the melting point of the glaze. The concomitant need to collect and to manage greater amounts of fuelwood might have had important impacts on local environments and allocations of time and labor.

It seems, however, all this was not a limitation as lead-glazed indigenous-style vessels were broadly distributed in early colonial times. This new style of decoration simplified the process of surface finishing, as it was not necessary to burnish the vessel in detail; it also made the surface impermeable. Still, the glossy finish was probably the most attractive quality for potters. Some early vessels were glazed in areas where water-resistance and reduced labor from burnishing were not important criteria. For example, tripod *molcajetes* had glazed supports.

Decoration with lead glaze was, without a doubt, the aspect of the Spanish ceramic technology most implemented by indigenous potters. Nevertheless, early colonial indigenous-style vessels also show other decorative elements taken from the Spanish tradition. The majority, however, were not copies of decorations of Spanish-style ceramics but reinterpretations of Spanish motifs. For example, a few colonial red wares were painted with black curvilinear motifs which had some resemblance to motifs painted on the earliest majolica wares in Mexico City (see Charlton et al. 2007, 449, 472–477). However, this kind of adornment was rather exceptional. Most of the new decorations with Spanish influence were not taken from Spanish-style vessels, but from other media. For example, a number of early colonial native black-on-orange vessels were painted with images of flowers, fishes, birds, leaves and ears of wheat (Figure 13.5). Most of these images were not new in Mesoamerican ceramics, but they were not painted before in this ware. Their style of representation was also a bit different from earlier figural depictions, and showed a little Spanish influence. In addition to decorative elements, indigenous potters also incorporated a few vessel shapes of the Spanish tradition. For example, extended plates or small cups with flat handles were the most common forms of majolica wares (Charlton et al. 2007, 463) while they, in that particular shape, were not made before in Mesoamerica. These vessel forms were incorporated into the wide, formal, early colonial repertoire of red wares. The rest of the morphological novelties were details for embellishing the vessels rather than for modifying their function. For example, red wares incorporated ring bases and cover lids that were characteristic of Spanish vessels at that time. In addition, as in the case of decoration, some morphological innovations were not imitation of European ceramics but were inspired by the new culture that arrived to Mesoamerica. This was clearly the case of vessels with supports modeled in new shapes inspired by colonial animals, like pig hoofs and lion claws.

Thus, central Mexican potters, after the encounters with the potter's wheel, the glazing and the majolica ware, openly and selectively incorporated or reinterpreted a number of elements of Spanish technology, in particular in vessel decoration and morphology. They were also inspired by the new colonial world

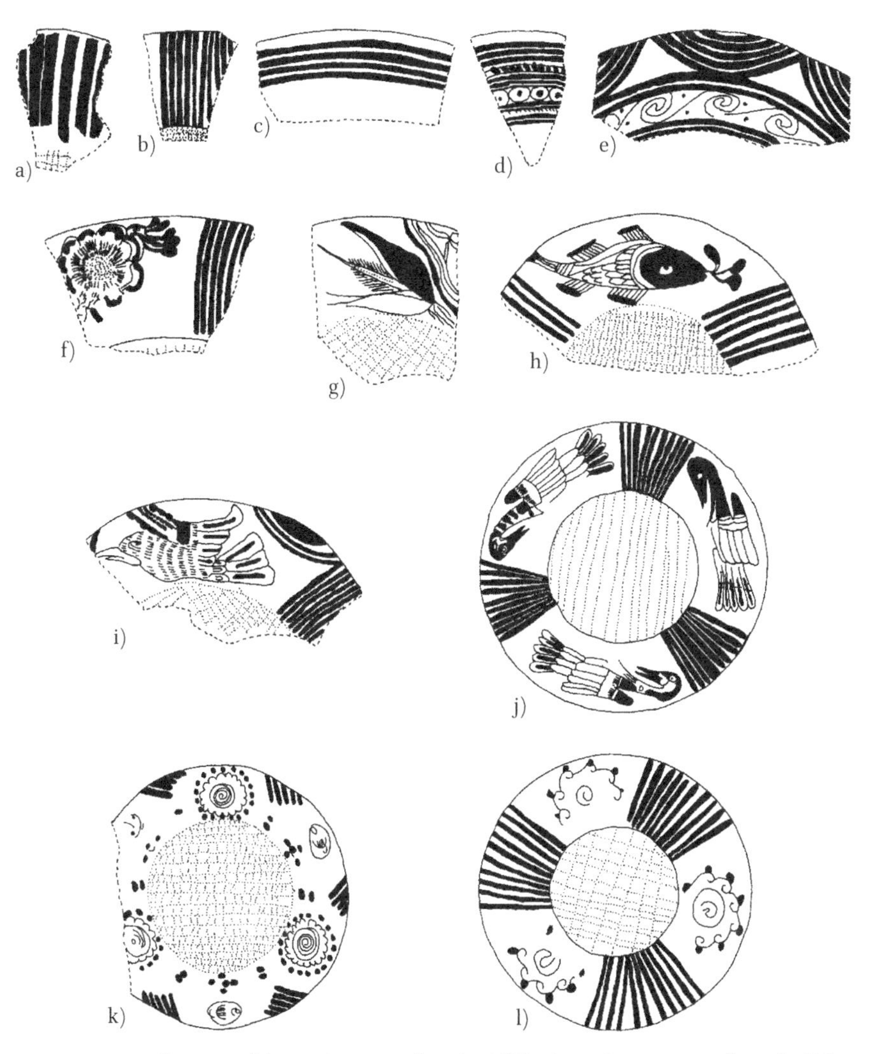

FIGURE 13.5 Patterns of decoration on early colonial Black-on-Orange wares from the Valley of Mexico: (a-d) *molcajetes* and tripod bowls (based on Charlton et al. 2007, Figure 12); (e-g, k) *molcajetes* and tripod bowls (based on Charlton et al. 2007, Figure 13); and (h-j, l) *molcajetes* (based on Charlton et al. 2007, Figure 14)

and created new decorative compositions, which were more figural and iconic than in pre-Hispanic times. In contrast, the Spanish ceramic technology adopted practically nothing from the indigenous tradition, neither in Seville nor in Mexico City. Peoples associated with the colonial rule in the Valley of Mexico used indigenous-style pottery. For example, many red wares have been found in the most prominent area of Mexico City (Rodríguez Alegría 2003), and in churches and convents elsewhere in central Mexico. However, Spanish-style workshops did not produce vessels of the indigenous-tradition. This is suggested by the lack of indigenous-style vessels made with the wheel. The reason for this rejection might be in part related to the colonial situation; namely, the need of the colonizers to maintain the cultural association with their fatherland or the conviction of European technological superiority. However, part of the reason was probably of technical nature. As was the case for indigenous-style workshops, Spanish-style workshops were attached to their own methods of manufacture.

In brief, early colonial central Mexican potters had various reactions to the encounter with Spanish ceramics. On one hand, they selectively incorporated and reinterpreted elements of decoration and morphology, and were inspired by the world brought by the Spaniards in the creation of new decorative motifs. On the other, they did not implement technical devices that they did not need, such as the potter's wheel. In my opinion, the inclusion of Spanish decorative elements in the vessels was not related to attitudes of submission, just as the rejection of Spanish devices was not related to attitudes of subversion. A potter's work followed the same dynamics as in ancient times, that is, they conserved their familiar methods of forming while they adapted the visible aspects of the vessels to the situation of the present time. These two basic characteristics, existing at the same time, are evident in the entire ceramic production of the pre-Hispanic history. Thus, in the early colonial period the incorporation, adaptation or rejection of Spanish elements was not politicized by potters. They just wanted to maintain their way of living and adapt to the new post-conquest society.

After the early colonial period, ceramic-making in central Mexico experienced more changes. Clay recipes, method of forming and firing technology were consistent with ancient times. Whereas morphology, finishing and decoration were so modified that vessels became gradually more differentiated from their pre-Hispanic antecedents. This trend continues to the present day (Hernández Sánchez 2012, 207). Thus, today vessels do not look like those of the precolonial past. In central Mexico, the majority of the production is now concentrated on lead-glazed wares, which are embellished with motifs not related to ancient decorations. The shapes continue to resemble ancient forms although potters have made many innovations in minor formal details. Nevertheless, the method of forming has been maintained. Vessels continue to be made with molds and coiling, not with the potter's wheel. This is intimately related to the core of this tradition, and therefore it is sign of the continuation of the pre-Hispanic ceramic culture until the present time.

Acknowledgments

This study presents results of the research project "Ceramics and Social Change. The Impact of the Spanish Conquest on Middle America's Material Culture" carried out from 2006 to 2010, under the generous support of a VENI grant from the Innovational Research Incentives Scheme of the Netherlands Foundation for Scientific Research (NWO) and the Faculty of Archaeology of Leiden University.

References

Blanton, Richard E., and Jeffrey R. Parsons. 1971. "Apendix I: ceramic markers used for period designations." In *Prehistoric settlement patterns in the Texcoco region, Mexico*, Memoirs of the Museum of Anthropology No. 3, edited by Jeffrey R. Parsons, 255–313. Ann Arbor: Museum of Anthropology.

Brumfiel, Elizabeth M. 2005. "Ceramic chronology at Xaltocan." In *Production and power at Postclassic Xaltocan*, edited by Elizabeth M. Brumfiel, 117–152. Pittsburgh: University of Pittsburgh.

Carrera de la Stampa, Manuel. 1954. *Los gremios mexicanos. La organización gremial en Nueva España, 1521–1861*. Mexico: Edición Iberoamericana de Publicaciones.

Cervantes, Enrique A. 1939. *Loza blanca y Azulejo de Puebla*, 2 vols. Mexico: Privately printed.

Cervantes, Juan and Patricia M. Fournier. 1995. "El complejo Azteca III Temprano de Tlatelolco: consideraciones acerca de sus variantes tipologías en la cuenca de México." In *Presencias y encuentros. Investigaciones arqueológicas de salvamento*, edited by Dirección de Salvamento Arqueológico, 83–110. México City: Dirección de Salvamento Arqueológico, INAH.

Cervantes, Juan, Patricia M. Fournier, and M. Carballal. 2007. "La cerámica del Posclásico en la cuenca de México." In *La producción alfarera en el México antiguo, volumen V: La alfarería en el Posclásico (1200–1521 d.C.), el intercambio cultural y las permanencias*, edited by Leticia Merino and Angél García Cook, 277–320. México, City: INAH.

Charlton, Thomas H. 2000. "The Aztecs and their contemporaries: the central and eastern Mexicanhighlands." In *The Cambridge history of the native peoples of the Americas, vol. II: Mesoamerican, Part 1*, edited by Richard E.W. Adams and Murdo J. MacLeod, 500–557. Cambridge: Cambridge University Press.

Charlton, Thomas H., Deborah L. Nichols, and C. Otis Charlton. 1992. "Aztec craft production and specialization: archaeological evidence from the city-state of Otumba, Mexico." *World Archaeology* 23: 98–113.

Charlton, Thomas H., C. Otis Charlton, and Patricia M. Fournier. 2005. "The Basin of Mexico A.D. 1450–1620. Archaeological dimensions." In *The Postclassic to Spanish-era transition in Mesoamerica. Archaeological perspectives*, edited by Susan Kepecs and Rani T. Alexander, 49–63. Albuquerque: University of New Mexico Press.

Charlton, Thomas H., Patricia M. Fournier, and Juan Cervantes. 1995. "La cerámica del periodo Colonial Temprano en Tlatelolco: el caso de la Loza Roja Bruñida." In *Presencias y encuentros. Investigaciones arqueológicas de salvamento*, edited by Dirección de Salvamento Arqueológico, 135–155. México City: Dirección de Salvamento Arqueológico, INAH.

Charlton, Thomas H., Patricia M. Fournier, and C. Otis Charlton. 2007. "La cerámica del periodo Colonial Temprano en la cuenca de México: permanencia y cambio en la cultura material." In *La producción alfarera en el México antiguo. La alfarería en el Posclásico (1200–1521 d.C.), el intercambio cultural y las permanencias,* edited by Beatriz L. Merino Carríon and Ángel García Cook, 429–496. México City: INAH.

Deagan, Kathleen A. 1987. *Artifacts of the Spanish Colonies of Florida and the Caribbean, 1500–1800, Volume 1: Ceramics, Glassware, and Beads*. Michigan: University of Michigan.

Deagan, Kathleen A. and José M. Cruxent. 2002. *Columbus's outpost among the Taínos. Spain and America at La Isabela, 1493–1498*. New Haven: Yale University Press.

D'Owler, Luis N. and Howard F. Cline. 1973. "Bernardino de Sahagún, 1499–1590." In *Handbook of Middle American Indians, Volume 13: Guide to ethnohistorical sources, part 2*, edited by Robert Wauchope, 186–206. Austin: University of Texas Press.

Duarte, Carlos. and María L. Fernández. 1980. *La cerámica durante la época colonial venezolana*. Caracas: Ernesto Armitano Editor.

Fairbanks, Charles H. 1966. "A feldspar-inlaid ceramic type from Spanish colonial sites." *American Antiquity* 31 (3): 430–432.

Foster, George M. 1960. *Culture and conquest: America's Spanish heritage*. New York: Viking Fund Publications in Anthropology 27, Wenner Gren Foundation.

Fournier, Patricia M. 1996. "Tendencias de consumo y diferencias socioétnicas en el valle de México. Contraste entre Tlatelolco y la ciudad de México durante los periodos colonial y republicano." In *Memoria del primer congreso nacional de arqueología histórica,* edited by E. Fernández and S. Gómez, 448–457. México City: Consejo Nacional para la Cultura y las Artes.

Fournier, Patricia M., M. James Blackman, and Ronald L. Bishop. 2007. "Los alfareros Purépecha de la Cuenca de Pátzcuaro: producción, intercambio y consumo de cerámica vidridada durante la época virreinal." In *Arqueología y complejidad social,* edited by Patricia M. Fournier, Walburga Wiesheu and Thomas H. Charlton, 195–221. México City: Instituto Nacional de Antropología e Historia.

Gámez Martínez, Ana P. 2003. "The forgotten potters of Mexico City." In *Cerámica y cultura. The story of Spanish and Mexican Mayolica*, edited by Robin F. Gavin, Donna P.

Piercel and Alfonso Pleguezuelo, 227–242. Albuquerque: University of New Mexico Press.

Garraty, Christopher P. 2006. "Aztec Teotihuacan: political processes at a Postclassic and Early Colonial city-state in the Basin of Mexico." *Latin American Antiquity* 17 (4): 363–387.

Goggin, John M. 1960. *The Spanish olive jar: an introductory study,* University Publications in Anthropology No. 62, Yale University. New Haven: Yale.

Goggin, John M. 1968. *Spanish majolica in the New World: types of the sixteenth to eighteenth centuries.* New Haven: Department of Anthropology, Yale University.

Gosselain, Olivier P. 2000. "Materializing identities: an African perspective." *Journal of Archaeological Method and Theory* 7 (3): 187–217.

Hare, Timothy S. and Michael E. Smith. 1996. "A new Postclassic chronology for Yautepec, Morelos." *Ancient Mesoamerica* 7 (2): 281–297.

Hernández, Carlos, Robert H. Cobean, Alba G. Mastache, and María E. Suárez. 1999. "Un taller de alfareros en la antigua ciudad de Tula." *Arqueología* 22: 69–88.

Hernández Sánchez, Gilda. 2005. "Vasijas para ceremonia. Iconografía de la cerámica tipo códice del estilo Mixteca-Puebla." PhD diss., Leiden University.

Hernández Sánchez, Gilda. 2008. "Indigenous pottery after the Spanish conquest of Mexico: potter's reactions to the new colonial society." *Leiden Journal of Pottery Studies* 24: 5–18.

Hernández Sánchez, Gilda. 2012. *Ceramics and the Spanish conquest. Response and continuity of Indigenous pottery technology in central Mexico.* Leiden: Brill.

Katz, Roberta R. 1977. "The potters and pottery of Tonalá, Jalisco, Mexico: a study in aesthetic anthropology." Unpublished PhD Diss., Columbia University.

Lind, Michael D. 1994. "Cholula and Mixteca polychromes: Two Mixteca-Puebla regional sub-styles." In *Mixteca-Puebla. Discoveries and research in Mesoamerican art and archaeology,* edited by Henry B. Nicholson and Keber E. Quiñones, 79–99. Culver City: Labyrinthos.

Lister, Florence C. and Robert H. Lister. 1978. "The first Mexican Maiolicas: imported and locally produced." *Historical Archaeology* 12 (1): 1–24.

Lister, Florence C. and Robert H. Lister. 1982. *Sixteenth Century Maiolica pottery in the valley of Mexico.* Tucson: University of Arizona Press.

Lister, Florence C. and Robert H. Lister. 1984. "The potter's quarter of colonial Puebla, Mexico." *Historical Archaeology* 18 (1): 87–102.

Lister, Florence C. and Robert H. Lister. 1987. *Andalusian ceramics in Spain and New Spain.* Tucson: The University of Arizona Press.

Lockhart, James. 1992. *The Nahuas after the Conquest.* Stanford: Stanford University Press.

López Cervantes, Gonzalo. 1976. *Cerámica colonial en la ciudad de México,* Colección Científica No. 38. México City: INAH.

Mendieta, Jerónimo de. 1980. *Historia eclesiástica Indiana*. México City: Editorial Porrúa.

Minc, Leah D., Mary G. Hodge and James Blackman. 1994. "Stylistic and spatial variability in Early Aztec ceramics: insights into pre-imperial exchange systems." In *Economies and polities in the Aztec realm*, edited by Mary G. Hodge and Michael E. Smith, 133–174. Albany: Institute for Mesoamerican Studies.

Noguera, Eduardo. 1954. *La cerámica arqueológica de Cholula*. México City: Editorial Guaranía.

Novelo, Victoria. ed. 2007. *Artesanos, artesanías y arte popular en México*. México City: Consejo Nacional para la Cultura y las Artes.

Pleguezuelo, Alfonso. 1999. "Cerámica de Sevilla (1248–1841)." In *Cerámica española. Summa Artis XLII*, edited by Trinidad Sánchez-Pacheco, 343–386. Madrid: Espasa Calpe.

Rice, Prudence M. 1987. *Pottery analysis. A sourcebook*. Chicago: The University of Chicago Press.

Rodríguez Alegría, Enrique. 2002. "Food, eating, and objects of power: class stratification and ceramic production and consumption in colonial Mexico." Unpublished PhD diss., University of Chicago.

Rodríguez Alegría, Enrique. 2003. "Indigenous ware or Spanish import? The case of Indígena Ware and approaches to power in colonial Mexico." *Latin American Antiquity* 14 (1): 67–81.

Roux, Valentine. 2003. "A dynamic systems framework for studying technological change: application to the emergence of the potter's wheel in the southern Levant." *Journal of Archaeological Method and Theory* 10 (1): 1–30.

Rye, Owen S. 1981. *Pottery technology: principles and reconstruction*. Washington, D.C.: Taraxacum.

Sahagún, Bernardino de. 1961. *The Florentine Codex. General history of the things of New Spain, Book 10*, edited and translated by Charles Dibble and Arthur J. Anderson. Santa Fé: The School of American Research.

Sánchez, José M. 1994. *El oficio de ollero en Sevilla en el siglo XVI*. Sevilla: Diputación Provincial de Sevilla.

Sánchez, José M. 1996. "La cerámica exportada a América en el siglo XVI a través de la documentación del Archivo General de Indias I. Materiales arquitectónicos y contenedores de mercancías." *Laboratorio de Arte* 9: 125–142.

Sánchez, José M. 1998. "La cerámica exportada a América en el siglo XVI a través de la documentación del Archivo General de Indias (II). Ajuares domésticos y cerámica cultual y laboral." *Laboratorio de Arte* 11: 121–133.

Sánchez Cortegana, José M. 1994. *El oficio de ollero en Sevilla en el siglo XVI*. Sevilla: Diputación Provincial de Sevilla.

Sanders, William T., Jeffrey R. Parsons and Robert S. Santley. 1979. *The basin of Mexico: ecological processes in the evolution of a civilization*. New York: Academic Press.

Shepard, Anna O. 1948. *Plumbate: a Mesoamerican trade ware*. Washington, D.C.: Publication No. 573, Carnegie Institution of Washington.

Tolstoy, Paul. 1958. "Surface survey of the northern valley of Mexico: the Classic and Post- Classic periods." *Transactions of the American Philosophical Society* 48 (5): 1–100.

Vega, Constanza. 1975. *Forma y decoración en las vasijas de tradición azteca*. México City: Instituto Nacional de Antropología e Historia.

Whalen, Michael E. and Jeffrey R. Parsons. 1982. "Ceramic markers used for period designations. Appendix I." In *Prehispanic settlement patterns in the southern valley of Mexico: The Chalco-Xochimilco regions*, Memoirs of the Museum of Anthropology, No. 14, edited by Jeffrey R. Parsons and Michael E. Whalen, 385–459. Ann Arbor: The University of Michigan Press.

CHAPTER 14

War and Peace in the Sixteenth-Century Southwest: Objected-Oriented Approaches to Native-European Encounters and Trajectories

Clay Mathers

1 Introduction

Although the Southwestern United States was the focus for the largest sixteenth-century *entrada* in North America, evidence for the indigenous use, modification, and consumption of early European objects in this region has been surprisingly modest.[1] In reviewing the archaeological record of sixteenth-century Southwestern *entradas* there is a notable scarcity of early European artifacts in indigenous domestic, mortuary, and other contexts. While sixteenth-century European objects are present in the Southwest, they are linked predominantly with sites associated with Spaniards and their indigenous Mexican allies, rather than indigenous Americans. More striking is that although large assemblages of European contact period items are found where Spanish-led expeditions spent the most time and encountered the greatest indigenous resistance, these same areas present limited evidence that early European objects were utilized in any significant way by indigenous communities – as tools, for display, or for ceremonial purposes. Elsewhere in the Southwest, where more peaceful relations prevailed, early contacts did result in materials being exchanged and incorporated into indigenous contexts, though these objects seldom bear signs of purposeful modification. This discussion argues that the clash and entanglement of material culture and ideational systems at the earliest phase of contact in the Southwest cannot be understood

1 In this discussion the physical and behavioral distinctions between these three are as follows: "use": refers to the employment of objects and materials without the purposeful redesign of their physical form – e.g., edge wear on a knife or axe resulting from repeated episodes of cutting; "modification": denotes a deliberate alteration in the physical form of an object to affect an aesthetic, functional, and/or symbolic transformation that differs from the object's original form, purpose, and/or cultural significance – e.g., converting a ceramic fragment into a perforated pendant; and "consumption": refers to the definitive act of taking an object out of circulation – e.g., by 'disposal' in a burial, ceremonial deposit, or trash midden where the object is unlikely to be retrieved.

without reference to the types of initial contacts between indigenous peoples and Europeans (antagonistic or peaceful), the frequency of contacts and the primary motivations behind them, as well as the political histories of different indigenous groups encountered by early *entradas*. One of the key questions here is why sustained indigenous contact with large Spanish-led expeditions, carrying plentiful supplies of European materials, did not result in large-scale indigenous use, modification, and consumption of European goods.

The chapter concludes by outlining specific episodes of war and peace, and material exchange, during the 1540–1542 *entrada* of Francisco Vázquez de Coronado. Although conflict and "conquista" campaigns characterized many early contacts between indigenous and European groups in New Spain, *La Florida*, and the interior Southeast, the transformation of objects, communities, and strategic policies in these areas was locally variable and changed dramatically by the close of the sixteenth century. Materials characteristic of these changes and variegated responses are found across the Southwest, but have seldom been explored for the insights they provide into broader anthropological themes such as resistance, exchange, and agency. While part of this study focuses on the contextual analysis of objects, its broader goal is to begin comparing cultural trajectories at an interregional scale, particularly the American Southwest and Southeast in the first century of New-Old World contact. Both areas transitioned from initial imperial strategies of resource acquisition and conflict, to policies of settlement and missionization by the end of the 1600s, and a similar suite of European objects was available in both areas. Nevertheless, the manner in which these objects were employed, modified, and consumed by indigenous groups and Europeans varies significantly, and in ways that reveal important aspects of the earliest colonial encounters in North America.

2 Macro-Regional Trajectories and Comparisons

Before turning to objects and assemblages, it is important to understand some of the broader contexts and histories in which these materials were embedded. Cultural and physical landscapes along the southern margins of North America in the sixteenth century held very different geographic, demographic, economic, and geopolitical potentialities for both indigenous polities and the bourgeoning Spanish Empire. Although early Southeastern *entradas* focused on the littoral margins of *La Florida* and on the indigenous chiefdoms nearby, later expeditions targeted larger, more aggregated, and hierarchical Mississippian chiefdoms in the interior. Compared with the Trans-Mississippi

West, the Southeast was more proximal to the trade routes linking Spain's prosperous colonial activities in the Americas to its operational home base in Iberia. Though less socially and politically stratified than state-level societies in Mexico and Peru, Mississippian groups did exhibit multi-tiered site hierarchies, well-defined social ranking, institutionalized disparities in resource access, and clear signs of local and regional tribute payments.

By contrast, indigenous communities in the sixteenth-century Southwest were more "nodal," with large uninhabited or thinly populated areas extending 50–100 miles between some primary settlement clusters – particularly in arid interfluvial areas. In addition, a large buffer zone of hostile sedentary and hunter-gatherer groups (known pejoratively by the Spaniards as "Chichimeca") populated large swaths of territory for many hundreds of miles between Spanish-occupied regions in central Mexico and autonomous Pueblo communities situated on New Spain's northern frontier. This barrier persisted until the early 1590s when a pan-regional megadrought ended, more benign Crown policies of gift exchange and resettlement of indigenous communities were introduced, and the nearly 50-year long Chichimec War was concluded (Powell 1967).

Throughout the sixteenth century, the American Southwest was extremely remote from primary centers of Spanish colonial activity in the Caribbean and the Valley of Mexico. Located on the northern edge of New Spain, it was largely landlocked, peripheral to many colonial trade routes, and of questionable economic value with respect to agricultural, ranching, craft production, trade, and mining activities. Compared with Spanish-led expeditions in the Southeast, Southwestern *entradas* were almost exclusively pedestrian affairs, involving little or no maritime travel, and often requiring overland journeys of 800–1000 miles before reaching their initial destinations. Within the Southwest, there were few signs of the vertically integrated regional and interregional polities seen in contemporary Southeastern communities. Equally difficult to discern in the sixteenth-century Southwest were the developed systems of tribute, warfare, social ranking, and wealth disparities akin to those found in many Mississippian and Peninsular Florida chiefdoms.

3 Comparing Material Trajectories

As a first step in assessing indigenous-European material encounters in the American Southeast and Southwest, it is important to emphasize the general similarity in materials and object types present in the archaeological record of both areas. A list of these early object and material types – dating largely to the early/mid-sixteenth century – is included in Table 14.1 (below). In it we

TABLE 14.1 Typical non-perishable artifact types and raw materials used as trade goods in the early Southwestern and Southeastern *entradas* (ca. AD 1500–1599)

Object type	Raw material	Southwest	Southeast
Adzes	Ferrous	✓	✓
Awls	Ferrous	✓	✓
Axes	Ferrous	✓	✓
Chisels	Ferrous	✓	✓
Knives	Ferrous	✓	✓
Clarksdale Bells	Cupreous	✓	✓
Tubular Beads	Cupreous		✓
Chevron-Faceted Beads	Glass	✓	✓
Nueva Cádiz Beads	Glass	✓	✓
Tumbled Blue Beads	Glass		✓

can see that iron objects such as flat axes, chisels, knives, and awls, as well as copper bells, and distinctive forms of glass beads were items frequently mentioned as "trade goods" or "gifts" in expedition documents. These same items appear in archaeological contexts in the Southeast and Southwest associated with indigenous communities known to have been in contact with Spanish-led *entradas*.

Despite some morphometric differences between early Spanish-European objects documented currently in these regions, the overall similarities are striking and are clearly important in helping to distinguish early *entradas* and contacts (ca. AD 1500–1550), from later expeditions and encounters (ca. AD 1551–1600) in both regions (Blanton 2018; Little 2008; Smith 1987). As Mathers (2013) has argued, some of the major distinctions between these assemblages, pre- and post-1550, relate to a fundamental shift in Crown policy from "conquista" campaigns to missions and colonization.

Other important variables in evaluating early indigenous-European material trajectories in the Southeast and Southwest include:

- the incidence of indigenous-European encounters;
- the chronology of these contacts, and
- the types of the activities they represent (e.g., what types of ethnic groups/ social classes were involved? And in what spatial-behavioral contexts?)

1. **Contact Frequency**. One stark difference in the cultural trajectories of *La Florida* and the Southeastern interior on the one hand, and northern New Spain on the other, is the incidence of indigenous-European

contacts. Documented encounters between indigenous communities and Europeans began in the Southeast in the second decade of the 1500s and continued with some regularity to the end of the century (Table 14.2). By contrast, the earliest documented indigenous-European contact in the Southwest is 26 years later, in 1539, when Esteban the Moor and his indigenous allies reached the Zuni village of Hawikku (Flint and Flint 2005, 59–64). Indigenous-European contacts in the Southeast are not only more frequent throughout the sixteenth century, they also represent efforts to establish colonies and missions earlier than in the Southwest (e.g., compare the 1526 colonization attempt by Vázquez de Ayllón in coastal Georgia (Hoffman 1994), to Oñate y Salazar's 1598 colonial efforts in northern New Mexico (Ellis 1989, 9–23), and the 1566 mission established at Santa Elena in Georgia by Menéndez de Áviles (Milanich 2006, 92–83), with the earliest Southwestern mission founded in 1598 by Oñate y Salazar in northern New Mexico (Barrett 2012, 36–37)). The relative lag in the Southwest with respect to both colonization and missionization, and the general dearth of Spanish expeditions in that area until the late 1500s, contributed significantly to the modest volume of European objects known to indigenous communities throughout the sixteenth century. The low frequency of indigenous-European contact in the sixteenth-century Southwest was due in large measure to: (1) its peripheral geopolitical/

TABLE 14.2 Comparison of the Spanish/European presence in the US Southeast and Southwest during the sixteenth century (Based on the number of years with documented indigenous-European contact)

Decade	SE	SW
1500–1509		
1510–1519	5	
1520–1529	4	
1530–1539	1	1
1540–1549	5	3
1550–1559	1	
1560–1569	11	
1570–1579	12	
1580–1589	10	3
1590–1599	10	5
Total No. of Years with Episodes of Contact	59	12

geographic position relative to other strategic assets; (2) the success of Chichimec groups in limiting Spanish activities north of Mexico City, (3) the general lack of large, aggregated villages with stored food reserves, and (4) the severe megadrought that dominated the region between the mid-1570s and mid-1590s (Van West et al. 2013, 88, 93).

2. **Contact Chronologies.** Given the major shift in emphasis of Spanish-led expeditions at approximately mid-century – transitioning from large, "conquista"-style military efforts to smaller, evangelical missions and settlement – some attention to the timing of indigenous-European encounters in the Southwest and Southeast is appropriate. As Mathers (2013) suggests, the disruptive effects, crippling costs, and unfavorable international visibility resulting from "conquista" campaigns and their associated *encomienda* rewards system, characterized the early sixteenth century. Indigenous uprisings were common in the Americas throughout the sixteenth century (Figure 14.1) and were expensive to remedy. The adverse, wide-ranging consequences of these conflicts encouraged the Spanish Crown into enacting new policies and legislation (notably the New Laws of 1542–1543), designed to reorient its New World priorities towards smaller, less overtly belligerent expeditions. In comparing the two regions, it is significant that only one "*conquista*"-style *entrada* is known

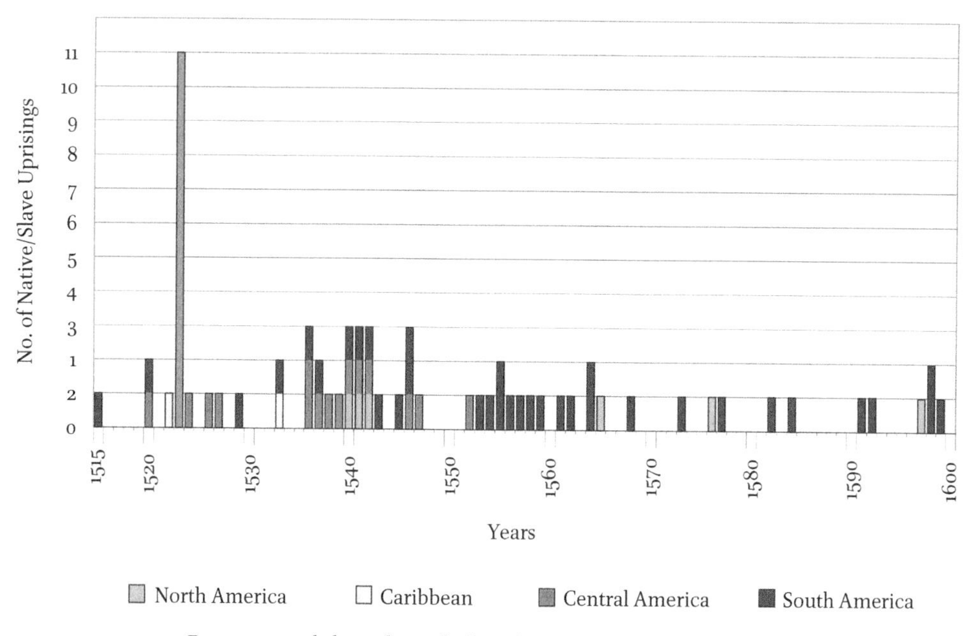

FIGURE 14.1 Frequency and chronology of selected sixteenth-century native and slave-related uprisings in the Americas. The source of these data are a range of published archaeological and ethnohistorical literature from the Americas

in the Southwest during the sixteenth century. Moreover, the number of years in which indigenous-European contact can be documented in the Southwest is only about a fifth of the contact years evidenced in the Southeast (Table 14.2). Encounters in the Southeast began earlier and were more frequent and varied than in the Southwest, involving a variety of terrestrial and maritime contacts, as well as an earlier shift from military-style expeditions to *entradas* involving a greater emphasis on settlement, gift exchange, and missionization.

3. **Nature and Context of Contacts.** In both the Southwest and Southeast, the character of sixteenth-century indigenous-European encounters varied from amiable and ephemeral, to protracted conflict and open warfare. Significantly perhaps, when terrestrial *entradas* remained for an extended period amongst indigenous communities in both areas the result was sustained conflict (e.g., Clayton et al. 1993a, 71–73; 1993b, 192–194; Mathers 2013). Smaller, more mobile maritime expeditions seldom faced such hostility, partly because they bulk transported their provisions, thereby reducing the pressures on local communities to supply them with food, shelter, and clothing. *Entradas* involving marine and riverine transport could often move rapidly from one indigenous area to another, with less effort, defend themselves more easily with large, well-armed water-borne vessels, and retreat more readily from danger, than their land-based counterparts.

Food shortages were the perpetual bane of sixteenth-century *entradas* and colonies from *La Florida* (Hudson 1997, 102–104, 167–171, 185–187, 378; Priestly 1928a, 139, 153, 203, 209; 1928b, 57) and *Tierra Nueva* (Flint and Flint 2005, 255–257, 291, 557–558) south to the Río de la Plata (Dominguez 2005, 9, 41, 107; García Loaeza and Garrett 2015, 6, 23, 56, 90) and Tierra del Fuego (Clissold 1954, 165–167). Provisioning large expeditions tended to strain indigenous-European relations the longer the latter made significant demands on the former. On the other hand, small expeditions were consistently in danger of attack and/or annihilation.

Smith and Hally (2019) have proposed that sixteenth-century European materials were obtained by indigenous Southeastern communities through mechanisms such as formal gift exchange, barter, battle trophies, theft/pilfering, scavenging, and shipwreck salvage. To this list, we could add transfers of European goods from indigenous guides and allies, who had obtained these items by various means, but often as a result of services rendered to European expeditionaries. The latter form of exchange is particularly significant since the "*chaîne opératoire*" for the transmission of European materials is made more complex in expeditions that included significant numbers of indigenous allies, such as the many Mexica and Tlaxcalans in Vázquez de Coronado's *entrada* in the Southwest (Flint 2009) and similar groups amongst Luna y Arellano's expedition in the Southeast (Bratten 2009, 109–110). While Spaniards engaged in early expeditions are known to have presented a variety of European-made goods to indigenous communities – especially as a means of establishing peaceful relations and strategic alliances (Flint 2002, 281–282; Hernández 2005, 110; Wilson 1990, 69) – conventional archaeological approaches have often seen such indigenous-European exchanges in rather simple linear terms – and with exchanges taking place primarily between elites (Figure 14.2). These simple models fail to account for indigenous allies as major actors in Spanish-led expeditions across the Americas and for their opportunities to participate actively in the exchange of European objects at a variety of social levels.

Because indigenous allies are mentioned infrequently in *entrada*-related narratives, examples of payments and rewards to indigenous expedition members are rare, and are generally restricted to higher social ranks. Examples include the right of the Tlaxcalan allies of Hernán Cortés to carry swords and

Model 1: Binary-Linear Transmission

Model 2: Tripartite-Linear Transmission

Model 3: Multi-Actor-Elliptical Transmission

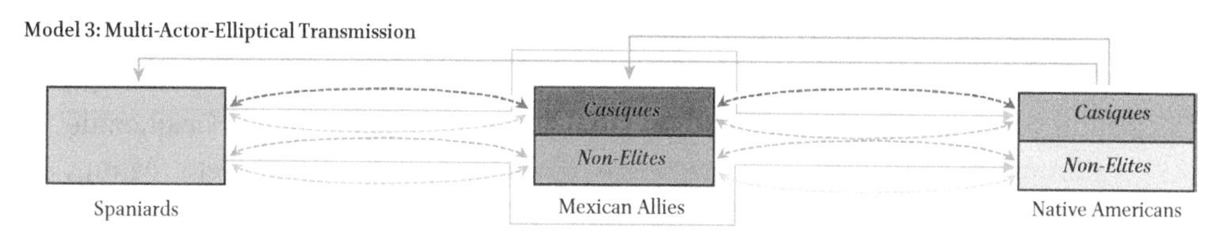

FIGURE 14.2 Schematic models of indigenous-European exchange in the American Southwest and Southeast

firearms, wear Spanish clothing, and ride horses (Flint 2009, 75; Gibson 1968, 155), the gifts given by Marcos de Niza to discourage the desertion of his Mexican allies (Flint and Flint 2005, 74), and the "provisions for gratuitous distribution" offered by Adelantado Álvar Núñez Cabeza de Vaca to his indigenous Paraguayan allies (Hernández 2005, 173–174). It also appears that not all allies were regarded equally and that Spanish recruitment of indigenous auxiliaries often recognized a distinction between well-trained warriors (such as Nahau), and others (such as Tabascans), and apportioned expedition tasks accordingly (Chuchiak 2007, 199).

Having received or otherwise procured European objects, some allies may have created their own rules concerning when, what, and with whom they might trade, including both mundane and exotic European items. Accounts of indigenous auxiliaries starving and freezing to death on various expeditions, such Vázquez de Coronado's in the Southwest (Flint and Flint 2005, 235–236), or Diego de Almagro's on the Argentine-Chilean border (Pocock 1967, 26–27), suggest that the hardships associated with Spanish-led *entradas* could be extreme and have lethal consequences for many indigenous allies – through combat, illness, exposure, poor diet, starvation, or some combination of these deprivations (e.g., Clayton et al. 1993a, 242; Flint 2009, 74; Francis 2007, 35; Restall and Asselbergs 2007, 16). Trading European objects for food in such resource-stressed or life-threatening situations may therefore have been commonplace – with respect to both indigenous allies and Europeans. Given the utilitarian nature of many of the sixteenth-century European objects that have emerged recently from the Stark Farm and Glass sites, in Mississippi and Georgia, respectively (Cobb and Legg 2017; Blanton 2019) – compared with traditional "gift items" such as beads and axes – we may be seeing material evidence of these more broadly-based trade relations between both indigenous and European non-elites. The complexities inherent in modeling indigenous-European exchange processes are compounded further in the Vázquez de Coronado *entrada* since after the expedition returned to Mexico City in 1542, a number of Mexican allies remained behind, choosing to live the remainder of their lives in indigenous communities such as the Zuni pueblo of Halona:wa (Hammond and Rey 1967, 89, 93).

With these complexities and challenges in mind, the following case studies focus on historically–, and/or archaeologically–, documented assemblages and objects, associated with the 1540–1542 *entrada* of Francisco Vázquez de Coronado. These four case studies highlight geographically discrete regions, ethnically distinct indigenous American communities, and generally follow the chronological trajectory of this expedition.

4 Case 1: Hernando de Alarcón Naval Contingent, Sea of Cortes/ Lower Colorado River, (May–September(?) 1540)

Although designed as a supportive supply link for the terrestrial portion of this *entrada*, and perhaps some general reconnaissance, the naval segment of this expedition never fulfilled its primary mission owing to the scarcity of navigable waterways in the Desert Southwest. Departing from Acapulco, and then Culiacán, in what is now western Mexico, Alarcón's maritime expedition was well supplied with food, clothing, and weapons, as well as an unspecified number of crew. Once Alarcón reached the Colorado River, he was forced to continue using two small launches that could better navigate the swift, changeable currents.

Judging by Alarcón's report of his encounters with indigenous Yuma communities in the Lower Colorado River (LCR) Basin (Flint and Flint 2005, 185–205), his contacts were characterized by:

- Generally peaceful relationships, with no signs of conflict in the extant archaeological record or historical documents;
- Amicable gift giving – involving the presentation of European goods such as clothing, beads, and food;
- And as further evidence of peaceful relations, Alarcón reports numerous large wooden crosses being erected by Spanish expeditionaries, and their veneration by local Yuma communities.

Alarcón's generally benign contacts with indigenous communities in the LCR were largely the result of the expedition's:

1. **Small Size**: the relative symmetry between the size of the expedition and the indigenous communities they encountered;
2. **Minimal Expedition Requirements**: because of their relative self-sufficiency with regard to clothing, food, and shelter, the expedition made minimal demands on local communities;
3. **Minimal Threats**: the potential menace posed by Alarcón's *entrada* was negligible, owing to the expedition's modest size and requirements, minimal periods of occupation, and the relatively modest investment of Yuman communities in institutionalized social ranks (i.e., positions that could have been impacted adversely by the Spaniards, their alternative cultural values, and the dissent created by differential acquisition of exotic European objects);
4. **Mutually Favorable Exchanges of Goods**: Alarcón's plentiful supplies and the manner in which he distributed them, helped promote friendly relations with Yuman communities;

To date there is no archaeological evidence from the LCR region to suggest how early trade items obtained from the Spaniards were used or consumed by indigenous communities.

5 Case 2: Zuni (July–December 1540)

Moving north from their initial departure point in western Mexico, and into what is now the border zone between New Mexico and Arizona, the Vázquez de Coronado expedition encountered a largely uninhabited area ("despoblado"), where provisions began to run dangerously low. Upon reaching the Zuni pueblo of Hawikku, Vázquez de Coronado was confronted by indigenous warriors and warned against entering their settlement. After a short engagement between Spaniards and their allies on the one hand, and Zuni warriors on the other, the former succeeded in defeating the Zunis and occupying Hawikku for the next four to five months. Shortly after the Battle of Hawikku, Vázquez de Coronado sought out the *caciques* from many Zuni pueblos to negotiate more peaceful relations. These contacts resulted in material exchanges between Spaniards and Zuni leaders that are confirmed in both sixteenth-century Spanish documents and in the archaeological record (Flint 2002, 281–282, Howell 2001; Mathers et al. 2011).

Since the size of the Vázquez de Coronado expedition at Zuni approached 3,000 individuals (~75% consisting of indigenous Mexican auxiliaries), and more than 7,000 animals, local indigenous communities were faced with an immediate logistical and military dilemma of considerable magnitude. Unable to defeat the Spaniards and their allies in direct combat, Zuni communities opted for a more tactical plan of passive resistance and active reconnaissance to ensure that the Spaniards would soon look elsewhere for labor, clothing, food, and shelter. As part of this plan, Zuni guides spent months directing the Vázquez de Coronado expedition north to the Grand Canyon, west to the Hopi Pueblos, south to the Piro area, and east to the Buffalo Plains.

Our best archaeological evidence for encounters between the Vázquez de Coronado *entrada* and the Zuni come from the large-scale, 1917–1923 excavations at Hawikku (Smith et al. 1966) and Kechiba:wa (Hodge 1920; Lothrop 1923). Investigations at both pueblos recovered Vázquez de Coronado objects from residential and mortuary contexts, though excavators failed to recognize both their date and significance.

Analyses of Vázquez de Coronado objects from mortuary contexts at Kechiba:wa (Mathers et al. 2011) indicate that only 8.4% of the 266 excavated burials included European objects. All of these burials were adults; the individual's gender was not recorded. And while no comprehensive analyses have been undertaken of Vázquez de Coronado materials from Hawikku, Howell's research (2001, 151) on the 955 burials excavated there indicates only 13 included iron objects (1.4%) and only 3 (0.3%) contained copper artifacts; children and adults are more or less equally represented in these totals. Only seven Hawikku burials (0.73%) contained European glass beads (Smith et al. 1966, 265), and only two excavated interments at Kechiba:wa contained Vázquez de Coronado-period glass beads – a nearly identical percentage (0.75%). What is clear from these data is that known trade objects, dating to the Vázquez de Coronado *entrada*:

- Were directed towards a small percentage of adults and children at Zuni, indicating the presence of well-defined social ranks there (evidence of such hierarchical status and ranking is present at Zuni in both the prehistoric and early historic periods);
- These findings are congruent with Spanish narratives describing trade with Zuni *caciques* following the Battle of Hawikku;
- The exchange of European objects at Zuni emphasized items with a ceremonial, rather than utilitarian, value (e.g., small hatchet-like axes too thin and fragile to have been functional for cutting);
- European trade items at Zuni appear unmodified and are not visibly altered to conform to Zuni cultural aesthetics/norms.

The relatively peaceful relations between Zuni communities and the Vázquez de Coronado expedition, following their initial confrontation at Hawikku, is further demonstrated by the construction of large wooden crosses – documented by members of the Espejo *entrada* who noted these large, well-built crosses throughout the Zuni area in 1583 (Hammond and Rey 1967, 89).

Another significant object from the Hawikku excavations (National Museum of the American Indian Ref. No. 085899.000 (1)) is a small, reshaped ceramic fragment with Mexican painted motifs on its interior surface. This fragment has a small annular ring base on its exterior surface and was included as part of Hawikku Burial 899 – an infant accompanied by a small, finely woven basket/tray, with this broken, decorated bowl fragment placed on its legs (Hodge 1918b, 140–141). The surviving decoration appears to represent a duck-like figure floating on water, with various additional symbols and glyphs along its margins. Discussions with a number of Mexican ceramic specialists in the US, Canada, and Mexico suggest that this fragment:

- Probably dates to the mid-late sixteenth century;
- Originates from Central Mexico – perhaps the Puebla-Tlaxcala region or more probably the Southern Basin of Mexico;
- Includes dot symbols (signifying numerals) and a "tepetl" (or place) glyph (Philip Arnold, Joseph Ball, Patricia Fournier, Geoffrey McCafferty, and Jeffrey Parsons, personal communications February 2017).

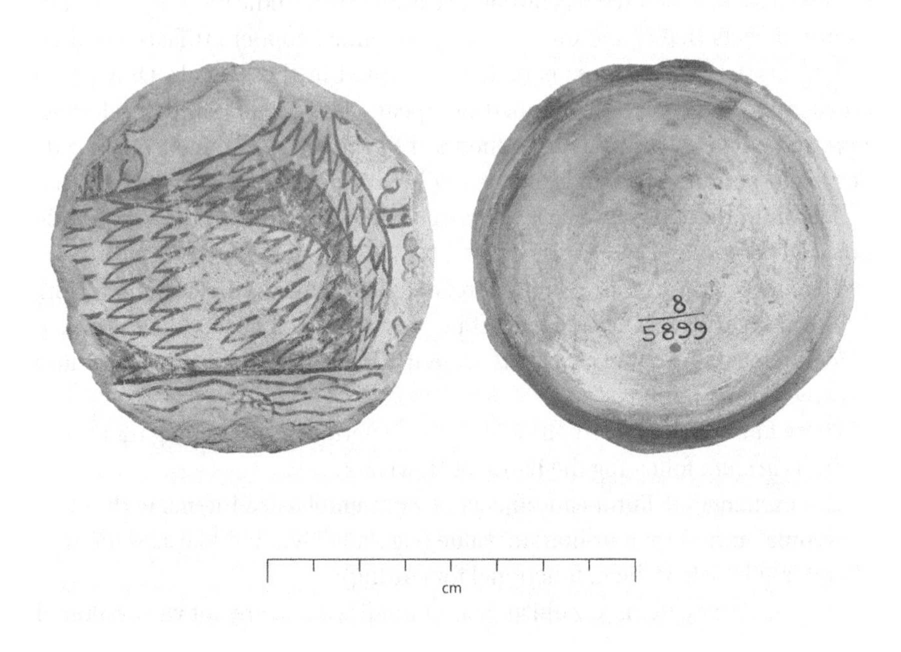

FIGURE 14.3 Shaped Mexican ceramic fragment with bird motif
NATIONAL MUSEUM OF THE AMERICAN INDIAN (REF. NO. 085899.000 (1))

Patricia Fournier (personal communication, April 2012) suggested the design motif "might be part of early colonial period traditions that show drastic deterioration and oversimplified patterns, surface treatment, and colors of the Postclassic Cholula polychrome wares." In addition, the surface treatment of this sherd, its iconography, and annular base, are consistent with a mid- to late sixteenth-century Mexican chronology, possibly corresponding to Aztec IV-style ceramic traditions (Griffin and Espejo 1950, Jeffrey Parsons personal communication, February 2017). Since the Vázquez de Coronado expedition spent four to five months encamped at Hawikku, with ~2000 Mexican auxiliaries, and at least three or four Mexican allies remained at Zuni after the expedition returned to Mexico City, the balance of evidence suggests this bowl is associated with the Vázquez de Coronado *entrada*.

The reshaping of this broken sherd appears to have been executed fairly carefully to produce a regularized edge that mirrors the annular base on its exterior surface. Significantly, a variety of other reshaped ceramic fragments appear at Hawikku (n=16) – all of them found in mortuary contexts (Smith et al. 1966, 237). These circular or rectilinear fragments are associated with both cremations and inhumations, and with adult, adolescent, and child burials. While the positioning of these sherds on the body was recorded in only three cases, one child burial had a single shaped sherd on its legs and another on its knee, in a manner similar to the placement of the shaped Mexican fragment from Burial 899. So, while there is evidence at Zuni for the reshaping of local ceramic fragments and incorporating them into burials, similar mortuary practices have not been documented to date in Central Mexico during the early historical period (Patricia Fournier personal communication, February 2017).

Whether the decorated Mexican fragment from Hawikku was associated with material left by Spaniards associated with the Vázquez de Coronado *entrada*, by Mexican allies during their 1540 occupation, or by indigenous auxiliaries remaining behind after the expedition returned to Mexico, it seems likely that this fragment:

- Constitutes one of the few material signatures of indigenous Mexicans known to date in the contact period Southwest;
- Was modified and then grafted into a set of existing Zuni mortuary practices that had previously been based largely, if not exclusively, on locally available indigenous American objects and materials; and
- Represents only a fraction of the extant indigenous Mexican material and behavioural repertoire from the contact period Southwest. Further evidence is suggested by a votive stone object from Hawikku Burial 59 with an appearance similar to the sword-like Mexica weapons known as "macanas" or "macuahuitls" and undecorated Mexican ceramic fragments – both found in the Western Cemetery (Hodge 1918a:376, fig. 8; 1918b, 371, 375).

6 Case 3: Tiguex (Summer 1540–Spring 1542)

Attempts by the Zuni to redirect Vázquez de Coronado to other areas of the Southwest, and to free themselves of this large occupying force, began bearing fruit in the autumn of 1540. Having scouted widely in regions surrounding Zuni, Vázquez de Coronado and his commanders were convinced that the Tiguex region (now the metropolitan area surrounding Albuquerque, New Mexico) was the best place to establish themselves for the winter of 1540–1541. Bitterly cold conditions, the absence of clothing and other resources expected from Alarcón's supply ships, and the difficulty of provisioning such a large expedition, all conspired to create problems and animosity. Shortly afterwards, these tensions erupted into open hostility and sustained warfare. The Tiguex War lasted for more than a year and half, resulted in large numbers of casualties, and saw every Tiguex pueblo burned and damaged.

Despite its scale and intensity, generations of scholars appear to have regarded this conflict as rather inconsequential, without either major or lasting consequences for the trajectory of indigenous-European relations in the Southwest. However, archaeological investigations are now providing evidence of the ferocity and consequences of the Tiguex War that have been missing from the rather limited perspective of Spanish documents (Mathers and Marshall 2014). These new analyses provide a fresh look at the material manifestations of combat, conflict, trade, and cultural transmission, and an opportunity to re-evaluate the Tiguex War using the rich detail of the archaeological record. A single European object from Santiago Pueblo, provides an example of these new perspectives.

At the time Santiago Pueblo was excavated, in 1934–1935, there was considerable enthusiasm to celebrate the 400th anniversary of the Vázquez de Coronado *entrada*. Hoping to uncover material proof of this expedition in time for the 1940 anniversary, excavations were undertaken at Santiago, and at the nearby Pueblo of Kuaua. Ironically, materials uncovered at Santiago demonstrated exactly what the excavators hoped for, but nevertheless failed to recognize.

Diagnostic Vázquez de Coronado objects found during excavations at Santiago include copper crossbow quarrels, a caret-headed nail, and long copper lace chapes (Ellis 1957; Museum of Indian Arts and Culture n.d., Tichy 1939, 161–162). Despite their distinctive shape, size, form of manufacture, and widespread presence on European sites, in museums, and in the historical/archaeological literature, crossbow quarrels at Santiago were regarded as possible "pen tips," as they had been in earlier investigations at Pecos Pueblo (Kidder 1932, 307, figure 251, i). Consequently, when a crossbow quarrel was found in the chest cavity of an indigenous burial at Santiago, it passed with little comment or analysis (Tichy 1939, 162).

Ellis (1957) was the first to identify crossbow quarrels in the American Southwest and specifically addressed his article to the examples found at Santiago Pueblo. In his pioneering paper, however, Ellis (1957, 213–214) suggested that crossbow quarrels could have been recovered and repurposed by indigenous communities for "re-use in Indian ways." Furthermore, Ellis (1957, 214) made specific reference to one crossbow quarrel-like object found in the Santiago excavations (Museum of Indian Arts and Culture, Catalog No. 44778/11, Bp 38/15). Ellis suggested it resembled the "tinklers" used by Pueblo and other Southwestern indigenous communities. The object described by Ellis is indeed a crossbow quarrel, though there is no evidence whatever to indicate that it was found in a context suggesting indigenous use, that there were modifications after it was distorted and broken following discharge, or that such objects had been modified and used by either Southern Tiwa communities or any other Pueblo/indigenous group in the Southwest.

Having predicted, and then discovered, a large Vázquez de Coronado battle assemblage in the vicinity of Santiago Pueblo (Mathers 2011, Mathers and Marshall 2014), questions arose about the function, and possible indigenous modification, of Vázquez de Coronado objects found in the 1930s excavations. In particular, whether crossbow quarrels were simply fired and unmodified or whether they were objects later repurposed by indigenous Pueblo communities into the form Ellis regarded as tinkler-like or ornamental.

Re-examination of the object Ellis suggested could be a crossbow quarrel transformed by indigenous groups into a tinkler-type ornament, suggests that:

- This object is heavy, thick, and an unsuitable size and shape for producing any audible resonance;
- Furthermore, its wall thickness (1.4 mm) would have made it difficult to reshape without a knowledge of annealing, and there are no tell-tale surface indications of the dimpling and impressions of stone tool working necessary to produce a more suitable tinkler-form;
- Instead, its maximum length, width, and thickness (37.5 mm, 9.6 mm, and 0.5 mm) are all close to the mean and standard deviation (SD) of 18 nearly complete crossbow quarrels from Santiago with length and thickness measurements, and 11 crossbow quarrels with meaningful width measurements – i.e., mean = 38.9 mm, 9.2 mm, and 0.7 mm, and SD = 13.8 mm, 1.80 mm, and 0.3 mm respectively.

Apart from the single burial at Santiago, neither crossbow quarrels nor metal tinklers of any kind are known from 394 excavated burials there, from the ~600 burials excavated at the Southern Tiwa Pueblo of Kuaua (Dutton 1963, 26), or the 42 and 23 burials with data from the Tiguex Pueblos of Alameda and Chamisal, respectively (Cordero 2013, 201–220). A similar absence of both object types is notable from ~2000 excavated burials at the Towa Pueblo of Pecos, and from the Zuni Pueblos of Hawikku and Kechiba:wa with 955 and 266 excavated burials, respectively. All of these Southern Tiwa, Towa, and Zuni communities are known to have been contacted by the Vázquez de Coronado *entrada*, but none have produced burials with metal tinklers or crossbow quarrels.

Several salient patterns emerge from this analysis of Vázquez de Coronado materials in the Tiguex area:

- Despite the expedition's size and protracted occupation in this region, there are no clear signs that early European objects were used, modified, or consumed by Southern Tiwa communities – despite major excavations at six Tiguex sites (Cordero 2013; Dutton 1963; Marshall 1982; Pooler 1940; Sargeant 1985; Tichy 1939).

- Early European objects found to date in and around Tiguex Pueblos appear to be the result of occupation, and/or attacks, by the Vázquez de Coronado *entrada*;
- Mortuary, ethnohistorical, and anthropological evidence from Tiguex suggests a history marked by little emphasis on the display of high value, exotic goods for marking rank and status (evidence of social 'leveling' or distributed authority, ranks, and status are marked features of Southern Tiwa mortuary practices throughout the prehistoric and early historic periods);
- Despite the unusual opportunities that Southern Tiwa groups had to acquire early European objects, very little evidence has emerged to suggest they did so;
- The overwhelming resource demands made by the Vázquez de Coronado expedition, and their arrival at the beginning of winter – when both indigenous communities and Europeans would be reliant on stored (indigenous) food supplies and clothing – created a lethal threat to indigenous communities that could not be finessed easily, as it was at Zuni;
- The antagonistic relationships between indigenous and European groups in Tiguex is underlined by only a single reference to the exchange of "a few small items" (involving Alvarado's initial encounter with Tiguex communities in 1540 (Flint and Flint 2005, 305)) and the lack of any clear evidence for the erection of crosses anywhere in Tiguex in all known historical narratives relating to this phase of the expedition.

7 Case 4: Southern Plains (Spring–Summer 1541)

Following their two-month siege of Moho, the expedition left Tiguex in the spring of AD 1541 in search of Quivira and major indigenous settlements in the Great Plains. Led by indigenous guides, they traveled east past Pecos Pueblo into the territory of Querecho and Teja groups in the Texas Panhandle, western Oklahoma, and Kansas (Habicht-Mauche 1992). Mobile buffalo hunters in these regions were represented by fairly small groups and mostly temporary settlements. Relations between the expedition and these indigenous communities appear to have been peaceful, despite the latter's reputation as fierce warriors (Flint and Flint 2005, 421, 423).

Notwithstanding the generally small, migratory nature of Querecho and Teja communities in the Southern Plains, and their often-ephemeral archaeological footprint, there are archaeological traces of European and Mexican objects from the Vázquez de Coronado *entrada* recorded in this region, as well as documentary evidence of wooden crosses being erected (Billeck 2009; Sudbury 1984; Bell 1959; Hoard et al. 2008, 223–225 and Flint and Flint 2005, 517, respectively). Artifacts include Nueva Cádiz and faceted chevron beads, a possible prismatic core of Mexican obsidian, and three examples of Pachuca obsidian.

It is likely that the overwhelming size and firepower of the Vázquez de Coronado expedition discouraged any hostile encounters with Querecho-Teja groups in the Southern Plains, and that the expedition's demands on these indigenous communities would have been minimal given the latter's modest size and resources. Objects linked with the Vázquez de Coronado expedition do not appear in any definitive contexts suggesting exchange targeted exclusively at indigenous Querecho-Teja leaders, so it is difficult, at this stage, to determine if such items were directed towards restricted social ranks. To date there is no sign that the original form of these objects was modified or reconfigured. And again, the erection of crosses and generally positive tone of expedition narratives for this area, suggests peaceful relations with the expedition throughout this region.

8 Conclusion

This chapter has emphasized the role of war and peace in influencing early encounters between indigenous and European communities in the American Southwest. It has also underlined a degree of congruence between the documentary and archaeological records on matters of indigenous-European material exchanges and relations. In addition, it has highlighted some of the major factors that impacted the trajectory of these exchanges and relations throughout the sixteenth century; the history and consequences of these interactions continued to resonate in the seventeenth century as well. A variety of geographic, climatic, historical, and geopolitical dynamics helped to maintain a relatively low level of indigenous-European contact in the Southwest, including its largely arid, landlocked landscapes; marginal productivity; peripheral position relative to other Crown assets; an historic mega-drought in the late 1500s; the paucity of large aggregated indigenous communities with significant stored food supplies; and a protracted, region-wide indigenous-European conflict lasting nearly fifty years.

Nevertheless, as the case studies above illustrate, the reaction of indigenous communities to encounters with Europeans in the sixteenth-century Southwest was far from uniform. Some groups – not overtaxed by demands for labor, food, and other resources – seem to have had reasonably amiable relations with Vázquez de Coronado and later *entradas*. And in those indigenous groups, like Zuni, where social ranking and material differentiation had some historical roots, European items obtained from Spanish-led expeditions were used to promote and sustain elevated socio-political and economic positions. Some key features of the more amiable indigenous-European encounters in the Southwest, appear to be:

- Limited exposure/European contact (involving relatively short expedition stays in the territories of any one indigenous community);
- Limited demands on indigenous resources; and
- Indigenous regions with limited to modest economic productivity (especially areas where stored food supplies and village sizes were restricted).

These conditions appear to have prevailed in the Lower Colorado River Valley, at Zuni (after some initial conflict), and in the Southern Plains. In Tiguex, however, conditions were badly aligned for any peaceful outcome – particularly with the arrival of the expedition at a time of year when the stored resources of Southern Tiwa communities were essential to their survival, and to the large occupying force of the Vázquez de Coronado expedition and their numerous livestock. Here local resources were insufficient to maintain both groups and their relations soon became lethal on a region-wide scale. The political histories of Southern Tiwa communities also played a key role in discouraging an interest in, as well as the acquisition and display of, exotic materials and objects.

Recent work in the Southeast to model *entrada* assemblages associated with Hernando de Soto (e.g., Blanton 2019; Mitchem 2014) have suggested that as this expedition moved inland and began to deplete its supplies and personnel, the resources available for distribution to indigenous communities began to diminish. While their resource reduction model is a new and important development in understanding early indigenous-European encounters, as well as the nature and trajectory of resource use through time, the Tiguex example above clearly suggests that the availability of "trade goods" (however defined) is only one element in a complex constellation of factors that need to be considered when attempting to understand issues of object use, modification, and consumption.

This discussion has attempted to consider many of the material exchange "vectors" involved in early indigenous-European relations and the assemblages of objects that resulted from them. One of the least tangible of these concern indigenous materials obtained by European expeditionaries – either from their indigenous allies (such as cotton-quilted armor and Mexica-indigenous weapons), or from indigenous American communities (such as blankets, ceramics, and food). Finding traces of these less visible items/exchanges remains a challenge for archaeologists addressing the early historical period. Nevertheless, the repurposed ceramic vessel from Hawikku, the Pachuca (Mexican) obsidian from the Vázquez de Coronado encampment near Santiago Pueblo in Tiguex (Vierra 1989, 119), the Mexica ceramics and obsidian from Luna y Arellano sites in the Pensacola area of Florida (Worth 2016), and the Coosawattee book/box plate from the Poarch Farm site in Georgia (Langford 1990), may point the way towards new possibilities and perspectives. Further material evidence of the Mexican presence in early Spanish-led *entradas* may help us uncover the breadth and depth of the multi-, rather than unidirectional, nature of indigenous-European exchanges and relations in the fifteenth and sixteenth centuries. In addition, it may help us understand the major organizational shift from Spanish-led expeditions that included indigenous allies recruited in Mexico (a sixteenth-century model) to indigenous American auxiliaries recruited locally (a practice more typical of the seventeenth century). Furthermore, as we recognize more of these objects, and the ethnic groups associated with them, we may begin to penetrate beneath the catch-all terminology of "*indios amigos*" and begin the more interesting analyses of the varied ethnic origins, roles, and cultural practices within different groups of indigenous allies during the first century of European contact.

Examples of indigenous Mexicans and Africans in the Southwest (Hammond and Rey 1967, 89; Flint and Flint 2005, 502), and Africans in the Southeast (e.g., Clayton et al. 1993b, 313), remaining behind when the expeditions they belonged to originally returned home, remind us that there are elements of the early historical period that remain silent and await thoughtful analysis. Zuni pueblos, for example, may have welcomed Mexica allies into their communities because of their military prowess and knowledge of Spanish customs and tactics, as Hopi villages were to incorporate Tano refugees for similar reasons following the First Pueblo Revolt in the late seventeenth century (Brooks 2016, 69–71, 83, 85). Such cases remind us that, as we grapple with episodes of war and peace in the Early Americas, our understanding of indigenous-European material encounters – and the multifarious cultural dynamics that underlie them – becomes richer and more variegated. It is hoped that this modest effort to untangle a specific web of motivations, history, agencies, and processes will begin to encourage similar comparative efforts elsewhere – including approaches that are both regional and interregional. In doing so, we will continue the challenging task of connecting the objects of microhistory (and the "lived lives" of indigenous and European actors) to the broader patterns of macrohistory in historically meaningful ways.

Acknowledgments

A number of colleagues were instrumental in providing support for the development of this chapter including: Rani Alexander, Philip Arnold, Joseph Ball, William Billeck, Tom Briones, Kathy Deagan, Tom Evans, Patricia Fournier, Jay Hart, Ron Kneebone, Alex Kurota, Shannon Mann, Michael Marshall, Geoffrey McCafferty, Peter McKenna, Jeffrey Mitchem, Patricia Nietfeld, Jeffrey Parsons, Helen Pollard, Dyane Sonier, Anthony Trujillo, and Tom Windes.

My profound thanks to Floris Keehnen, Corinne Hofman, Penelope Drooker, Robbie Ethridge, and Chris Rodning for their thoughtful comments on earlier versions of the manuscript.

A special thanks to the New Mexico Archaeological Council for their financial support in helping me to carry out work on the Hawikku/Kechiba:wa collections at the National Museum of the American Indian (NMAI), Suitland, MD.

References

Barrett, Elinore M. 2012. *The Spanish Colonial Settlement Landscapes of New Mexico, 1598–1680.* Albuquerque: University of New Mexico Press.

Bell, Robert E. 1959. "Obsidian Core Found in Western Oklahoma." *El Palacio* 66 (2): 72.

Billeck, William. 2009. "Traces of Coronado: Spanish Glass Beads in the Southwest and the Plains." Poster presented at the *74th Annual Meeting of the Society for American Archaeology, Atlanta, 22–26 April.*

Blanton, Dennis B. 2019. "Modeling Entrada-related Material Culture by Comparative Analysis of Sixteenth-Century Archaeological Assemblages from the Southeast." In *The Destiny of Their Manifests: Modeling Sixteenth-Century Entradas in North America,* edited by Clay Mathers. Gainesville: University Press of Florida.

Bratton, John. 2009. "Mesoamerican Component of the Emanuel Point Ships: Obsidian, Ceramics, and Projectile Points." *The Florida Anthropologist* 62 (3/4): 109–114.

Brooks, James F. 2016. *Mesa of Sorrows: A History of the Awat'ovi Massacre.* New York: W.W. Norton and Company.

Chuchiak, John F. IV. 2007. "Forgotten Allies: The Origins and Roles of Native Mesoamerican Auxiliaries and Indios Conquistadores in the Conquest of Yucatan, 1526–1550." In *Indian Conquistadors: Indigenous Allies in the Conquest of Mesoamerica,* edited by Laura E. Matthew and Michel R. Oudijk, 175–225. Norman: University of Oklahoma Press.

Clayton, Lawrence A., Vernon J. Knight, Jr., and Edward C. Moore (eds). 1993a. *The De Soto Chronicles: The Expedition of Hernando de Soto to North America in 1539–1543 Volume 1*. Tuscaloosa: University of Alabama Press.

Clayton, Lawrence A., Vernon J. Knight, Jr., and Edward C. Moore (eds). 1993b. *The De Soto Chronicles: The Expedition of Hernando de Soto to North America in 1539–1543 Volume 2*. Tuscaloosa: University of Alabama Press.

Clissold, Stephen. 1954. *Conquistador: The Life of Don Pedro Sarmiento de Gamboa*. London: Derek Verschoyle.

Cobb, Charles and Tony Legg. 2017. "Indigenous Appropriations of Spanish Metal Goods in Southeastern North America." Paper presented at the *82nd Annual Meeting of the Society for American Archaeology, Vancouver, 29 March–2 April*.

Cordero, Robin M. 2013. *Final Report on Excavations of the Alameda School Site (LA 421): A Classic Period Pueblo of the Tiguex Province. OCA/UNM Report no. 185–969. Office of Contract Archaeology*. Albuquerque: University of New Mexico.

Dominguez, Luis L. (translator). 2005. *The Conquest of the River Plate (1535–1555). I. Voyage of Ulrich Schmidt to the Rivers of La Plata and Paraguai from the Original German Edition, 1567. II. The Commentaries of Álvar Núñez Cabeza de Vaca from the Original Spanish Edition, 1555*. Reprinted Elibron Classics, www.elibronclassics.com. Originally published in 1891. London: Hakluyt Society.

Dutton, Bertha P. 1963. *Sun Father's Way: The Kiva Murals of Kuaua, a Pueblo Ruin, Coronado State Monument, New Mexico*. Santa Fe: Museum of New Mexico Press.

Ellis, Bruce T. 1957. "Crossbow Boltheads from Historic Pueblo Sites." *El Palacio* 64 (7/8): 209–214.

Ellis, Florence H. 1989. *San Gabriel del Yungue as Seen by an Archaeologist*. Santa Fe: Sunstone Press.

Flint, Richard. 2002. *Great Cruelties Have Been Reported. The 1544 Investigation of the Coronado Expedition*. Dallas: Southern Methodist University Press.

Flint, Richard. 2009. "Without Them, Nothing Was Possible: The Coronado Expedition's Indian Allies." *New Mexico Historical Review* 84 (1): 65–118.

Flint, Richard, and Shirley Cushing Flint, eds. and trans. 2005. *Documents of the Coronado Expedition, 1539–1542. "They Were Not Familiar with His Majesty, nor Did They Wish to Be His Subjects."* Dallas: Southern Methodist University Press.

Francis, J. Michael. 2007. *Invading Columbia: Spanish Accounts of the Gonzalo Jiménez de Quesada Expedition of Conquest*. University Park: Pennsylvania State University.

García Loaeza, Pablo and Victoria L. Garrett. 2015. *The Improbable Conquest: Sixteenth-Century Letters from the Río de la Plata*. University Park: Pennsylvania State University.

Gibson, Charles. 1968. *The Aztecs Under Spanish Rule: A History of the Indians of the Valley of Mexico*. Stanford: Stanford University Press.

Griffin, James B., and Antonieta Espejo. 1950. "La alfaería correspondiente al último período de ocupación nahua del valle de México, II." *Tlatelolco a través de los tiempos* 11: 15–66.

Habicht-Mauche, Judith A. 1992. "Coronado's Querechos and Teyas in the Archaeological Record of the Texas Panhandle." *Plains Anthropologist* 140: 247–259.

Hammond, George P., and Agapito Rey, eds. and trans. 1967. *Expedition into New Mexico Made by Antonio de Espejo 1582–1583 as Revealed in the Journal of Diego Pérez de Luxán, A Member of the Party.* Reprinted Arno Press: New York. Originally published 1929, Quivira Society Publications, Volume 1, Los Angeles: The Quivira Society.

Hernández, Pero. 2005. *The Commentaries of Álvar Núñez Cabeza de Vaca from the Original Spanish Edition, 1555.* Reprinted Elibron Classics, www.elibronclassics.com. Originally published 1891. London: Hakluyt Society.

Hoard, Robert J., C. Tod Bevitt, and Janice McLean. 2008. "Source Determination of Obsidian from Kansas Archaeological Sites Using Compositional Analysis." *Transactions of the Kansas Academy of Science* 111 (3/4): 219–229.

Hodge, Frederick W. 1918a. "Excavations at the Zuni Pueblo of Hawikuh in 1917." *Art and Archaeology* 7 (9): 367–379.

Hodge, Frederick W. 1918b. "Hawikuh. III." *Catalogue of Collections.* Unpublished fieldwork notes. Carl A. Kroch Library, University of Microfilms, Ithaca.

Hodge, Frederick W. 1920. "The Age of the Zuni Pueblo of Kechipauan." *Indian Notes and Monographs* 3 (2): 41–60.

Hoffman, Paul E. 1994. "Lucas Vázquez de Ayllón's Discovery and Colony." In *The Forgotten Centuries: Indians and Europeans in the American South, 1521–1704*, edited by Charles Hudson and Carmen Chaves Tesser, 36–49. Athens: University of Georgia Press.

Howell, Todd. 2001. "Foundations of Political Power in Ancestral Zuni Society." In *Ancient Burial Practices in the American Southwest: Archaeology, Physical Anthropology, and Native American Perspectives*, edited by Douglas R. Mitchell and Judy L. Brunson-Hadley, 149–166. Albuquerque: University of New Mexico Press.

Hudson, Charles. 1997. *Knights of Spain, Warriors of the Sun: Hernando de Soto and the South's Ancient Chiefdoms.* Athens: University of Georgia.

Kidder, Alfred V. 1932. *The Artifacts of Pecos.* New Haven: Yale University Press.

Langford, James B. Jr. 1990. "The Coosawattee Plate: a Sixteenth-Century Catholic/Aztec Artifact from Northwest Georgia." In *Columbian Consequences, Vol. 2: Archaeological and Historical Perspectives on the Spanish Borderlands East*, edited by David Hurst Thomas, 139–151. Washington, D.C.: Smithsonian Institution Press.

Little, Keith J. 2008. "European Artifact Chronology and Impacts of Spanish Contact in the Sixteenth-Century Coosa Valley." Unpublished PhD diss., Department of Anthropology, University of Alabama.

Lothrop, Samuel K. 1923. "Ketchipauan – General Notes and Diary." Transcription by Keith Kintigh. Manuscript on file at the University Museum of Archaeology and Anthropology, University of Cambridge, England.

Marshall, Michael P. 1982. *Excavations at Nuestra Señora de Dolores Pueblo (LA 6777), A Prehistoric Settlement in the Tiguex Province.* Office of Contract Archaeology, Albuquerque: University of New Mexico.

Mathers, Clay. 2011. "Tangled Threads, Loose Ends, and Knotty Problems: The Place of Moho in Tiguex Archaeology, Geography, and History." In *The Latest News from 1540. People, Places and Portrayals of the Coronado Expedition*, edited by Richard Flint and Shirley Cushing Flint, 367–397. Albuquerque: University of New Mexico Press.

Mathers, Clay. 2013. "Conquest and Violence on the Northern Borderlands Frontier: Patterns of Native-European Conflict in the Sixteenth-Century Southwest." In *Native and Spanish New Worlds: Sixteenth-Century Entradas in the American Southwest and Southeast*, edited by Clay Mathers, Jeffrey M. Mitchem, and Charles M. Haecker, 225–255. Tucson: University of Arizona Press.

Mathers, Clay, and Michael P. Marshall. 2014. "'Missing Links' and the War of the Worlds in Tiguex (1540–1542)." Paper presented at the *79th Annual Meeting of the Society for American Archaeology, Austin, TX, 23–27 April.*

Mathers, Clay, Dan Simplicio, and Tom Kennedy. 2011. "'Everywhere They Told Us He Had Been There': Evidence of the Vázquez de Coronado Entrada at the Ancestral Zuni Pueblo of Kechiba:wa, New Mexico." In *The Latest Word from 1540: People, Places, and Portrayals of the Coronado Expedition*, edited by Richard Flint and Shirley Cushing Flint, 262–285. Albuquerque: University of Mexico Press.

Milanich, Jerald T. 2006. *Laboring in the Fields of the Lord: Spanish Missions and Southeastern Indians.* Gainesville: University Press of Florida.

Mitchem, Jeffrey M. 2014. "Archaeological Evidence of the Hernando de Soto Expedition in the Southeastern United States." *Lecture presented for the Archaeological Conservancy, Albuquerque, 17 September.*

Museum of Indian Arts and Culture. n.d. Collections for Puaray or Santiago Pueblo (Laboratory of Anthropology Site Number 326). Object Catalog Numbers: 44778/11, Bp 38/15, Object Name: "Nails"; 49925/11a-c, Bp 38/5, Object Name: "Tinkler." Santa Fe, NM.

Pooler, Lolita M. 1940. "Alameda Pueblo Ruins." *El Palacio* 47 (4): 84–89.

Powell, Phillip W. 1967. *Mexico's Miguel Caldera: The Taming of America's First Frontier (1548–1597).* Tucson: University of Arizona Press.

Priestley, Herbert Ingram, ed. and trans. 1928a. *The Luna Papers, Documents Relating to the Expedition of Don Tristán de Luna y Arellano for the Conquest of Florida in 1559–1561. Volume 1.* Deland: Florida State Historical Society.

Priestley, Herbert Ingram, ed. and trans. 1928b. *The Luna Papers, Documents Relating to the Expedition of Don Tristán de Luna y Arellano for the Conquest of Florida in 1559–1561. Volume 2.* Deland: Florida State Historical Society.

Restall, Matthew, and Florine Asselbergs. 2007. *Invading Guatemala: Spanish, Nahua, and Maya Accounts of the Conquest Wars.* University Park: Pennsylvania State University.

Sargeant, Kathryn. 1985. *An Archaeological and Historical Survey of the Village of Los Ranchos.* Santa Fe: New Mexico Historic Preservation Division.

Smith, Marvin T. 1987. *Archaeology of Aboriginal Culture Change in the Interior Southeast: Depopulation during the Early Historic Period.* Gainesville: University Press of Florida.

Smith, Marvin T. and David J. Hally. 2019. "The Acquisition of Sixteenth-Century European Objects by Native Americans in the Southeastern United States." In *The Destiny of Their Manifests: Modeling Sixteenth-Century Entradas in North America*, edited by Clay Mathers. Gainesville: University Press of Florida.

Smith, Watson, Richard B. Woodbury, and Nathalie F.S. Woodbury. 1966. *The Excavation of Hawikuh by Frederick Webb Hodge. Report of the Hendricks-Hodge Expedition 1917–1923. Contributions from the Museum of the American Indian, Heye Foundation, Volume 20.* New York: Museum of the American Indian, Heye Foundation.

Sudbury, Byron. 1984. "A Sixteenth-Century Spanish Colonial Trade Bead from Western Oklahoma." *Bulletin of the Oklahoma Anthropological Society* 33: 31–36.

Tichy, Marjorie. 1939. "The Archaeology of Puaray." *El Palacio* 46(7): 145–163.

Van West, Carla R., Thomas C. Windes, Frances Levine, Henri D. Grissino-Mayer, and Matthew W. Salzer. 2013. "The Role of Climate in Early Spanish-Native American Interactions in the U.S. Southwest." In *Native and Spanish New Worlds: Sixteenth-Century Entradas in the American Southwest and Southeast*, edited by Clay Mathers, Jeffrey M. Mitchem, and Charles M. Haecker, 81–98. Tucson: University of Arizona Press.

Vierra, Bradley J. ed. 1989. "A Sixteenth-Century Spanish Campsite in the Tiguex Province." Laboratory of Anthropology Note 475. Santa Fe: Museum of New Mexico.

Worth, John E. 2016. "Interpreting Spanish Artifact Assemblages in the Mid-Sixteenth-Century Southeast: The View from the 1559–1561 Tristán de Luna Settlement on Pensacola Bay." Paper presented at the *73rd Annual Meeting of the Southeastern Archaeological Conference, Athens, GA, 26–29 October.*

Wilson, Samuel M. 1990. *Hispaniola: Caribbean Chiefdoms in the Age of Columbus.* Tuscaloosa: University of Alabama Press.

CHAPTER 15

'Beyond the Falls': Amerindian Stance towards New Encounters along the Wild Coast (AD 1595–1627)

Martijn van den Bel and Gérard Collomb

According to the Treaty of Tordesillas, the Guianas represent the border region of the Demarcation Line that is approximately situated to the East of the mouth of the Amazon River. Both Iberian nations, firmly implanted on both sides of this line by the second half of the sixteenth century (i.e. Margarita and Pernambuco) do not establish themselves in this area. It is possible they do this to avoid confrontations between themselves but this settlement pattern also leaves an opening for the English, Dutch and French to intervene in the area and barter with the indigenous population (Figure 15.1).

After the discovery of pearls off the Coast of Paria in 1508 and their depletion by the 1530s, the focus of the Spanish upon the Aztec, Inca and Muisca gold seemed to diminish their interest in these new parts of the Americas.

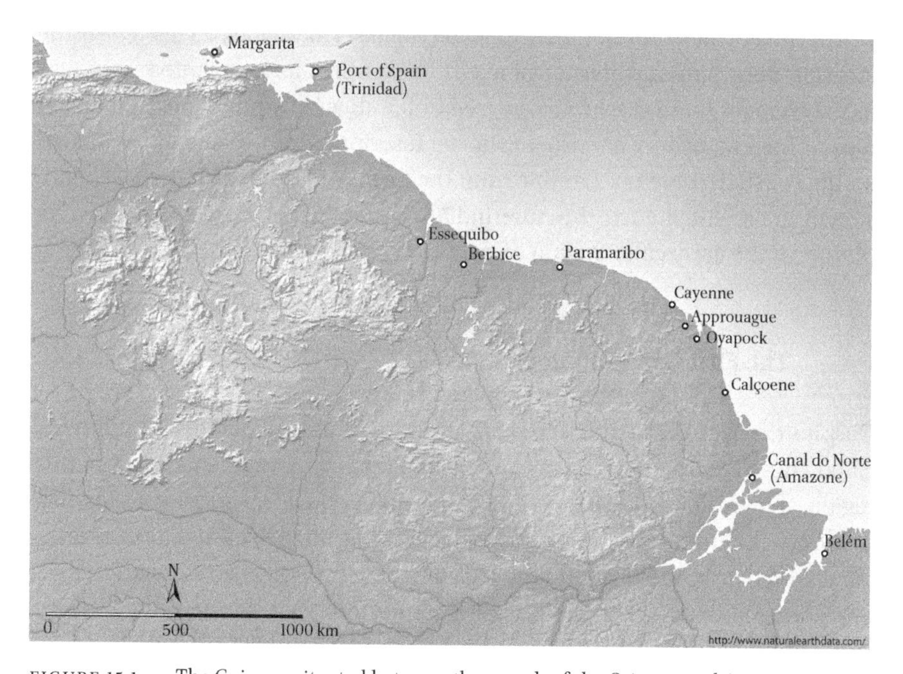

FIGURE 15.1 The Guianas, situated between the mouth of the Orinoco and Amazon Rivers

However, once these empires were controlled and plundered, the quest for new El Dorados pointed them back to the unexplored areas of the interior of the Guianas as witnessed by the numerous Spanish explorations upon the Orinoco River (Hemming 1978).

On the other side of the Demarcation Line, the Portuguese had firmly established numerous sugar plantations in Pernambuco since 1534, also leaving a large geographical margin with the Amazon River. From the second half of the sixteenth century onwards, this area was soon invested in by French entrepreneurs such as Jacques Riffault, who established themselves at Saint Louis de Maranhao and who were eventually chased out by the Portuguese in 1614 who, in turn, founded Belém two years later.

This is a pivotal moment, at the turn of the sixteenth and seventeenth centuries, when small coastal trade (dye wood, annatto, tobacco, hammocks, gums in exchange for iron implements, beads…) reaches its peak, with the regular passage of Dutch, English and French ships, while, at the same time, giving gradually way to colonial settlement projects for production of cash crops, i.e., tobacco, annatto, sugar, which will take shape by the middle of the century. This European intrusion on the Guiana or Wild Coast (coastline inhabited by those who are called Savages or *Wilden*) as well as the demographic collapse that it has caused, adversely affected the native populations throughout this region, upsetting the ancient social, economic and also warlike relationships which linked these Amerindian groups (Gallois Tilkin 2005). This led to the creation and reinforcement of new ethnic frontiers as an adaptive response to the changes that were occurring (see Collomb and Dupuy 2009; Whitehead 1992). Thus the history and the territorial inscription of these peoples, settled on the coast, became inseparable from the forms of European colonial expansion in the region; hence, the Amerindians were – to a certain extent – active players in the early colonization of the Guianas as traders and middlemen.

1 The First Confrontations

Despite the fact that Francisco de Orellana had already almost circumnavigated the Guianas in AD 1542, only a few Spanish documents discuss this region that formed the eastern limit of their official possessions. Spanish intelligence about the Guianas, and especially the allied *Aruaca* nation, is principally known through the writings of Figueroa (*ca.* 1520) and Navarrete (*ca.* 1570) (see Whitehead 2011). Although they did not visit this area themselves, they gathered information about the Amerindian population from Aruacas that resided in or visited Trinidad or Margarita. As a result, Figueroa reported to the Spanish Crown a polarized image of friends (*guatiao*) and enemies (*caribe*). Much like the Lower Orinoco River, the eastern and western Guianas were seen by the Spanish as supply areas of victuals and slaves, which could be exploited through Spanish/Aruaca combined raiding parties.

According to Navarrete's sources, the Aruacas inhabited the rivers of *Bermeji* [Berbice], *Curetuy* [Corantyne], *Dumaruni* [Demerara], *Desguixo* [Essequibo], *Baorome* [Pomeroon], and *Moraca* [Moruca], which were previously inhabited by the *Caribes*. At first, they lived in peace together but eventually they found themselves fighting each other, which is evidenced by seasonal warfare of large raiding parties to procure slaves, as described by Navarrete. When English explorers, such as John Burgh, Jacob Whiddon, Robert Dudley and Walter Ralegh, arrived by the end of the sixteenth century, they were thus confronted with an indigenous population that was trying to establish a confederation to fight the allied forces of the Aruacas and Spanish.[1]

It is evident that the socio-political situation of pre-Columbian times, although already infused with warfare between distinct nations, had been altered by the impetus of another important player, the Spanish, to whom the leaders had to adapt eventually. Ralegh (1596, 5–6) suggests, for example, that this new political situation had changed the status of the Amerindian lords or *cacique*, "called Acarewana in their own language" because they adopted the European military term "Capitaynes", to acquire, in a way, elite European status.[2]

2 The Second Confrontation

Arriving at Trinidad, Ralegh instantly took San José de Oruña and its commander Antonio de Berrio. His Virginian experiences coupled with the information gathered from de Berrio made it clear to Ralegh that the key to success in locating *El Dorado* was to have the local Amerindian elite on his side. However, he also recognized that the Spanish were too strong to be conquered, and that he must look for other ways to get to El Dorado rather than sailing up the Orinoco and the *Caroli* River, to reach the lake of *Parimé*. Once back in England, he procured *A Second Voyage to Guiana* under the command of Lawrence Keymis who, in company of William Downe, searched the Guiana coast

1 "The sea coast is nowhere populous, for they have much wasted themselves, in mutuall warres. But now in all parts so farre as Orenoque, they live in league and peace" (Keymis 1596, Gr).

2 Derived from the latin *caput*, head, this term appears in the Middle Ages in its military sense, pointing to one who is head, a leader of war.

for possible Spanish colonies and to obtain intelligence about other rivers that would lead to El Dorado. This voyage, which took approximately six months, was followed by a third voyage under the command of Thomas Masham, accompanied by Leonard Berry, serving the same interests as the second voyage of Keymis but this time carefully exploring the Courantyne River.

The 20th of April 1597, they encountered the *John of London* under the command of John Leigh [Ley] with whom they teamed up to explore the upper reaches of this river, which would join the *Desekebe* River "within a dayes journey of the lake called *Perima*, whereupon *Manoa* is supposed to stand" (Masham 1890, 190). Along with English merchants, such as John Ley, Dutch merchants, lured by the writings of Ralegh about mines, began to appear along the Guiana coast (Netscher 1888, 32).[3] But, instead of investing in the construction of expensive (gold) mines in the hostile regions of the Lower Orinoco, private Dutch merchant companies started to install *'leggers'* or traders near Amerindian villages situated upon the lower reaches of various rivers to procure local goods in exchange for European ware (Hulsman 2010, 2011).

In fact, the English assistance along the Guianas is also accounted for in the French voyage to the *Wiapoco* and *Caliana* Rivers, as written by Jean Mocquet (1617). In 1604, this expedition, under the command of Daniel de la Ravardière, set sail towards the Guianas to check for goods to be procured among the Amerindians of this region and possibly to pinpoint a location for a colony. It is noteworthy that a large part of this crew and even the pilot were English. (Mocquet 1617, 148). They were accompanied by the "Indian" guide that once belonged to "milord *Ralle*[*gh*]" who was "the son of a King from the Island of Trinidad" (*ibidem* p. 97).[4]

3 *Wiapogo*: a Place of Trade

Since the first Spanish attempts, the Island of Cayenne as well as the Oyapock River ["Wiapogo"], were quickly landmarked as the most important ports of call for trade with the Amerindians (Collomb and van den Bel 2014). In fact, the embouchures of these rivers are easily recognizable along the Guiana coast because of their large table mountains that stand directly in the Atlantic

3 Although the Dutch were already trafficking salt from Punta de Araia within Spanish territory, they relied on English intelligence when calling on ports for barter upon the Guiana coast (Goslinga 1971, 486).

4 Earlier French visits to Guiana have been noted by Keymis (1968 [1596], F4r) but may be related to the French colony at Maranhão, Brazil.

Ocean (Figure 15.2). The Oyapock River, in particular, was soon targeted as an important river for European settlement as attested by the (failed) colonies and eye-witness reports of the Leigh brothers [1604–1606], Robert Harcourt [1608–1613, 1630], Jan Pietersz [1615], Lourens Lourensz [1618–1626], Jesse de Forest [1624–1625] and Jan van Ryen [1627]. Their accounts allow us to establish a list of the numerous groups dwelling here (Table 15.1).

When he entered the Oyapock estuary, Keymis found only empty houses at Mount *Caripo* [Montagne Bruyère]. Similarly, Masham did not encounter any

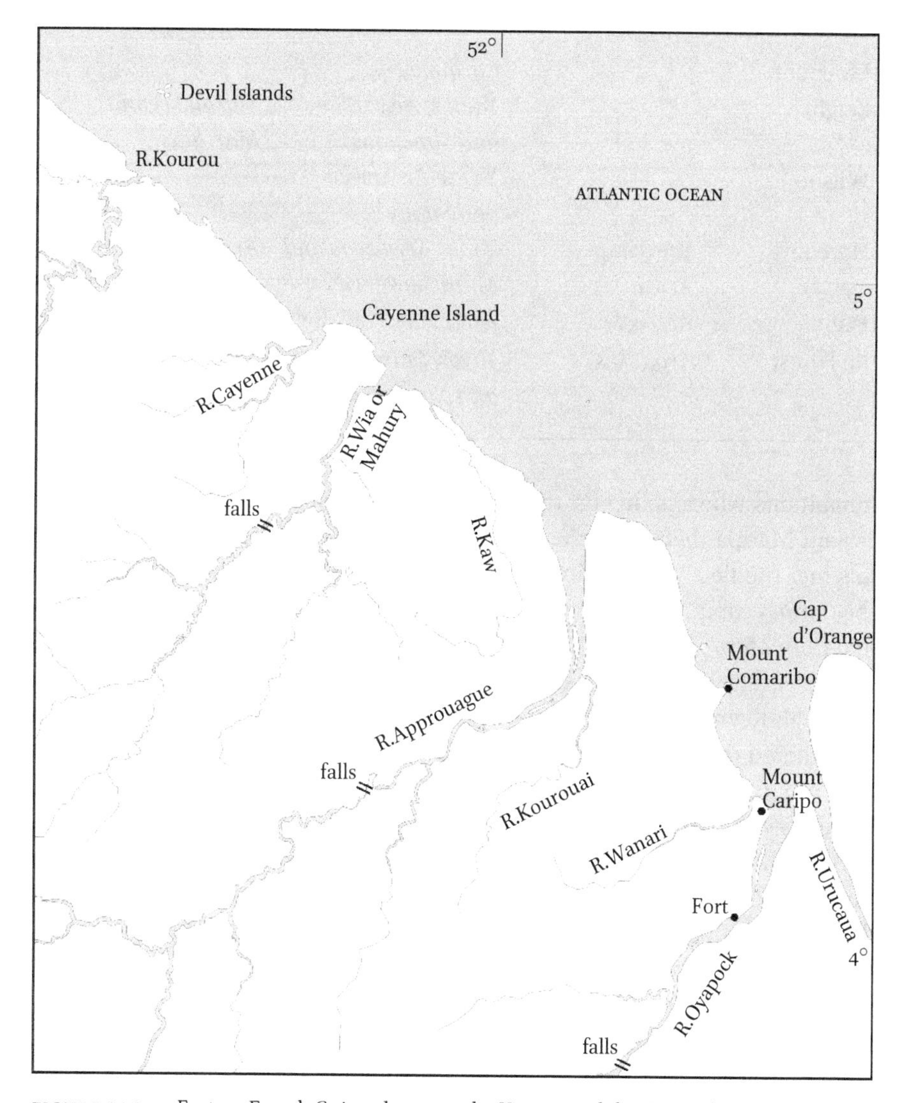

FIGURE 15.2 Eastern French Guiana between the Kourou and the Oyapock Rivers

TABLE 15.1 Overview of Nations dwelling in the Oyapock basin during the early seventeenth century. All names can be found in the bibliography

	Wanari	Wiapogo	Arracow
Keymis	*Charibes*	*Coonoracki, Wacacoia, Wariseaco*	*Marowanas*
Ley		*Iayos*	*Morowonow and Marawen*
Mocquet		*Caripous*	
Leigh		*Yaioas, Arwarkas and Sapayoas*	*Marauvas and Marraias*
Wilson		*Yayes, Arwackes and Supayes*	
Harcourt	*Yaios and Arwaccas*	*Yaios, Arwaccas and Marashewaccas*	*Arracoories*
Fisher	*Areecole*	*Areecola*	*Yaios*
de Forest	*Arouakes*	*Yayos, Maraons and Nourakes*	

inhabitants when sailing up this river until the first waterfalls (today known as Saut Maripa) because he was likely preceded by John Ley. Eventually, Keymis met the fled *Yao* captain *Wareo* at the *Cawo* River whereas Masham met the *Caribes* captain *Ritimo* at *Chiana*. When meeting Keymis, the Yao, who had fled from *Moruga*, feared that they were Spanish whereas Masham was asked by the *Caribes* to join forces in order to attack the Spanish on the Orinoco. Although Keymis did not meet any people on the Oyapock, it is possible that he gathered this information from Downe who told him that he sailed up the *Wiapoco* until the first falls when they met eventually at large of the Orinoco Delta (Keymis 1968 [1596], D4v).

According to John Ley, the Yao are firmly installed on the Kaw and Oyapock Rivers at the beginning of the seventeenth century (Lorimer 2006, 326). Ley does not mention any other nations upon this river such as those suggested by Keymis and, later on, by Leigh. It appears that the Yao had clearly promoted themselves as the one nation with whom the English had to trade upon this part of the Guiana coast. Ley does not mention any other nations upon this river, which may have arrived some time later, but neither remarks the presence of any (autochthonous) local groups except for the *Morowonow* of the

Arowcoa or modern Urucauá River, affluent of the Oyapock in its embouchure, facing Mount Caripo.

After this first English survey and their publications in English and subsequent translations in other languages, a large number of English, Dutch and French entrepreneurs frequented the Guiana coast for trade with the Amerindians. As mentioned before Dutch private companies, such as the *Compagnie op Guiane en de Wiapoco* based in Amsterdam (established in the early 1600s), issued contracts for a two-year stay upon the Oyapock for a trader in service for that particular company (Hulsman 2009, 61). These traders may have lived among the Amerindians but they also built warehouses with stockades in order to secure the goods that were coming and going. They were sustained by the local population for they provided them with European objects, notably iron tools, in exchange for Amerindian wares, such as tobacco, dye wood, annatto, cotton and feathers, as well as with victuals such as fish, game and local bread or cassava. Frequently, perhaps every six months or every year, the European companies sent a ship to the trader to bring back the traded goods and to render another stock of barter goods (Figure 15.3).

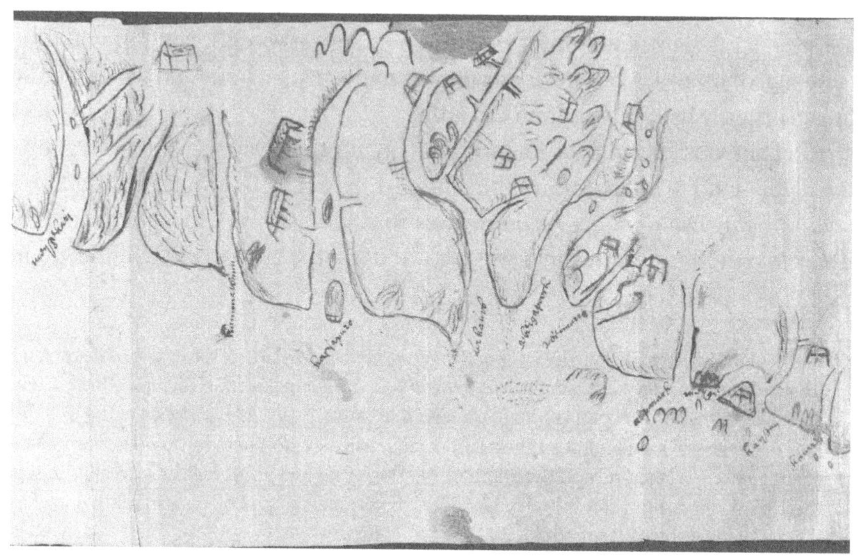

FIGURE 15.3 Detail of an anonymous Dutch map of the Eastern Guianas between the rivers Kaurro and Mayhekarj, showing the different Amerindian villages upon this coast to be visited for barter by the Dutch (Nationaal Archief, Den Haag 4.VEL652, ca. 1615). One observes the Oyapock Bay in the middle of this picture where the Urucauá, Oyapock and Wanari Rivers all come together. It is noteworthy that nearly all Amerindian villages, except for Cayenne, are situated a little distance up river.

By 1610, there were various companies trading along the Guiana Coast as well as in the Lower Amazon River, with a fairly large number of European ships calling on the numerous traders, as witnessed by the published journals of Jean Mocquet (1617), Charles Leigh (1625), John Wilson (1625),[5] Robert Harcourt (Harcourt [1625] 1928) and Unton Fisher (Harcourt [1625] 1928). According to Wilson, the Dutch traders were very well equipped for their tasks which appeared small in comparison to their own failing colony: "Neither had we any store of commodities to trade up in the Maine, as the two Hollanders hath which are there, and were left there at our comming from thence by John Sims, Master of a Ship called the Hope of Amsterdam, of the burthen of one hundred tuns Fraughted by the Merchants of Amsterdam, and by their Charter partie was bound to lye in the River of Wiapoco, and of Caliane six moneths time" (Wilson 1625, 1264).[6]

It may be likely that these traders were completely dependent on their relationship with the Amerindians, not only for their work but also for their life. An intimate relationship may have existed between the local population and various traders who certainly lived among them and married indigenous women, e.g., the post-scriptum of the journal of Lourens Lourensz explains that the wife of an Amerindian *Captain* bore a daughter from a Dutchman (Wassenaer 1627, 64v).[7] Subsequently, if the local population was not content with the presence of a certain colony or any person in particular, a sudden death would most certainly be brought upon the latter.

After the observations of Keymis and Ley up the Oyapock, the French entered this land (Mocquet 1617), but interestingly they only refer to the *Caripous*[8] of *Yapoco*. Mocquet further states that the King of Yapoco is a certain *Anacaioury* (Figure 15.4) who is waging war upon the *Caribes* of Cayenne, their

5 The Zealander that sailed up the Oyapock River wanted to sell African slaves to Charles Leigh: "His comming unto us to Wiapoco, was to have sold unto our Generall Negroes, whose kindnesse we did requite in helping him to such commodities as wee had, and did get the Indians to provide Cassavi and Guinea Wheate for bread, with Potato Roots for his Negroes to eat, who departed on the one and twentieth of May (after he had bin some three weekes in the River of Wiapoco) for Point de Ray [La Brea], where he shipped of our company into his Countrimens ships" (Wilson 1625, 1262).

6 Here we refer to the well documented dissertation of the late Lodewijk Hulsman (1950–2016) for further readings and archival research on Dutch factors and their different companies, later to be fused into *De Guiaansche Compagnie*, upon the Guiana coast for this early period (Hulsman 2009, 2010).

7 A first English translation of the Lourensz journal is published by M. van den Bel (2009).

8 According to F. Grenand and P. Grenand (1987, 10) *Caripous* is considered "a new bourgeon for the old word Charib-Karipuna." Whitehead (Ralegh 1997, 62) states that the term *caripou* is "a garbled attempt to render 'Palicour' since the substitution of 'p' for 'b' and 'r' for 'l' is common in European transcriptions of native American languages," and that the Caripou

FIGURE 15.4 "Comment les Caribes tirent le poisson (CC). Amazone allant à la guerre (D). Forme de danse des Caribes DD." Engraving taken from Jean Mocquet, Voyages en Afrique, Asie, Indes orientales & occidentales [...], Livre II, Paris, 1617, p. 157 © BNF, RÉSERVE DES LIVRES RARES

eternal enemies (Mocquet 1617, 81). Although Caripous are not mentioned by other Europeans, Anacaioury is met by them and represents an emblematic personality of the Oyapock River and chief of a larger geographical area, as esteemed by Robert Harcourt (Harcourt 1625, 1271): "Beyond the Country of *Morrownia* to the Southward bordering the River of *Arwy*, is the Province of *Norrak*; the people thereof are *Charibes*, and enemies both to the *Morrowinnes*, the inhabitants of *Morrownia*, and to the *Wiapocoories* ; who are also under the subjection of *Anaky-u-ry*, the Principall and greatest Lord, or *Cassique* of all the *Yaios* in those Provinces, bordering upon the Sea betwixt the *Amazones*, South-eastward, and *Dessequebe* North-westward."

Anacaioury, a Yao far away from his homeland in Trinidad, was probably able to obtain an important socio-political position in a newly conquered and occupied country (Oyapock) depending heavily upon European interaction in this region. In this manner, the "King" Anacaioury would be able to control access and distribution of "new" iron tools among the important leading groups as well as among their alliances; thus reshaping and establishing a new political balance in the region. The historical and famous warfare between the Caribs of Cayenne and the Palikur of the Oyapock River must therefore be seen within the light of this Yao arrival and subsequent new balance of regional power.

4 Indigenous Networks

Despite some archaeological evidence, the interior of the Guianas remains virtually unknown, just as it was at the beginning of the seventeenth century, compared to the better explored coastal zone. Nevertheless, some observations shed light on complex processes which linked the native societies throughout the Guianas and Greater Amazonia, embedded in large social, economic, cultural spaces. In this regard, it is worthwhile to note the social and commercial networks, which operated beyond the region where the local group lived. These societies were integrated into these networks, in which alliance, war and

described by Mocquet are actually the Yao described in the English documents. It must be noted here that Mocquet (1617, 133) remarks that "*il y en a de plusieurs sortes, et celle des Caripous est aucunement différente de celle des Caribes, et ont assez de peine à s'entendre, encorequ'ils ne soient pas fort éloignés les uns des autres*," suggesting that the Caripous and Charibe language were actually different languages. However, if Caripou are Yao, as suggested by various authors, they must have understood each other better since the Yao language is supposed to be of Cariban stock (Taylor 1977). This argument favours an interpretation of Caripou as an Amerindian idiom for social or political status instead of a group name.

trade were opportunities to periodically reactivate social relations between different groups.[9] These trade relations were based upon ties forged between trade partners, i.e. *pawana, banare*, sometimes separated by great distances, along trade routes that connected peoples, step by step, throughout the interior of the Guianas, from the Orinoco and the Rio Negro to the Amazon and the Atlantic coast (Butt Colson 1973; Dreyfus 1992; Dupuy 2008). The European goods that arrived in large quantities on this coast were exchanged for local goods by the Amerindian middlemen who traded them again for other goods in these extensive networks.

It is pertinent to note that Wilson ([1625] 1906) is in awe that the Amerindians of the Oyapock were already aware of the future arrival of a (Dutch) ship upon their river since three ships, according to Amerindian intelligence, had already sailed up the Amazon River, revealing a possible land route between these river basins. One can recall here the complete crossing of the Guiana interior accompanied by Amerindians (probably by some Aruã) from the Lower Amazon, by ascending the Rio Paru (tentatively) and descending the Suriname River towards the ocean, accomplished by the Irishman Bernard O'Brian in 1625 (Mathews 1970, 92). Moreover, there are many testimonies of the antiquity of an important native commercial route connecting Essequibo and deeper Guiana through the 'Pirara portage' in the Rupununi area (Edmundson 1904, 10–13). Thus, Acuña reported the presence of iron arms and tools amongst groups living near the Rio Branco and said these goods were obtained by trade with peoples living near the sea, who themselves bartered them with Dutch merchants settled in the Essequibo (Acuña 1641, 30v–31r).

In such networks, of which some remained active until at least the nineteenth century,[10] circulated objects (and information) in which different

9 War not only allowed the consumption of enemies, through which is achieved the "social production of persons by persons," as pointed out by Carlos Fausto (2001). It was, as well, a means to capture wives, and finally to renew social relations through peace and spouses exchange (Collomb and Dupuy, 2009). See also Harcourt (1928, 86): "But with the Charibes inhabiting the in-land parts upon the Mountaines, they have as yet no peace at all; for they doe often times come downe upon them in great numbers, spoile and burne their houses, kill their men, and carry away their women, which is the greatest cause of warre and hatred amongst them." Also see Santos-Granero (2009) for Amerindian warfare during late pre-Columbian and early historic times.

10 As evidenced, for example, by Im Thurn at the end of the nineteenth century: "To interchange their manufacture the Indians make long journeys. The Wapianas visit the countries of the Tarumas and the Woyowais, carrying with them canoes, cotton hammocks, and now very frequently knives, beads, and other European goods; and, leaving their canoes and other merchandise, they walk back, carrying with them a supply of cassava-graters, and leading hunting dogs-all which things they have received in exchange for the things

groups were specialized such as trained dogs for hunting, cassava graters, ceramics and manufactured products traded on the coast, but also valuable assets, such as greenstone frog-pendants which came from various production centers in Guiana or the Lower Amazon River (Boomert 1987) as well as the gold ornaments were brought up on the coast from the interior of Guiana or Andean foothills (Whitehead 1990). The sometimes long trips comprising these exchange networks were often ended in lengthy visits and celebrations. However, the networks were eventually transformed as a result of the contact and the arrival of European goods, especially by shifting perpendicular to the coast, where new goods were arriving. These exchange networks constituted a chain of different populations in which the value of the goods depended on their provenance, manufacturers and cultural interest per group.

"At our returne to Wyapoco we gave to the Indians for their paines, and providing of us victuals in our journey an Axe, for which they would have travelled with us two or three moneths time if occasion had required. And for an Axe they found us victuals two moneths time at our houses, as Bread, and Drinke, and Crabbes, and Fish, and all such kinde of flesh as they killed for themselves, for the same price: but if we desired any Hennes or Cockes of them, then we were to have given them some small trifles, as Beades ; so likewise if they brought us in our travell to any of their friend Indians houses, we must doe the like as at our departure, to give them some trifles, as Knives and Beades. So that we lived very good cheape" (Wilson 1625, 1263). John Wilson's comment also shows how the exchanges with the Amerindians were unfavorable to them, and how what economists designate as 'terms of trade' were unbalanced. The novelty of European manufactured products, the attraction towards objects that Native Americans could consider as desirables goods because of their exoticism, the technological advantage that metal tools provided in the exercise of most daily tasks, gave the traders an important advantage in negotiating. This allowed them the opportunity to make huge profits that warranted the necessary down-payment and risk-taking that the transatlantic trip represented.

Similarly, the coastal indigenous population, who were controlling the trade and alliances with the European merchants, were now able to manipulate to their advantage the value of the goods they sent to the interior along the indigenous routes, as Keymis (1968 [1596], Gv) explained: "Some images of golde, spleenestones, and others may bee gotten on this coast, but they doe somewhat extraordinarily esteeme of them, because every where they

which they took. [...]. In this way, travellers with goods and with news constantly pass from district to district" (Im Thurn 1883, 273).

are current money. They get their Moones, & other pieces of gold by exchange, taking for each one of their greater Canoas, one piece or image of golde, with three heades, and after that rate for their lesser Canoas, they receive pieces of golde of lesse value. One hatchet is the ordinarie price for a Canoa."

5 New Tools

The circulation of European manufactured goods in the indigenous networks signified important technological changes for the peoples who adopted them, but these changes remain difficult to assess and to describe with precision, insofar as the effects of the use of these new tools were gradual and likely different depending on the groups concerned. We know, for example, that the introduction of the axe allowed an increase of up to ten times the efficiency of tree cutting for slash and burn cultivation. This technological leap had consequences far beyond the technical sphere, leading perhaps to a deep transformation of methods of cultivation and ways of life, by allowing the transition from intensive agriculture to itinerant agriculture, associated with an increasing mobility of villages which had to follow the moving of agricultural parcels in the forest (Denevan 2001).

The case of a tool such as the metal canoe adze (*canoodissel*) is less known but just as significant. Its introduction has gradually transformed the manufacturing technique of the canoes, which were previously made by controlled burning of wood, and likely had effects on the of navigation and fishing activities of the people who had adopted it. In fact, specific tools were in high esteem and demand and were necessary to keep good faith in the existing alliances between Europeans and Amerindians. A good example is given by the translator Jan Andries, a Dane and trader on Cayenne since the early 1630s for the Dutch WIC. He stated in 1640 that the cargo transported by Cornelis IJsbrantsz van des Sluijs, master of the *St Jan* which arrived in Cayenne was not good enough for the Amerindians, notably the straight adzes, which were not curbed and grooved; hence these were considered useless by the Amerindians and bad for trade according to Andries (Table 15.2).[11] For the contemporary Kali'na who dwell in eastern Surinam and western French-Guiana and are

11 West Frisian Archives, Hoorn, Notariele akten 1685 975, f.175r (16401020). Furthermore, Andries also stated that the color of the beads was not wanted by the Amerindians and the axes of the "noordhoeck" type were also useless for trade.

TABLE 15.2 Overview of trade goods shipped to the Guianas. The journal of Paulus van Caerden is found in the University Library of Göttingen (Germany), Cod. Ms 837, folio 5 ; the Fortuin cargo is found in the Commission Books of the States General in the National Archives of The Hague (The Netherlands), 1.01.02_12561, n° 32 ; the statement of trader Jan Andries can be found in the Notary Acts of the West-Frisian Archives in Hoorn (The Netherlands), 1685 975, folios 175–176 ; the listing of the Argyn can be found in the archives of the Old West Indian Company (OWIC) stored in the National Archives in the Hague (The Netherlands), 1.05.01.01, n° 25 but have also been translated for the British Guiana Boundary Commission 1898, p. 129–130 ; the delivery for Berbice by Van Pere is found in the same archive but another register 1.05.01.01, n° 42, folio 82 (Transcription and translation by M. van den Bel)

Ship	**Den Hollantschen Tuijn**	**Fortuijn**	**St Jan**	**Argyn**		**Middelburgh**
Capitain	**Paulus van Caerden, Antilles / Wild Coast, 1604**	**Pieter Swaeroogh, Amazone, 1616**	**Factor Jan Andries for Cornelis Ijs-brantsz der Sluijs, Cayenne, 1640**	**Essequibo, 1642**		**Cornelis van Pere, Berbice, 1647**
	16 gross dog-bells	63 lb *terwasge*?	200 *noortheckse* axes	400 large and small axes	Parcel of presents	100 large axes
	2 lb cut *margrieten* [glass beads]	60 lb *cakoenen*? with and without straps [*stroppen*]	150 large and small knives	400 small axes	500 large shark-knives? [*haeymessen*]	40 small axes
	168 bondles of *quispel-greijnen* [bead strings]	10 assorted *quispelgrauwen* white, blue and lavender	All kinds of small beads	300 medium cut-lasses with curved points	4 gross large and small copper bells	236 *deekens caniseros*
	half of 500 Spanish	200 violet *magariten* [glass beads]	100 canoe chisels	300 small cutlasses with curved points	3 dozen black flat male-hats	50 dozen black straight knives

the [other] half of 500 cheap needles	29 lb blue and white with eyes [beads]	cutlasses	60 cassava chisels	30 pairs of shoes	50 boatmen knives
4 gross trumps [mouth harp]	56 lb blauwe ende witte met ogen [pen]	small handaxes	3000 *carnaseros* [butcher knives?]	30 costumes of canvas clothes	6 dozen metal mirrors
20 bunch of coloured *teijckenen*	250 *deecken carniseros* [butcher knife?]	large red and blue beads	1500 large boatmen knives	1 bundle canvas *petris*	3 dozen scissors
24 *deeckens carniceros* [butchers knife?]	80 *decken taruse-ros* [knife?]	large and small harpoon points	800 small boatman knives	50 shirts	1 dozen large scissors
40 *deeckens* Bohemian [knives]	24 bushels trumps	small pins and needles	40 dozen razor-blades, large and small	4 dozen stone pitchers	10 dozen trumps
8 lb margarita beads [*margriettijnen*]	50 dozen metal mirrors	cleijne en grote visch hoecken	20 dozen razorblades	50 lb thin copper sheets	1250 fish-hooks
14 dozen mirrors	20 dozen scissors	wooden combs	10 dozen framed mirrors	100 lb coarse gunpowder	140 lb beads or *quispelg*[*rijn*]
	diverse long rectangular, en-graved and striped *pullenarij*	small rectangular mirrors	10 large black mirros	50 lb fine gunpowder	41 lb *cassoeren* [beads, see Biet 1664, p. 427 ; Renault-Lescure 1999, p. 276]

TABLE 15.2 Overview of trade goods shipped to the Guianas. (*cont.*)

Ship	**Den Hollantschen Tuijn**	**Fortuijn**	**St Jan**	**Argyn**		**Middelburgh**
Capitain	**Paulus van Caerden, Antilles / Wild Coast, 1604**	**Pieter Swaeroogh, Amazone, 1616**	**Factor Jan Andries for Cornelis Ijs-brantsz der Sluijs, Cayenne, 1640**	**Essequibo, 1642**		**Cornelis van Pere, Berbice, 1647**
		1812 lb axes and cutlasses	box of Venetian ear-rings	40 dozen double thimbles	50 lb sharp [shot] of which 1 lb iron [shot]	4 lb coloured cassoeren
		200 hand-axes	glass beads	10 gross trumps [mouth harps]	12 carabyns	1 dozen rasorblades
		200 plane chisels		2 cases wooden combs	6 pistols	20 [*v. poluer?*]
		220 cassavi-adzes		300 large fish-hooks	200 flints for arquebus and pistols	2 [*loat roers?*] [arquebus]
		200 chisels		200 large tinned fish-hooks	12 swords	20 [*potte slessen?*]
		angies, hooks, mirrors, *kissers*		400 cod-fish hooks	[List continues with carpenter tools]	1400 assorted nails
		250 large axes		30 dozen flintlocks		

divers merchandise	2 dozen letter pins
	200 large needles
	100 lb yellow *quispelgreyn* [bead strings]
	200 lb violet *quispelgreyn*
	50 lb green *quispelgreyn*
	60 lb white *quispelgreyn*
	60 lb tubular beads [*tierenarisen*]

respected navigators, these small adzes have been, until today, among the most wanted items of trade with Europeans.[12]

Other examples of particular iron tools for the Amerindian market are the cassava-picks (*cassavisteeckes*) and / or cassava-adzes (*cassavibeytels*) (Hulsman 2011, 188). Apparently, these cassava irons were frequently exported to the Guianas and were needed for daily work, such as manioc harvesting and / or tuber-peeling. Another example is the metal grater, or a copper plate pierced with holes and attached to a wooden plank in order to grate manioc tubers. It is believed that these metal graters were traded in order to facilitate the production of cassava and have replaced the use of stone and wooden graters (van den Bel 2015).[13]

Therefore, it can be concluded that the introduction of these new iron tools was indeed a highly interesting novelty for the Amerindians, but they soon demanded some changes in this hardware to better meet their daily needs. European smiths responded by producing iron tools which were more in line with Amerindian preferences.

6 Given to the Dead…

But beyond the technical benefits that these goods provided, their use and access represented an important symbolic and also political issue that justified the implementation of strategies to control access to trade and to the redistribution of the goods inward. An indication can be found in the Amerindian urn burials where they are most often discovered. Amerindian groups inhabiting the coastal fringe between the Lower Amazon and the Oyapock Rivers buried their dead in anthropomorphic urns such as those known from Late Aristé, Mazagão, Maracá and Aruã archaeological complexes. These urns have been attributed to the Late Ceramic Age and contact period since European trade items were found in them (Goeldi 1900; Meggers and Evans 1957; Nimuendajú 2004; Petitjean Roget 1995) (Figure 15.5). These objects consist primarily of glass beads whereas other items such as iron tools are less common. It is perhaps difficult to imagine but since the end of the sixteenth century literally tons of European barter goods, i.e., glass beads, trumps, bells, fish hooks, axes,

12 The contemporary Kali'na, amongst others, are the descendents of the historic *Caribbees* (English) or *Galibis* (French). See Collomb and Tiouka 2000.

13 An exemplary shipping list or 'cargasoen' for the Essequibo colony can be found in the proceedings of the WIC Zeeland Chamber dated 30 June 1642, revealing the presence of '50 lb thin, yellow [colored] copper plates' (British Guiana Boundary Commission 1898, 129–130). See also Table 15.2.

FIGURE 15.5 Glass beads found in Aristé urns on the Lower Oyapock at the archaeological sites of Trou Réliquaire and Trou Delft (Courtesy Jean-Paul Gess, IRAMAT-CRP2A, UMR 5060 CNRS-University Bordeaux-Montaigne). Karlis Karklins (pers. comm. January 2017) commented that these beads are rather common and have a large time-span ranging between the sixteenth and nineteenth centuries. Facetted amber and blue beads are probably made in Bavaria after 1650. Chemical analysis of different blue beads suggests a tentative date after 1600 due to the high arsenic-cobalt ratio (Ollagnier et al. 2011). A Delft faience which accompanied these glass beads is also dated after 1650 (Le Roux 1994) suggesting a late seventeenth-century collection for these funerary sites. This corresponds well with the trade beads found on Amerindian burial sites in the Brazilian State of Amapá as identified by Kenneth Kidd (Meggers and Evans 1957, 588)

hatchets, knives, needles, pins, mirrors, nails, etc., were dispatched and distributed in the Guianas of which only a handful have been retrieved by archaeological research. It is suspected that most of these goods have been exchanged within the interior since so few have been found in the first regions of contact, such as Cayenne, despite the fact that considerable archaeological research has been done there.[14]

The fact that these items are found in burial contexts also suggests that they were sufficiently important to be 'given to the dead' which, according to Amerindian burial practices, may reveal political hierarchy and/or social stratification (see Guapindaia 2001). This is confirmed by the first historic documents of the Eastern Guianas and Lower Amazon River.

Beyond their technical use, European objects were therefore also a symbolic object and probably represented a source of prestige for groups or individuals who possessed them, confirming trade alliances. In this context, they could sometimes travel through time, as suggested by the example of an archaeological site, near Kourou: here, objects imported during the seventeenth and eighteenth centuries seem to have acquired a genuine heritage value (heirloom) and have been transmitted between generations, to finally be placed in an Amerindian tomb during the second half of the nineteenth century (van den Bel 2010b; van den Bel et al. 2015). These items were considered valuable and integrated into the local trade system, already containing spleen stones (*muiraquitãs*), crescent-shaped golden plates (*caracolis*), and strings of shell beads (*quiripás* or *ocayes*) which were used as gifts, status symbols and heirlooms among the multiple groups of the entire Guiana Plateau. The vastness of this early colonial trading network is evidenced by the geographical distribution of stylistically similar objects such as green stone (nephrite) frog-shaped pendants or *muiraquitãs* as well as particular techniques of decoration on specific vessel shapes which have been attributed to Late Ceramic Age Koriabo pottery (van den Bel 2010a; Boomert 1987, 2004; Cabral 2011; Rostain 1994). The latter ceramic ware is found between the mouths of the Orinoco and Amazon Rivers as well as in the Guiana Highlands and the Tumuc Humac Mountains, showing stylistic homogeneity within this vast area. However, further research is needed to locate a possible origin or to discern various regional production centers, which apparently, shared interregional production methods.

14 Conversely, little is left of the large quantities of tobacco and annatto balls, precious wood, golden objects and spleen stones that were shipped to the other side of the Atlantic Ocean, either. Only a few (personal) trade objects are still present in European museums.

7 Final Remarks on the 'Yao Connection'

The stories of the late sixteenth and seventeenth centuries tell us, indirectly, about the nature of the trade between Native Americans and Europeans but also about the political strategies that each developed to ensure the maximum benefit from these exchanges. Furthermore, they also allow us to better understand the way in which Amerindians sought to draw political and warlike profits from the relationships they established with the newcomers.

The Yao presented themselves as the absolute partners of the English, relying upon their early contacts with Ralegh and their role as guides.[15] With this partnership the Yao controlled the flow of European and Amerindian goods in the Oyapock basin. The Yao controlled the local production of tobacco, annatto and cotton, and also kept an abundance of precious hard wood and victuals, which were supplied by their allies or the Amerindians from 'above the falls.' A similar operation can be suggested for the Charibes of Cayenne who represent the premier trading partner for the different European Nations in this part of the Guianas. In this manner, these Amerindian powers created a 'zone franche,' or mutual trading ground, which they ruled and controlled the access of trade goods. Hence, it can be argued that the deadlock war between Cayenne and the Oyapock was an elite-war for prestige of who will eventually possess all trading privileges with the Europeans. By the end of the 1620s the trade was abandoned on the Oyapock where numerous skirmishes between traders and Amerindians took place (de Laet [1644] 1932 t. ii, 16–18; van Ryen 1924). This social pressure was probably exacerbated by the European demand to have larger and more permanent settlements. This led to more aggressive intrusions into indigenous politics and created a less controllable situation for the ruling indigenous group. The English and Dutch traders who fled the Portuguese attacks upon the Lower Amazon River settled in other regions such as the Lesser Antilles, leaving the Oyapock vacant of European powers (Lorimer 1989, 1993; Williamson 1923). Although the Island of Cayenne was still being considered for settlement by the Dutch and French by 1630 (De la Roncière 1910, t. iv, 668–669; de Vries 1655), more permanent colonies such as Berbice (van Pere family) and Surinam (Marshal) provided new and more successful territories on this part of the coast. Again, this was based on trade with the Amerindians in which the Aruacas and Charibes played important political and economic roles.

15 For a wider discussion on the Indian guides of Walter Ralegh, see Vaughan (2002).

Eventually, during the 1630s, the Dutch had finally swept the Caribbean free from Spanish ships, which made the Europeans turn to the Lesser Antilles for permanent settlement (Lorimer 1989, 101). By this time a long stretch of coast, roughly between Cayenne and the Greater Antilles, was to be explored further and exploited by the Europeans. At the beginning of the second half of the seventeenth century, they introduced larger settlements and a sugar economy reliant on the labor of African slaves, thus resulting in Amerindian warfare over the loss of their land.

References

Acuña, Cristoval, d'. 1641. *Descubrimiento del Gran Rio de las Amazonas*. Madrid: Imprenta del Reyno.

Bel, Martijn, van den. 2009. "The journal of Lourens Lourenszoon (1618–1625) and his stay among the Arocouros on the Lower Cassiporé River, Amapá (Brazil)." *Boletim Musei Paraense Emílio Goeldi, Ciencas Humanas* 4 (2): 303–317.

Bel, Martijn, van den. 2010a. "A Koriabo site on the Lower Maroni River, Results of the preventive archaeological -excavation at Crique Sparouine, French-Guiana." In *Arquéologia Amazônica*, edited by Edithe Pereira and Vera Guapindaia, 61–93, Belém: MPEG/IPHAN/SECULT.

Bel, Martijn, van den. 2010b. "Grès rhénan dans une tombe guyanaise." *Archéopages* 28: 98–99.

Bel, Martijn, van den. 2015. "Uma nota sobre a introdução de raladores de metal e sobre a produção e consumo da mandioca e do milho na zona costeira das Guianas, durante o século XVII." *Revista Amazônica* 7 (1): 100–132.

Bel, Martijn, van den, Thomas Romon, Christian Vallet, and Sandrine Delpech. 2015. "Un village 'galibi' en Guyane française: le cas du site archéologique d'Eva 2." *Recherches amérindiennes au Québec* 44 (2/3): 127–141.

Biet, Antoine. 1664. *Voyage de la France Eqvinoxiale en l'Isle de Cayenne, entrepis par les françois en l'année MDCLII*. Paris: François Clovzier.

Boomert, Arie. 1987. "Gifts of the Amazons: 'green stone' pendants and beads as items of ceremonial exchange in Amazonia and the Carribean." *Antropológica* 67: 33–54.

Boomert, Arie. 2004. "Koriabo and the Polychrome Tradition: The Late-Prehistric era between the Orinoco and Amazon mouths." In *Late Ceramic Age Societies in the Eastern Caribbean*, edited by André Delpuech and Corinne L. Hofman, 251–266. London: Archaeopress, British Archaeological Reports, International Series 1273.

British Guiana Boundary Commission [BGBC]. 1898. *Arbitration with the United States of Venezuela, Appendix to the case on behalf of the Government of Her Britannic Majesty*, Vol. I. (1593–1723). Foreign Office.

Butt Colson, Audrey. 1973. "Inter-tribal trade in the Guiana Highlands." *Antropológica* 34: 1–70.

Cabral, Mariana, P. 2011. "Juntando Cacos: uma reflexão sobre a classificação da Fase Koriabo no Amapá." *Revista Amazônica* 3 (1): 88–106.

Collomb, Gérard and Felix Tiouka. 2000, *Na'na Kali'na. Une histoire des Kali'na en Guyane*. Cayenne: Ibis Rouge Édition.

Collomb, Gérard and Francis Dupuy. 2009. "Imagining group, living territory: a Kali'na and Wayana view of history." In *Anthropologies of Guayana: Cultural Spaces in Northeastern Amazonia*, edited by Neil L. Whitehead and Stephanie Aleman, 113–123. Boulder: University of Arizona Press.

Collomb, Gérard and Martijn van den Bel, eds. 2014. *Entre deux mondes. Amérindiens et Européens sur les côtes de Guyane, avant la colonie (1560–1627)*. Paris: CTHS.

Denevan, W. 2001. *Cultivated Landscapes of Native Amazonia and the Andes*. Oxford: Oxford University Press.

Dreyfus, Simone. 1992. "Les réseaux politiques indigènes en Guyane occidentale et leurs transformations aux XVIIe et XVIIIe siècles." *L'Homme* 122/124: 75–98.

Dupuy, Francis. 2008. "Wayana et Aluku: les jeux de l'altérité dans le haut Maroni." In *Histoires, identités et logiques ethniques. Amérindiens, Créoles et Noirs Marrons en Guyane*, edited by Gérard Collomb and Marie-José Jolivet, 165–201. Paris: CTHS.

Edmundson, George, Reverend. 1904. "The Dutch on the Amazon and Negro in the Seventeenth Century, Part 2: Dutch trade in the basin of the Rio Negro." *The English Historical Review* 73: 1–25.

Fausto, Carlos. 2001. *Inimigos fiéis: história, guerra e xamanismo na Amazônia*, São Paulo: Ed. USP.

Gallois Tilkin, Dominique, ed. 2005. *Redes de relações nas Guianas*. São Paulo: Associação Editorial Humanitas.

Goeldi, Emílio. 1900. *Excavações archeologicas em 1895. Executadas pelo Museu Paraense no Littoral da Guyana Brazileira entre Oyapock e Amazonas. 1ª Parte: As cavernas funerárias artificiaes de Índios hoje extinctos no Rio Cunany (Goanany) e sua ceramica*. Belém: Museu Paraense de História Natural e Ethnographia, (Memórias do Museu Goeldi, I).

Goslinga, Cornelis. 1971. *The Dutch in the Caribbean and on the Wild-Coast (1580–1680)*. Groningen: Van Gorcum.

Grenand, Françoise and Pierre Grenand. 1987. "La côte d'Amapá, de la bouche de l'Amazone à la baie d'Oyapock, à travers la tradition orale Palikur." *Boletim do Museu Paraense Emílio Goeldi Série Anthropologia* 3 (1): 1–77.

Guapindaia, Vera. 2001. "Encountering the Ancestors: The Maracá urns." In *Unknown Amazon. Culture and Nature in Ancient Brazil*, edited by Colin McEwan, Cristiana Barreto, and Eduardo Goes Neves, 156–173. London: British Museum Press.

Harcourt, Robert. [1625] 1928. *A relation of a voyage to Guiana*, second series LX. London: Hakluyt Society.

Hemming, John M. 1978. *The search for El Dorado*. London: Phoenix Books.

Hulsman, Lodewijk A.H.C. 2009. "Nederlands Amazonia. Handel met indianen tussen 1580 en 1680." Unpublished PhD diss., University of Amsterdam.

Hulsman, Lodewijk A.H.C. 2010. "De Guiaansche Compagnie: Nederlanders in Suriname in de periode 1604–1617." *OSO Tijdschrift voor Surinamistiek en het Caraïbisch gebied* 29 (2): 300–314

Hulsman, Lodewijk A.H.C. 2011. "Swaerooch: o comércio holandês com índios no Amapá." *Revista Estudos Amazônicas* 6 (1): 178–202.

Im Thurn, Everard F. 1883. *Among the Indians of Guiana, Being sketches chiefly anthropologic from the interior of British Guiana*. London: Kegan Paul, Trench & Company.

Keymis, Lawrence. 1968 [1596]. "A Relation of the second Voyage to Guiana." In *The English Experience* 65. Amsterdam: Theaturm Orbis Terrarum and New York: Da Capo Press.

Laet, Johannes, de [1644] 1931–1937. "*Historie of iaerlijck verhael van de verrichtinghen der geoctroyeerde West-Indische Compagnie, zedert haer begin, tot het eynde van 't jaer sesttien-hondert ses-en dertich* [...]." In: Werken van de Linschoten Vereniging 35, 4 Vols. Den Haag: Martinus Nijhoff.

Le Roux, Yannick. 1994. "Note à propos de la faience découverte sur le site de 'Trou Delft'." *Bilan scientifique de la Région Guyane* 93: 23–24.

Leigh, Charles. 1625. "Captaine Charles Leigh his voyage to Guiana and plantation there." In *Purchas his Pilgrimes: or Relations of the world and the religions observed in all ages and places discovered, from the creation unto the present*, edited by Samuel Purchas, 309–323. London: Henrie Fetherstone.

Lorimer, Joyce. 1989. *English and Irish Settlement on the River Amazon 1550–1646*. Second Series 171. London: Hakluyt Society.

Lorimer, Joyce. 1993. "The failure of the English Guiana ventures 1595–1667 and James I's foreign policy." *The Journal of Imperial and Commonwealth History* 21 (1): 1–30.

Lorimer, Joyce. 2006. *Sir Walter Ralegh's Discoverie of Guiana*. Cambridge: Hakluyt Society.

Masham, Thomas. 1890. "The thirde voyage set forth by Sir Walter Ralegh to Guiana, with a pinnesse called the Watte, in the yeere 1596, written by M. Thomas Masham a gentleman of the companie." In *The Voyages of the English Nation to America, Volume 4*, edited by Richard Hakluyt, 182–194. Edinburgh: E. & G. Goldsmid.

Mathews, Timothy G. 1970. "Memorial Autobiografico de Bernardo O'Brian." *Caribbean Studies* 10: 89–106.

Meggers, Betty J. and Clifford Evans. 1957. "Archaeological Investigations at the Mouth of the Amazon." *Bulletin of the Bureau of American Ethnology* 167, Washington D.C.: United States Government Printing Office.

Mocquet, Jean. 1617. *Voyage en Afrique, Asie, Indes Orientales et Occidentales. Fait par Iean Mocquet, Garde du Cabinet des Singularitez du Roi, aux Tuilleries*. Paris: Jean de Heucqueville.

Nimuendajú, Curt U. 2004. *In Pursuit of a Past Amazon. Archaeological Researches in the Brazilian Guyana and in the Amazon Region*, a posthumous work compiled and translated by Stig Rydén & Per Stenborg and edited by Per Stenborg & Jette Sandahl. Etnologiska Studier 45, Göteborg.

Netscher, Pieter M. 1888. *Geschiedenis van de Koloniën Essequebo, Demerary en Berbice, van de vestiging der Nederlanders aldaar tot op onze tijd*. Den Haag: Martinus Nijhoff.

Ollagnier, Céline, Max Schvoerer, Claude Ney, Virginie Mosca, and Nathalie Groué. 2011. "Contribution aux recherches de provenance et de chronologie, de perles en verre bleu trouvées en Guyane française, dans des sépultures à Ouanary (Trou Reliquaire et Trou Delft): Etude physique préliminaire." Paper presented at the *24th Congress of the International Association for Caribbean Archaeology, Fort-de-France, Martinique, 25–30 July*.

Petitjean Roget, Henri. 1995. "Fouille de sauvetage urgent, Site n° 97112314.16, Trou Delft: un site funéraire post colombien sur l'Oyapock en Guyane française." In *Proceedings of the 15th Congress of the International Association for Caribbean Archaeology*, edited by Ricardo Alégria and Miguel Rodríguez, 377–392. San Juan de Porto Rico: Centro de Estudios Avanzados de Puerto Rico y el Caribe.

Ralegh, Walter. (1596) 1968. *The Discoverie of the Large, Rich, and Bewtiful Empyre of Guiana*. In *The English Experience 3*. Amsterdam: Theaturm Orbis Terrarum, and New York: Da Capo Press.

Ralegh, Walter. 1997. *The Discoverie of the Large, Rich, and Bewtiful Empyre of Guiana*, edited by Neil L. Whitehead. Norman: University of Oklahoma Press.

Renault-Lescure, O. 1999. "Glossaire ethnolinguistique." In *Dictionnaire caraïbe-français, Révérend Père Raymond Breton 1665*, edited by Marina Besada Paisa, 267–303. Paris: IRD / Karthala.

Roncière, Charles B., de la. 1910. *Histoire de la marine française. En quête d'un empire colonial, Richelieu*. Paris: Plon.

Rostain, Stéphen. 1994. "Archéologie du littoral de Guyane. Une région charnière entre les influences culturelles de l'Orénoque et de l'Amazone." *Journal de la Société des Américanistes* 80: 9–46.

Ryen, Jan, van. 1924. "Brief van Jan van Ryen aan de Directeuren van de Kamer van Zeeland van de WIC Fort Nassau, Wiapoco, April 25, 1625 [1627?]." In *Documents relating to New Netherland 1624–1626*, In: The Henry E. Huntington Library, Translated and Edited by Arnold J.F. van Laer, Document B, pp. 21–33 and 256–258, San Marino: The Henry E. Huntington library and art gallery.

Santos-Granero, Fernando. 2009. *Vital Enemies: Slavery, Predation, and the American Political Economy of Life*. Austin: University of Texas Press.

Taylor, Douglas. 1977. *Languages of the West Indie.* Baltimore: The John Hopkins University Press.

Vaughan, Alden T. 2002. "Sir Walter Ralegh's Indian Interpreters, 1584–1618." *The William and Mary Quaterly* 59 (2): 341–376.

Vries, David Pietersz de. 1655. *Korte Historiael ende Journaels Aenteykeninge van verscheyden Voyagiens in de Vier Deeles des Werelds-ronde, als Europa, Africa, Asia, ende Amerika gedaen.* Brekegeest: Symon Cornelisz.

Wassenaer, Nicolaes. 1627. "Twaelfde deel of 't vervolvolgh van het Historisch Verhael aller gedenckwaerdiger geschiedenissen die in Europa […]." *Historisch Verhael aller gedenckwaerdiger geschiedenissen*, 17 Vols. (1622–1630). Amsterdam: Jan Jansen.

Whitehead, Neil L. 1990. "The Mazaruni pectoral: A golden artefact discovered in Guyana and the historical sources concerning native metallurgy in the Caribbean, Orinoco, and Amazonia." *Archaeology and Anthropology* 7: 19–38.

Whitehead, Neil L. 1992. "Tribes Make States and States Make Tribes. Warfare and the Creation of Colonial Tribes and States in Northeastern South America." In *War in the Tribal Zone: Expanding States and Indigenous Warfare*, edited by Ferguson R. Brian and Neil L. Whitehead, 127–150. Washington: University of Washington Press.

Whitehead, Neil L. 2011. *Of Cannibals and Kings: Primal Anthropology in the Americas*, Latin American Originals 7. Pensylvania: The Pennsylvania State University Press.

Williamson, James A. 1923. *English colonies in Guiana and on the Amazon, 1604–1668.* Oxford: Clarendon Press.

Wilson, John. [1625] 1906. *The Relation of Master John Wilson of Wansteed in Essex, one of the last ten that returned into England from Wiapoco in Guiana 1606*, in *Hakluytus Posthumus or Purchas His Pilgrimes, Contayning a History of the World in Sea Voyages and Lande Travells by Englishmen and others*, edited by Samuel Purchas, Volume 16, 338–352. Glasgow: James MacLehose & Sons.

CHAPTER 16

Colonial Encounters in the Southern Lesser Antilles: Indigenous Resistance, Material Transformations, and Diversity in an Ever-Globalizing World

Corinne L. Hofman, Menno L.P. Hoogland, Arie Boomert, and John Angus Martin

1 Introduction

The Lesser Antilles (Figure 16.1) represent one of the major regions in the world in which the lasting effects of the encounters between Europe and indigenous cultures with dramatically different ideological, social, technological, and economic frameworks are still very apparent. The small islands, which are located to the east of the Caribbean Sea, were linked through a vast web of social relationships in which Amerindians, Europeans, and Africans became entangled during the first centuries of European invasion and colonization. The intercultural dynamics which materialized during the early colonial period likely built upon local and regional networks of peoples, goods, and ideas that had developed in the insular Caribbean over the previous 6000 years (Hofman and Bright 2010; Hofman et al. 2011). By AD 1000, different island societies had developed in both the Greater and Lesser Antilles, and by 1492 a web of interlocking networks had spread across the Caribbean Sea, crossing local, regional, and pan-Caribbean boundaries (Hofman and Hoogland 2011). At the time of contact, these networks, which were flexible, robust, inclusive, and outward-looking systems, echoed the overarching patterns of human migration and mobility, and the intercultural dynamics among the communities of both islands and mainland(s) (Hofman et al. 2014). The Lesser Antilles were the last set of islands in the circum-Caribbean to be officially and permanently settled by Europeans in the course of the seventeenth and eighteenth centuries. Their occupation of these islands was fiercely contested by the Island Carib (*Kalinago*) and their mixed descendants, the Black Carib (*Garifuna*).[1]

1 The term Island Carib used throughout this chapter refers to the indigenous peoples often designated as Carib, who represented the Arawakan-speaking inhabitants of the Lesser Antilles during the historic period, but also lived on the South American mainland, notably

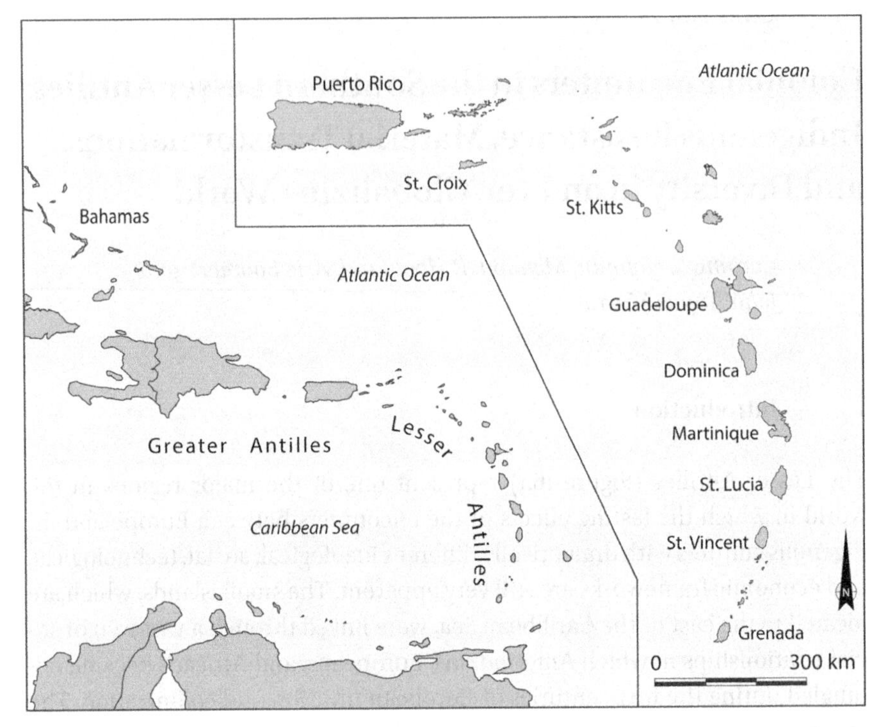

FIGURE 16.1 Map of the insular Caribbean with detail of Puerto Rico, and the Leeward and Windward Islands of the Lesser Antilles
MAP BY MENNO L.P. HOOGLAND

This chapter focuses on the impacts of the colonial encounters on Island Carib/*Kalinago* culture and society in the southern part of the Lesser Antilles, i.e. the Windward Islands, by looking at the village layout and material culture repertoires at two recently excavated early colonial sites on St. Vincent and Grenada. The research presented here offers the unique possibility of studying continuity and change of inter-community social relationships, in the advent

in the Orinoco Valley and the coastal zone of the Guianas, in parts of which these Cariban-speakers are known as *Galibis*. In the seventeenth century Island Carib and Mainland Carib jointly inhabited Grenada. Though the designation may not express the cultural/ethnic diversity that existed at either contact and/or colonization, it broadly defines the Amerindian population in the Lesser Antilles. 'Island Carib' is used throughout because of its historical acceptance and to avoid confusion, though the authors recognize the importance of *Kalinago* as a self-ascribed name for this people and its widespread acceptance, especially in the Caribbean. Actually, this name originally represented the self-denomination of the Island Carib men (Breton 1665/1666).

of European colonialism. We intend to recast *Kalinago* archaeology in a nuanced, inclusive manner, dissipating colonial documentary biases, and placing the transformations of *Kalinago* culture and society within the wider context of the European encounters and the globalizing world. The archaeological data that we present are embedded in a critical (re-)reading of the early (Spanish, Dutch, French, and English) documentary sources, involving the extraction of ethnographic information on *Kalinago* society that is compatible with the archaeological data. This line of inquiry thus integrates material and textual sources to provide a new and conjoined perspective on the transformations in *Kalinago* culture and society during early colonial times. The present-day indigenous peoples in the Lesser Antilles are the direct successors of the historic 'Island Carib' cultural traditions, with a considerable stake in the archaeological heritage (Hofman and Hoogland 2012, 2018; Honychurch 2000; Twinn 2006). Collaborations with the Carib/*Kalinago* communities in the Windward Islands, notably on Dominica, St. Vincent, and Trinidad, have been crucial to interpret our recent findings.

2 First Encounters with a "Phantastic Insular World"

The islands of the Lesser Antilles initially gained fame through Christopher Columbus' reports mentioning man-eating *Caribes*, who were allegedly raiding settlements to the north, i.e. the islands of the Greater Antilles (Curet 2005; Keegan and Hofman 2017; Oliver 2009; Petitjean Roget 2015; Rouse 1992). *Caribes*[2] rapidly became a generic term for the Spanish to denounce supposedly anthropophagous, fierce, and hostile Amerindians. When, at the end of his first voyage, Columbus was attacked at the Golfo de las Flechas in the area of Samaná in northeastern Hispaniola, the aggressors were identified as *Caribes*. The same happened when he was attacked by the indigenous peoples of the islands of Guadeloupe and St. Croix during his second voyage in 1493 and on his

2 The Spanish are responsible for spreading the term *Caribe* and other notions such as *Calino, Camballi, Caniba, Canima*, that were changed to *Cannibales* and *Caribales* and later on to *Caribes*, to indicate the pugnacious and man-eating Indians that were notorious for their resistance against the other Indians of the region and Spanish colonists. Las Casas, who uses the information from Columbus, mentions for 26th November 1492: '...toda la gente que hasta hoy ha hallado diz que tiene grandisimo temor de los de *Caribe* o *Canima*...' (Las Casas 1927). A month later, on the 26th of December, he mentions the term *Caribe* for the first time to indicate man-eating Indians. The islands where these *Caribes* lived were supposed to lay south of Hispaniola and half-way to all the other islands.

return to Spain in 1496 (Keegan and Hofman 2017). The Spanish colonizers were fueled with biases and misconceptions regarding the idea of cannibalism among these distant, unfamiliar *Caribes*, based on preconceived, late-medieval ideas about a "phantastic insular world" (Hulme 1986; Hofman et al. 2008).

Scholarly knowledge of the early colonial period in the Lesser Antilles is based primarily on the descriptions provided by early Western European chroniclers (e.g., Anonymous [1659] 1975; Anonyme de Carpentras 2002; Breton 1665/1666, 1978; Chanca 1988; Coppier 1645; Du Tertre 1667–1671; Labat [1722] 2005; Nicholl 1605; Pinchon 1961; and Rochefort 1658). As reported in these chronicles, the Lesser Antilles, especially the southern islands, were an ongoing contested space among the various Amerindian peoples of northern South America, Margarita and Cubagua, Trinidad, Tobago, and the islands of the Lesser Antilles, especially Grenada, St. Vincent, and Dominica (Anonymous [1659] 1975; Espinosa [1622] 1942; de Laet ([1630] 1988); Oviedo y Váldez 1959; Pelleprat 1965; Rochefort 1658). The arrival of the Spanish, followed by the French, English, and Dutch, added a new dynamic of contestation as they made allies with or fought against these indigenous populations for control of the region, its people, and resources, while the latter defended their way of life and homelands. Thus dawned the sixteenth century in the Lesser Antilles, changing landscapes and seascapes after the European invasion of the archipelago and the adjacent mainland(s), creating a multiplicity of social interactions and patterns of exchange in the ensuing centuries.

The early Amerindian-European relationship in the Lesser Antilles was one of originally amicable encounters and trade activities next to often violent clashes (Boomert 2002). It was the Spanish occupation of the Greater Antilles and that of the pearl islands of Cubagua and Margarita offshore Venezuela that determined their early historic interaction patterns with the Amerindians in the Lesser Antilles in which peaceful exchanges of trade goods alternated with violent meetings and endeavors at slave taking. There are few reports of exchanges between the *Kalinago* of the Windward Islands and the Spanish in this early period. The latter occasionally stopped to trade or refresh on their journeys, including those of the annual return fleets (*armadas*) from Spain, which generally entered the Caribbean by sailing through the Lesser Antilles (Boucher 1992; Breton 1665/1666; Moreau 1992). Though the Island Carib often traded with the Spanish for ironware and manufactured goods, which they had come to desire, they sometimes attacked the Spanish as retaliation for assaults committed against them previously. Before long, an atmosphere of open hostility became the norm between the two, as the one fought for the preservation of their own way of life, and the other for a foothold and dominance in the region.

Throughout the period of Spanish colonization, the Lesser Antilles were known for slave raiding, but also functioned as a refuge for Amerindians who wished to escape from the Spanish oppression of their communities under the *encomienda* system in the Greater Antilles. The Spanish failure at settlement in the Lesser Antilles, notably Guadeloupe, and their subsequent lack of interest in the archipelago except for its use as a realm for slave taking, allowed other European nations to become involved with the Lesser Antilles. While St. Christopher (St. Kitts) was the first island to be settled by the Europeans, jointly by the English and the French in 1624, the islands of Guadeloupe and Martinique were not permanently settled until 1635, and Grenada in 1649. St. Vincent and Dominica were never officially colonized in the seventeenth century, as in 1660 the English and French decided to designate these islands as 'neutral' and to be left in the possession of the *Kalinago*. The French, however, established settlements on both islands, unofficially colonizing them. A pattern of exchange developed in the late sixteenth and early seventeenth centuries between the European nations and the *Kalinago*. While some of the Lesser Antilles participated in the last phase of indigenous resistance to the colonial powers, from early on, the *Kalinago* communities of St. Vincent absorbed increasing numbers of escaped African slaves, leading to the formation of a Black Carib identity on the island alongside the communities that remained purely *Kalinago*. Following the eighteenth-century Carib-English Wars, many Black Carib were deported from St. Vincent to Central America in 1791, where they self-identify as *Garifuna* (Palacio 2005).[3] By 1700, a major demographic collapse among the Amerindian populations had dramatically reduced their presence in much of the archipelago and throughout the eighteenth century they became largely marginalized in the various islands. Nowadays, the Greggs community of St. Vincent feels itself to belong to the *Garifuna* ethnic group and actively connects with its kin from Central America. Descendants of the *Kalinago* live in Dominica (Kalinago Territory), St. Vincent (New Sandy Bay), and Trinidad (Santa Rosa community, Arima), where they actively claim their Amerindian ancestry within the multi-ethnic and multi-cultural Caribbean society of these islands (Boomert 2016; Hofman and Hoogland 2012; Honychurch 2000; Lenik 2012; Reid 2009; Sued Badillo 2003; Whitehead 1995).

3 This name is derived from *Kalipuna*, the original self-denomination of the Island Carib women (Breton 1665/1666).

3 *Kalinago* Strongholds

From the first decades of the sixteenth century, the indigenous populations of Trinidad and the Windward Islands were greatly harassed by Spanish slave raiders from the Greater Antilles and the Bahamas, the 'pearl islands' of Margarita and Cubagua, and the new town of Cumaná on the east Venezuelan coast. Slave taking of the *Caribes* was authorized by various royal decrees, starting in 1503. There are few details to shed light on the extent and impact of these raids on the various Amerindian populations, but by 1520 several of the islands in the Lesser Antilles, e.g. Barbados, were probably depopulated as a result (Boomert 2002; Boucher 1992; Watts 1987). Only the densely populated, mountainous and strategically located islands of Dominica, St. Vincent, Guadeloupe, Martinique, St. Lucia, and Grenada were able to resist these devastating raids to a certain extent. Exactly how the raids affected their populations is unknown, but they must have suffered tremendously from the relentless assault. One direct consequence must have been the depletion of the indigenous population through deaths due to resistance to capture and the raids themselves, as the Carib later related stories of Spanish massacres on various islands to the early French missionaries (Boucher 1992; Breton 1665; Gullick 1985). Another was the retreat along the inaccessible Atlantic coasts by the Island Carib to better protect and defend themselves (Breton 1665). As Breton notes, when the Island Carib constructed a settlement, they did not cut many trees, "so as to remain hidden from the Europeans" (Breton 1665, 279).

Spanish sources indicate that the Amerindians in the Lesser Antilles were a formidable force and resisted Spanish attempts to take control of any of the islands or even traverse the seas (Moreau 1992). The 1525 Spanish attempt to colonize Guadeloupe, then occupied by the Carib, failed, as did efforts to colonize Trinidad in the 1530s due to resistance by a multi-ethnic alliance of Amerindians from Trinidad and the Paria peninsula (Boomert 2016; Moreau 1992). The Spanish settlements on Puerto Rico were often targeted by the Carib who raided and destroyed plantations, and abducted Spaniards and enslaved Africans. The Spanish, with their array of weaponry and ships, were sometimes defeated or suffered heavy human and material losses at the hands of the Carib, who employed ambushes, raids, sabotage, and hit-and-run tactics, equipped only with dugout canoes which they skillfully maneuvered, poison-tipped arrows, and the feared *boutou* or war club.

By the second half of the sixteenth century, as more Spanish colonists entered the region along with an increase in maritime traffic, there was an escalation in the confrontations between the Spanish and the Carib who staged coordinated attacks on Spanish settlements. One such raid by the Carib of the southern Lesser Antilles is reported to have taken place in 1569 against the Spanish settlement of Carabelleda on the central Venezuelan coast (Oviedo y Baños 1987). Some 300 Carib in fourteen canoes landed during the night to prepare for an early morning assault on both the town and port. Though the Spanish were informed of the impending attack by Amerindian allies, they discounted it, only placing a guard to watch (Oviedo y Baños 1987). The Island Carib' attack on the city, with its armed Spaniards, proved unsuccessful when they were confronted with armed opposition and were forced to retreat (Oviedo y Baños 1987). Similarly, the islands of Cubagua and Margarita were raided repeatedly by Island Carib fleets from the Windwards.

The Island Carib attacks on Spanish shipping as well as settlements (Cody 1995; Shafie et al. 2017), and the wrecking of Spanish vessels on the shores of the islands led to the capture of European sailors and enslaved Africans. The *Kalinago* put many of them to work in their tobacco fields and food gardens, contrary to the Spanish belief that they were eaten (Espinosa [1622] 1942). In 1561 alone, the Island Carib on Grenada held at least 30 Spaniards, mainly women and children, following the wreck of a Spanish ship along the coast of the island (Moreau 1992). Fifty Spanish colonists from Margarita, aided by cooperative Amerindians, failed in their attempt to free the Spanish prisoners on Grenada (Moreau 1992, 74). The total number of these captives and the extent of their treatment are unknown, but they may have accounted for a sizeable population. This is substantiated by Francisco de Vides' 1592 royal contract to colonize Trinidad, wherein it is stated that to populate that island he should pacify the Island Carib on Grenada and liberate their Spanish and African captives (Moreau 1992). The many captives held by the Island Carib included Spanish, Portuguese, and enslaved Africans. Some were able, through various means, to escape as did the "three Christians (a Portuguese prisoner of five years, and two Spaniards who had been prisoners for two years)" in 1567, while in 1578 a Spaniard tricked his captors into releasing him, and in 1593 an enslaved African was able to escape (Moreau 1992). These tales of capture and escape of Europeans and Africans were quite common throughout the region until the mid-1600s. This is illustrated by the well-known capture in Trinidad of García Troche Ponce de León, in 1569, who still lived as a prisoner in Dominica ten years afterwards, and the escape of the free black Luisa Navarrete from Dominica in 1580 (Baromé 1966; Hulme and Whitehead 1992).

Contacts with the Carib produced both favorable and unfavorable accounts, the latter being the most often recorded. André Thévet, in the 1550s, described Grenada as unapproachable because of the large numbers of Carib settled there (Moreau 1992). In 1565, the infamous John Hawkins related the tale of the French privateer and slaver Captain Jean Bontemps of the ship *Dragon Vert* of Le Havre who in March 1565 "came to one of those Islands, called Granada, and being driven to water, could not doe the same for the Canybals, who fought with him very desperatly two dayes" (Hulme and Whitehead 1992, 49).

Beginning in the 1520s, the French entered the Caribbean as usurpers of the Spanish trade monopoly in the region; the English followed in the 1560s and the Dutch by the 1590s (Andrews 1978). The legendary riches of Spain's American colonies and the homeward-bound fleets enticed those Europeans who were excluded from this trade. The following century is a litany of exploits by privateers and contraband traders like Charles Fleury, Jambe de Bois, Jacques de Sors, François Le Clerc, Jean Bontemps, Francis Drake, John Hawkins, and Piet Heyn, who plundered Spanish ports and shipping or traded with their inhabitants. Many came in search of riches as merchants invested in what appeared to be very lucrative ventures (Andrews 1978). During their explorations, the northern Europeans encountered the Island Carib and traded with them, ultimately replacing the Spanish. The Amerindians often allowed the careening of vessels, the taking in of fresh water, and the resting of sick crew members while trading European ironware, notably nails, knives, needles, hooks, bills, sickles, hoes, hatchets, saws, iron griddles, colored glass beads, trinkets, mirrors next to combs, spirits, and, rarely, firearms in exchange for foodstuffs including plantains, sweet potatoes, cassava bread, hens, pineapples, and bananas as well as tobacco, cotton, turtle carapaces, hammocks, and *kalikulis*, i.e. ornaments made of a gold-copper alloy which the Island Carib obtained from the South American mainland (Boomert 2002). Some of the French, English, and Dutch sailors established temporary shelters in the Lesser Antilles that would subsequently pave the way for permanent settlements in the mid-1600s by the northern Europeans.

Spanish attempts to establish themselves in the Lesser Antilles proved futile, as their failure to settle Guadeloupe illustrates. Yet, it is quite evident that relative to the rest of Spain's empire in the Americas, the Lesser Antilles offered very little of value, except possibly a defensive one. As a matter of fact, when the Spanish were finally able to obtain a permanent foothold on Trinidad (1592), it was only as a base for El Dorado expeditions to the continent (Lorimer 2006). Though the Spanish colony at Trinidad remained neglected for over a century, its tobacco trade attracted northern Europeans, especially the English and Dutch, and the Spanish colonists engaged them in contraband trade (*rescate*) between the 1590s and around 1612 (Lorimer 1978). In order to suppress the contraband trade, the Spanish decided to prohibit the cultivation of tobacco on Trinidad and in the mainland coastal zone altogether. By this time the Island Carib had begun to grow tobacco for the foreign market, thus intensifying their interaction with the northern European contraband traders and privateers. It was the tobacco trade that created the impetus for the first attempt at non-Spanish settlement in the Lesser Antilles, when in 1609 an English-Dutch consortium of merchants sent colonists to establish an outpost on Grenada. Being attacked by the Island Carib, they were forced to evacuate the island before the end of the year (Andrews 1978; Martin 2013).

For over a century, the Island Carib had successfully defended their islands against Spanish and other European aggression. They had retained most of them by the time the northern Europeans entered the region as colonizers. The first six decades of the 1600s would prove to be most difficult for the Island Carib as the French, English and Dutch descended on the Lesser Antilles and challenged them for possession. The unsuccessful attempt to settle Grenada in 1609 was an early indication that the northern Europeans were serious about occupying these islands, but the final struggle began in earnest in the 1620s when the English and the French successfully established settlements on St. Kitts, followed by English colonies on Barbados and Nevis, while the Dutch occupied Tobago.

4 *Island Carib/Kalinago* Archaeology

The so-called Suazoid (or Suazan Troumassoid) and Cayoid ceramic series represent the latest precolonial and earliest indigenous colonial period pottery developments in the Windward Islands (Allaire 2013; Bright 2011; Hofman 2013; Keegan and Hofman 2017). Initial archaeological links between the Cayo complex and the Island Carib were made in the 1980s and 1990s (Boomert 1986, 1995), based on similarities between Cayo pottery and the Koriabo ceramics from the Guianas. The Koriabo ceramic series emerged on the South American mainland around AD 1100–1250. It is distributed in the coastal areas and on the river banks of Guyana, Suriname, French Guiana, the interior of the Guianas, and northeastern Brazil, as far south as the lower reaches of the Xingú River. Koriabo forms part of the Polychrome Tradition or Marajoaroid series of Amazonia (Boomert 2004; Rostain 2009; Van den Bel 2015). Its pottery complex continues to exist well into the early colonial period and represents the ancestral ceramic tradition of the contemporary *Kaliña* or Mainland Carib of the Guianas and Orinoco Valley (Boomert 1986). Specific Cayo elements may have been inspired by *Kaliña* carrying out raids in the Lesser Antilles, exchanging marriage partners with the local inhabitants (Boomert 1995; Davis and Goodwin 1990), or just by movements of peoples up the islands. Descriptions in the historic documents have led to the suggestion that there existed a ceramic repertoire of well-finished vessels with Cariban affiliated names and related to the men's realm in contrast to lesser finished domestic pottery and griddles with Maipuran Arawakan or European names and related to the female sphere of activities (Breton 1665; Boomert 1986, 1995, 2011). This linguistic dichotomy would concur with the male and female registers within the Island Carib language that were recorded by the seventeenth-century French chroniclers, notably a male register of Maipuran Arawakan character with an extensive *Kaliña* or *Kaliña*-derived vocabulary, and an entirely Arawakan female register (Hoff 1994, 1995). The male's 'language' shows elements suggesting that it was used as a pidgin during trade contacts between the Island and Mainland Carib (Taylor and Hoff 1980).

4.1 *Cayo Sites in the Lesser Antilles*

At present, more than twenty archaeological sites with Cayo remains have been documented throughout the Windward Islands (Bright 2011; Hofman and Hoogland 2012; 2018). These sites are located between Grenada and Basse-Terre, Guadeloupe, on the Grenadines (Ile de Ronde), St. Vincent, and Dominica (Allaire 1994; Boomert 1986, 2009, 2010; Bullen and Bullen 1972; Holdren 1998; Hofman 2016; Honychurch 2000; Kirby 1973; Petitjean Roget 2001/2002; Richard 2002). Besides, related materials are known from Grande-Terre, Guadeloupe, La Désirade (Hofman 1995; Hofman et al. 2004, 2014; de Waal 2006). Isolated Cayo vessels and sherds have been reported from St. Lucia, Martinique (Bright 2011), and St. Croix (Hardy 2008, Figure 57; Martijn van den Bel pers. communication 2016; Corinne Hofman pers. observation 2016).

Recent investigations on St Vincent and Grenada have provided unique new insights into the archaeology of the *Kalinago*, especially with regard to their village layout and associated material culture repertoires, including European trade wares (Hofman and Hoogland 2012; Hofman et al. 2015; Keegan and Hofman 2017). A series of radiocarbon samples from the sites of Argyle and La Poterie, on St. Vincent and Grenada respectively, have provided dates for these sites between the late fifteenth and the early seventeenth centuries, and are thus consistent with the documentary information presented in the foregoing paragraphs and the European materials encountered at these sites as shown below (Figure 16.2).

4.2 *Village Location and Layout*

The chroniclers describe hamlets or single households dispersed across the landscape (Labat [1722] 2005). Blondel's map of 1666, shows the distribution of Carib sites on the island of Grenada to be located particularly in the northern and eastern parts of the island (Figure 16.3). The French missionary Raymond Breton (1665) mentions that settlements were usually located close to the sea

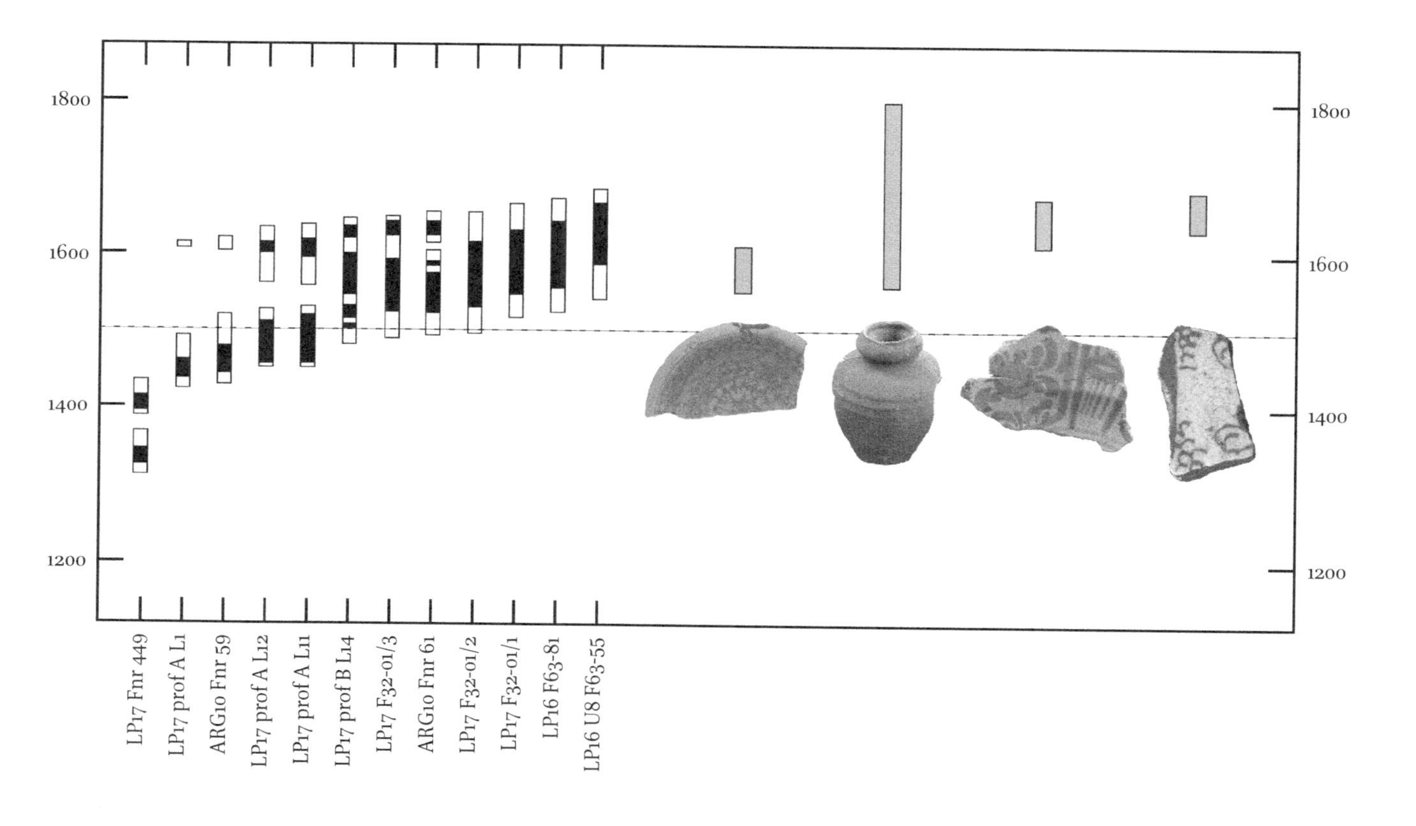

FIGURE 16.2 Radiocarbon dates for Argyle, St. Vincent and La Poterie, Grenada. On the right chronology based on late sixteenth-early seventeenth-century European earthenwares

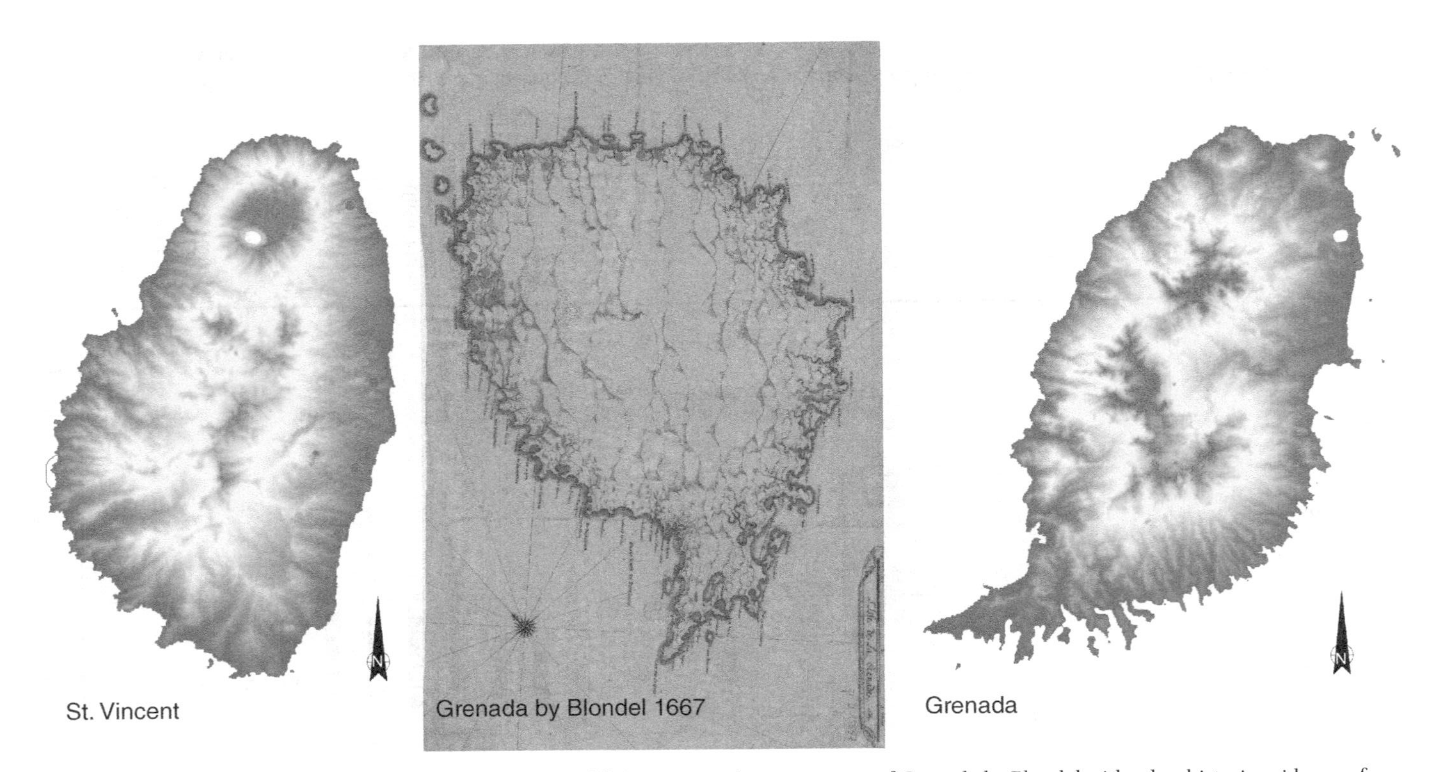

FIGURE 16.3 (left) Map of St. Vincent with Cayo sites; (middle) seventeenth-century map of Grenada by Blondel with ethnohistoric evidence of Carib/*Kalinago* settlements; (right) map of Grenada with Cayo sites
IMAGE BY MENNO L.P. HOOGLAND

and a river mouth (Breton 1665, 1978) on the rugged sides of the islands, facing the Atlantic Ocean. The distribution of Cayo sites on Grenada and other Windward islands, so far concur with these descriptions as the majority of sites has been found on the Windward side of the islands close to the sea and a river mouth (Hofman and Hoogland 2012) (Figure 16.3). According to Breton, a typical *Kalinago* village would include a rectangular to oval men's (assembly) house (*táboüi*) and a number of round family dwellings (*mánna*) around a *plaza* (e.g. Breton 1665/1666, 1978). The assembly house served as a meeting place for the men of the village, as an arms depot, a place to receive and accommodate guests, to hold communal feasts, and to bury deceased (male) members of the community. Several small rectangular structures such as racks and sheds were scattered between the houses. The *barbakot* or *boucan* is mentioned, i.e. a wooden rack consisting of four forked wooden sticks on which thin, straight branches were placed. Cooking places consisted of three stones on which wood and wood pulp were burned (Breton 1665; Hofman et al. 2015).

These descriptions concur with what has been recorded at the site of Argyle on St. Vincent where open-area excavations have revealed the postholes of eleven domestic structures located around two plazas (10x15 m and 15x25 m). The houses and the plazas are probably related to two, partially overlapping, building phases. During the second phase the village was rebuilt on the same location (Hofman and Hoogland 2012, 2018; Hofman et al. 2015). Nine of the eleven structures at Argyle harmonize with the descriptions for the mánna, i.e. small round to oval family houses (between 4.5x5 and 6x8 m), while the two largest structures can be interpreted as táboüi or oval men's houses. The largest Argyle táboüi measures 12x4 m (Hofman et al. 2015). At least 20 structures, all round to oval, were (re-)constructed from ca. 500 posthole features at the site of La Poterie in Grenada, of which some belong to the Cayo component of the site. All of them have a double row of posts, while some have two central posts and measure between 3 to 8 m in diameter. At La Poterie no plaza areas have been identified to date, obscuring the layout of the settlement and hampering full comparison to Argyle.

Burial pits were found in two of the round houses at Argyle. The skeletal material was not preserved due to the high acidity of the site's clayish soil, but two of the three burial pits yielded fragmented teeth of two adult individuals under the age of 25 years (Hofman and Hoogland 2012). The practice of burying the deceased under house floors is described by Father Breton (1665, 237–238) and other early chroniclers. Breton notes that the Island Carib "would dig a round pit three feet deep in the house for the dead person to be covered," continuing that the deceased was placed in a prepared grave and wrapped in a brand-new hammock, "in almost the same posture as a child in a mother's womb, neither turned upside down nor flat faced in the dirt, but upright, the feet below, the head above supported on the knees, and the grave covered with a plank." Indeed, the grave pit was sometimes covered by reed mats or boards/planks; additionally, clay pots were buried over the head. Burial examples from late precolonial components of sites in the Lesser Antilles suggest that this way of burying the dead was a widespread and long-lasting custom in the region (Hoogland and Hofman 2013).

4.3 *Material Culture Repertoire*

4.3.1 Cayo Pottery

Apart from the obviously South American origin of various stylistic and morphological elements as well as manufacturing techniques, Cayo ceramics exhibit links with the Greater Antillean Chicoid and Meillacoid ceramic series, particularly in decorative elements and specific vessel shapes. The latter include medium-sized biconical bowls with concave necks which are often decorated with punctuated or nicked small knobs at the corner point (e.g. Cayo Vessel Form 4 in Boomert 1986 Figures. 3:4, 5:4, 10:4). Form as well as ornamentation of this vessel shape show close similarities to the Late Chicoid bowls of Hispaniola, Puerto Rico and the Virgin Islands, especially those discussed by García Arévalo (1978, Figure 2, Lam. III:b-c). The decorative elements relating Cayo especially to the Koriabo complex of the mainland include painted or slipped designs, incised and grooved motifs, punctuations, lobed rims, and outward bossed wall sections (Boomert 1986, 27). Unrestricted bowls (so-called 'flower bowls') with carinated or indented ('lobed') rims are characteristic, often showing white-slipped interior surfaces. Some fragments suggest that originally they may have had red and/or black-painted designs on these white backgrounds. This way of decorating bowls is typical of the Koriabo pottery in the Guianas and occurs as far as northern Brazil (Boomert 2004; pers. communication Stephen Rostain and Christiana Barreto 2016). Another characteristic vessel shape is a large restricted jar made of reddish clay. It likely served as a container for the fermenting of cassava beer (*ouïcou*), and is mentioned as such by the early seventeenth-century chroniclers. In some instances, these jars have modeled decorations of animal/human faces (Figure 16.4).

4.3.2 European Tradewares

European earthenwares and merchandise were found intermingled with indigenous objects at these Cayo sites. Most of these trade wares come from the cliff side behind the houses, which was supposedly the area where the garbage was thrown. At La Poterie a fair amount was also found in the house area, but without clear association to a particular house or feature. European objects include

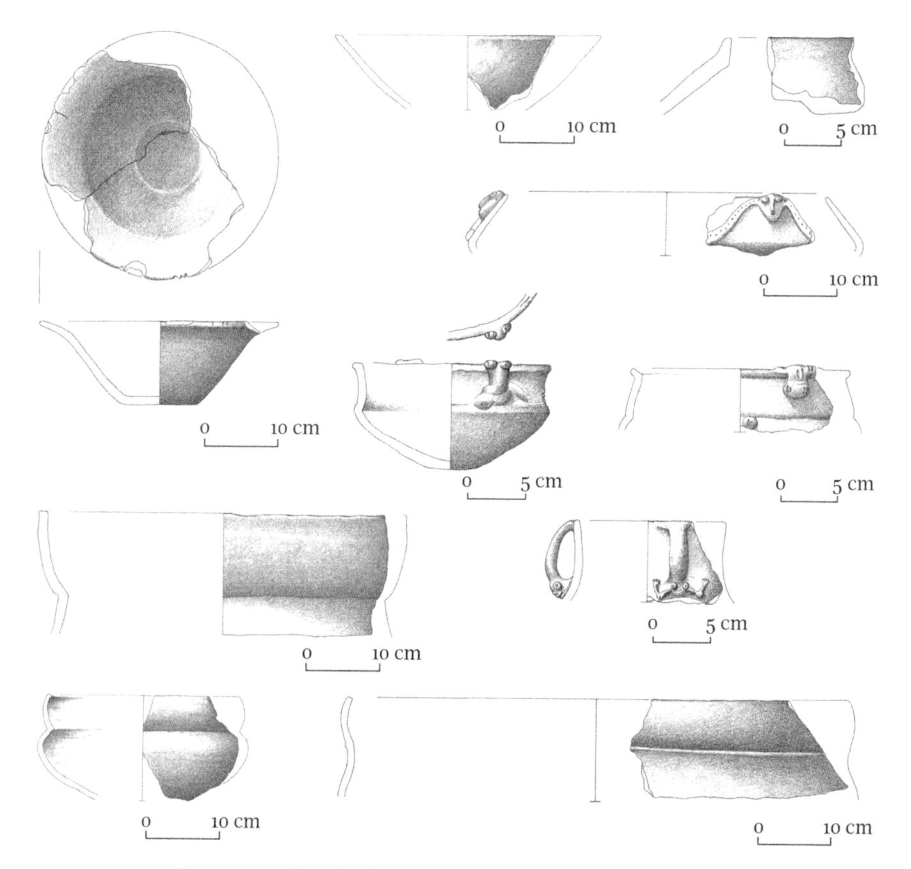

FIGURE 16.4 Cayo pottery from La Poterie, Grenada
IMAGE BY MENNO L.P. HOOGLAND

Spanish *maravedí* coins (https://finds.org.uk/database/artefacts/record/id/233375), pieces of iron, lead and glass, late sixteenth- and early seventeenth-century Spanish, Italian, and Dutch earthenwares, middle-style Iberian olive jars, as well as a series of glass beads (chevron and seed beads), and gun flints (Figure 16.5a-r). Some of the olive jar fragments have translucent green glaze on the exterior surface. An Argyle specimen has a mark on the rim that was stamped into the wet clay and represents the jar's ownership rather than its manufacturer's sign (Figure 16.5i; Marken 1994, 16). The first documented olive jars with rim marks of this type are from three securely dated Spanish wrecks from the first half of the seventeenth century (ca. 1622) (cf. Goggin 1960, 1968; Marken 1994, 50–51). The La Poterie assemblage also includes a tin-glazed fragment of a Beretino Ligurian Blue-on-Blue rounded plate with outwardly

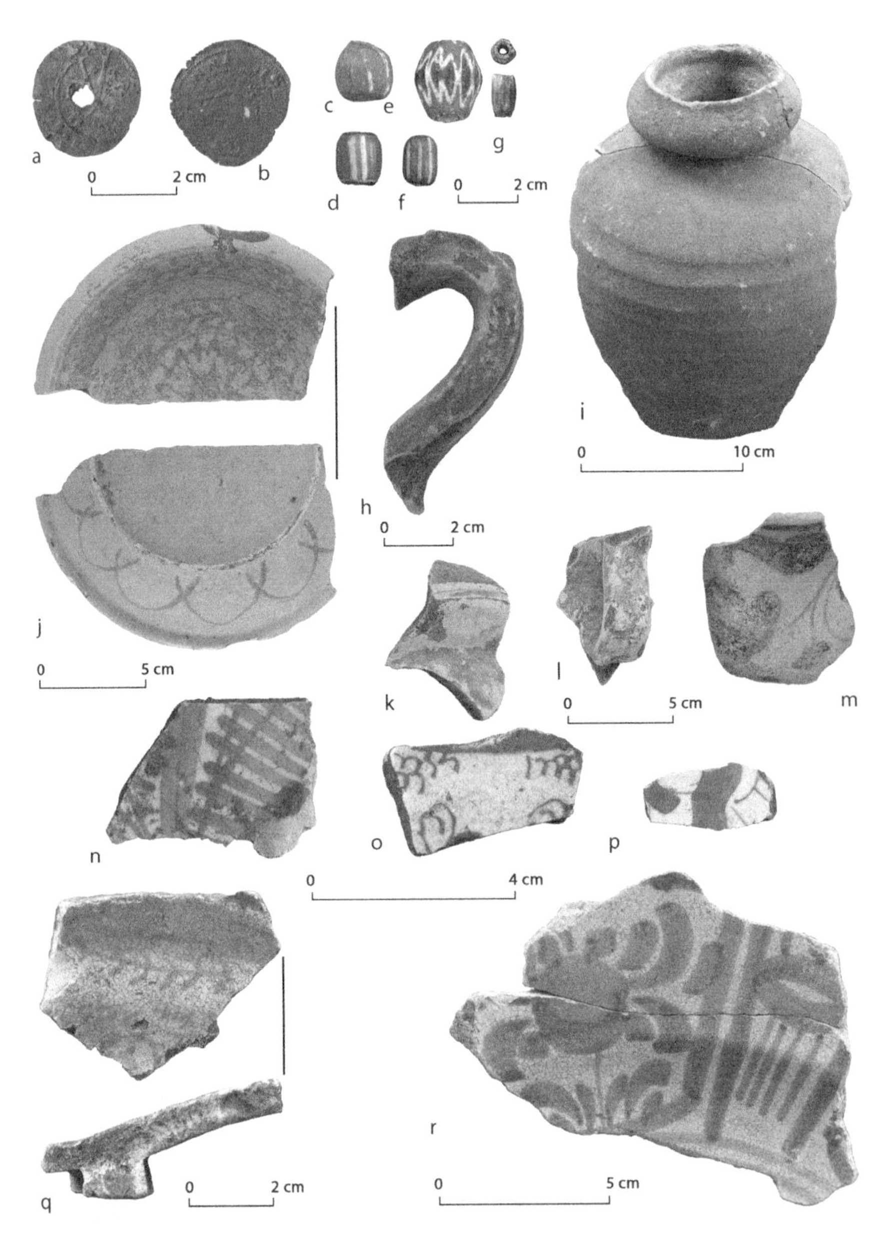

FIGURE 16.5 European trade wares from Grenada and St. Vincent: (a-b) Spanish copper *maravedí* coins; (c-g) sixteenth- and seventeenth-century beads; (h) fifteenth/sixteenth-century Spanish majolica; (i) Spanish olive jar; (j) sixteenth-century Ligurian *berettino* plate; (k-m) sixteenth-century Spanish majolica; (n-r) early seventeenth-century Dutch majolica and faience.
PHOTOS BY CORINNE L. HOFMAN AND MENNO L.P. HOOGLAND

curved rim and footring. This is a faience dating to 1550–1610. It has a buff fabric, it is blue tin-glazed (Figure 16.5j) on the front and back, with a decoration of a rosette with vegetable garland on the front and crossing arches on the back. The Ligurian Blue-on-Blue is the Italian prototype of the Seville Blue-on-Blue which was manufactured in Spain. On Seville Blue-and-Blue the motifs are simpler and less carefully executed than those of Ligurian Blue-on-Blue (www.floridamuseum.ufl/histarch/gallery). In addition, the La Poterie assemblage is composed of several pieces of Dutch faience and majolica dating from the late sixteenth to early seventeenth century, such as a fragment of a Dutch majolica (*'kraakporcelein'*) plate, with a buff fabric and a Chinese (Wanli)-motif of flowers and ribbons on the flange (Figure 16.5r). It is a very common motif and similar to a dish representing Mother Mary with Christ, from northern Netherlands, with similar floral motifs on the flange (Museum Boymans van Beuningen, Rotterdam; Scholten 1993, 76–77). It dates to the early seventeenth century (1610–1660). This is most likely the first evidence of Dutch trade ware in indigenous archaeological context from the Lesser Antilles.

5 Discussion and Concluding Remarks

The colonization of the Lesser Antilles was a lengthy process that spanned over 300 years. Much of that story was told solely from the perspective of Western European chroniclers until the beginning of the study of Caribbean archaeology in the twentieth century. Though the chronicles include biased information, they nonetheless provide important details that continue to shed light on the lifeways of the indigenous peoples, their encounters with Europeans and Africans, and their responses to European colonization. As we have demonstrated, archaeology of Island Carib/*Kalinago* sites in the Windward Islands of the Lesser Antilles have revealed important information on the lifeways, deathways and material culture of the early colonial Island Carib, placing the chronicles in perspective to better understand the vast network of social relationships in the southern Caribbean.

During the fifteenth to seventeenth centuries, the *Kalinago* strongholds participated in a complex trans-Atlantic system that emerged from the combination of new colonial and trade strategies with preexisting indigenous exchange and alliance networks. The *Kalinago* communities were evidently encapsulated within the expanding European territories, but also enjoyed a great amount of local autonomy and the capability to re-negotiate the new colonial realities and inflow of peoples, goods and ideas. They were clearly not only enemies, but also trading partners of the Europeans (Hofman et al. 2014).

The archaeological investigations in St. Vincent and Grenada have provided important new insights into the Amerindian settlement structure, burial practices and associated material culture repertoires of the fifteenth to seventeenth centuries. The blending of local, South American mainland and Greater Antillean ceramic traditions in what has been labelled the Cayo complex, or more recently the Cayoid series, evidences the role of the Lesser Antilles as a new conglomerate of peoples fleeing the Spanish threat, as well as the complex social relationships and intercultural dynamics that existed during the early colonial era.

The presence of locally produced Koriabo ceramics in the Lesser Antilles suggests that mainland Carib communities moved to the islands, probably during the Late Ceramic Age, taking with them their ancestral homeland traditions. This concurs with the oral history documented by the European chroniclers. It also emphasizes the possible role that Greater Antillean refugees or Carib raids on Greater Antillean settlements may have played in the transmission of stylistic traits from the larger islands to the Lesser Antillean ceramic assemblages. The presence of European trade wares (Spanish, Italian, Dutch) and the circulation of European beads and other adornments in the Lesser Antilles reflect the early negotiations between the indigenous peoples and Europeans in the initial years of their encounters around the Caribbean Sea (Hofman et al. 2014). Though it is difficult to identify the Spanish and other European finds at Grenada and St. Vincent due to direct trade between them and the Carib, the first exchange encounters in the sixteenth century and afterwards were referenced by the Spanish who stopped to refresh and trade (Cardona [1632] 1974; Martin 2013). Besides, by the early seventeenth century the interaction between Island Carib and Spanish became much less violent than previously, partially due to the sending of missionaries to the islands. It is recorded that the adoption of the spritsail by the Island Carib can be attributed to a Franciscan friar who stayed in Dominica for sixteen months in 1605–1606 (Boomert 2002). Early colonial sources also mention that indigenous goods, often made of perishable materials, had a lasting impact on the European-Amerindian exchange networks during the early-colonial period. They mention the use of hammocks that revolutionized the seventeenth and eighteenth-century maritime logistics, and the popularity of the indigenously domesticated tobacco. The differential absorption of escaped enslaved Africans by the Island Carib communities as the colonial period progressed undoubtedly had an impact on the material cultural objects produced by both peoples. The intercultural dynamics resulted in new and unique social formations influenced by Amerindian, European and African cultural elements.

Amerindian-African-European intercultural dynamics are still mirrored in today's locally produced earthenware (known as the Afro-Caribbean ware or folk pottery), which represents the last in a long line of local pottery manufacturing traditions in some of the islands of the Lesser Antilles, notably St. Lucia, Martinique, Antigua, and Nevis (Hauser and Handler 2009; Hofman and Bright 2004; Hofman et al. 2004). Indigenous elements are also visible throughout the Caribbean in the extensive use of forms of traditional agriculture (slash-and-burn cultivation, *conucos*) and other subsistence practices, techniques of food preparation, house building, an array of cultural and linguistic traditions, ritual practices as well as ecological knowledge, the intensive use of indigenous plant species for economic and curative purposes, all aspects that constitute an important part of everyday life in the Caribbean today (Hofman et al. 2018; Pesoutova and Hofman 2016).

The present-day *Kalinago* in the Lesser Antilles are the direct inheritors of the Island Carib cultural traditions, with a considerable stake in the archaeological cultural heritage. The archaeological findings represent a source of considerable historical interest for the *Kalinago* and *Garifuna* communities in St. Vincent, Trinidad and Dominica, as well as throughout the wider Caribbean area and Central America as their origin has long been contested due to a lack of firm archaeological evidence. In the context of the construction of the Argyle International Airport, St. Vincent, the government of Saint Vincent and The Grenadines requested the rebuilding of the Cayo village at its original location. In March 2015, Leiden University presented a model of the village to the National Public Library at Kingstown during the Garifuna Conference held on the island. In January 2016 we started the building of the first experimental *Kalinago* house at Argyle in collaboration with the local stakeholders. The village with one táboüi and four manná has just been completed, and will be one of the key interest locations on the island for the region's indigenous peoples, Vincentians and visitors to the island.

Acknowledgments

The archaeological/historical research discussed in this chapter is part of the CARIB project that received funding from the Humanities in the European Research Area (HERA) (grant agreement n° 1133). This work has also been supported by the NEXUS1492 project funded by the European Research Council under the European Union's Seventh Framework Programme (FP7/2007-2013) / ERC grant agreement n° 319209 and by NWO Island Networks grant n°360-62-060.

Locally, the project at Argyle has been facilitated by the National Trust of St. Vincent and The Grenadines and the St. Vincent and The Grenadines International Airport Construction Company Ltd. In this context we particularly would like to acknowledge Kathy Martin and Henry Petitjean Roget. The 2016 investigations at La Poterie were carried out under the Memorandum of Understanding, signed between the Ministry of Tourism, Civil Aviation and Culture of Grenada and Leiden University in 2015, and legalized in 2016. Permissions were granted by Minister of Culture Senator Brenda Hood, landowners Mrs. Cleopatrice Daniel Andrew and Cheo Christopher. We would like to express our gratitude to the community of La Poterie for having us in their village. The excavation teams included researchers and students of Leiden University, Northwestern University (Chicago), St. George's University, Grenada, community members and *Kalinago* and *Garifuna* from St. Vincent. Finally, we would like to thank Mark W. Hauser, Pauline Kulstad, Konrad Antczak and Nina Jansen (Terra Cotta Incognita) for their identification of the European trade wares, and Emma de Mooij for her help with the editorial work.

References

Allaire, Louis. 1994. "Historic Carib site discovered!" *University of Manitoba St. Vincent Archaeological Project Newsletter* 1: 1–3.

Allaire, Louis. 2013. "Ethnohistory of the Caribs." In *The Oxford Handbook of Caribbean Archaeology*, edited by William F. Keegan, Corinne L. Hofman, and Reniel Rodríguez Ramos, 97–108. New York: Oxford University Press.

Andrews, Kenneth R. 1978. *The Spanish Caribbean: trade and plunder, 1530–1630*. New Haven: Yale University Press.

Anonymous. (1659) 1975. "L'Histoire de l'isle de Grenade en Amerique, manuscript anonyme de 1659." Presented and annotated by Jacques Petitjean Roget. Text established by Élisabeth Crosnier. Montreal: University of Montreal Press.

Anonyme de Carpentras. 2002. *Un flibusiter français dans la Mer des Antilles, 1618–1620: relation d'un voyage infortuné fait aux Indes occidentales par le capitaine Fleury avec la description de quelques îles qu'on y rencontre, recueillie par l'un de ceux de la compagnie que fit le voyage, présenté par Jean-Pierre*. Paris: Éditions Payot & Rivages.

Baromé, Joseph 1966. "Spain and Dominica 1493–1647." *Caribbean Quarterly* 12 (4): 30–46.

Bel, Martijn van den. 2015. *Archaeological Investigations between Cayenne Island and the Maroni River: A cultural sequence of western coastal French Guiana from 5000 BP to the present.* Leiden: Sidestone Press.

Boomert, Arie. 1986. "The Cayo Complex of St. Vincent: Ethnohistorical and Archaeological Aspects of the Island-Carib problem." *Antropológica* 66: 3–68.

Boomert, Arie. 1995. "Island Carib archaeology." In *Wolves from the Sea: Readings in the anthropology of the native Caribbean*, edited by Neil L. Whitehead, 23–35. Leiden: KITLV Press.

Boomert, Arie. 2002. "Amerindian-European encounters on and around Tobago (1498-ca. 1810)." Antropológica 97/98: 71–207.

Boomert, Arie. 2004. "Koriabo and the Polychrome tradition: The late-prehistoric era between the Orinoco and Amazon mouths." In *Late Ceramic age societies in the eastern Caribbean*, edited by André Delpuech and Corinne L. Hofman, 251–266. Oxford: British Archaeological Reports, International Series 1273.

Boomert, Arie. 2009. "Between the mainland and the islands: The Amerindian cultural geography of Trinidad." *Bulletin of the Peabody Museum of Natural History* 50 (1): 63–73.

Boomert, Arie. 2010. "Searching for Cayo in Dominica," In *Proceedings of the 23rd Congress of the International Association for Caribbean Archaeology*, edited by Samantha A. Rebovich, 655–677. St. John's: Dockyard Museum.

Boomert, Arie. 2011. "From Cayo to Kalinago." In *Communities in contact: Essays in archaeology, ethnohistory and ethnography of the Amerindian circum-Caribbean*, edited by Corinne L. Hofman and Anne van Duijvenbode, 291–306. Leiden: Sidestone Press.

Boomert, Arie. 2016. *The Indigenous Peoples of Trinidad and Tobago: From the First Settlers Until Today*. Leiden: Sidestone Press.

Boucher, Philip P. 1992. *Cannibal Encounters: Europeans and Island Caribs, 1492–1763*. Baltimore: The Johns Hopkins University Press.

Breton, Raymond. 1665. *Dictionnaire caraibe-françois*. Auxerre: Gilles Bouquet.

Breton, Raymond. 1666. *Dictionnaire françois-caraibe*. Auxerre: Gilles Bouquet.

Breton, Raymond. 1978. *Relation de l'île de la Guadeloupe*. Basse-Terre: Société d'Histoire de la Guadeloupe.

Bright, A.J. 2011. *Blood is thicker than water: Amerindian intra- and inter-insular relationships and social organization in the pre-Colonial Windward Islands*. Leiden: Sidestone Press.

Bullen, Ripley P. and Adelaide K. Bullen. 1972. *Archaeological Investigations on St. Vincent and the Grenadines, West Indies*. Orlando: William L. Bryant Foundation.

Cardona, Nicolás de. (1632) 1974. *Geographic and Hydrographic Descriptions of Many Northern and Southern Lands and Seas in the Indies, Specifically of the Discovery of the Kingdom of California*. Translated and edited by W. Michael Mathes. LosAngeles: Dawson's Book Shop.

Chanca, Diego Alavarez. (1494) 1988. "The report of Dr. Chanca." In *The Four Voyayes of Columbus* Vol. 7, edited by Cecil Jane, 20–73. New York: Dover.

Cody, Annie. 1995. "Kalinago Alliance Networks." In *Proceedings of the 15th Congress of the International Association for Caribbean Archaeology*, edited by Ricardo A. Alegria and Miguel Rodriguez, 311–326. San Juan: Centro de Estudios Avanzados de Puerto Rico y el Caribe/Fundacion Puertorriquena de las Humanidades/Universidad del Turabo.

Coppier, Guillaume. 1645. *Histoire et voyages aux Indes Occidentales, et de plusieurs autres régions maritimes et esloignée*. Lyon.

Curet, L. Antonio. 2005. *Caribbean Paleodemography. Population, Culture History, and Sociopolitical Processes in Ancient Puerto Rico*. Tuscaloosa: University of Alabama Press.

Davis, Dave D., and R. Christopher Goodwin. 1990. "Island Carib origins: Evidence and Nonevidence." *American Antiquity* 55 (1): 37–48.

Du Tertre, Jean-Baptiste. 1667–1671. *Histoire générale des Antilles habitées par les François*. Paris: Thomas Jolly.

Espinosa, Antonio V., de. (1622) 1942. *Compendium and description of the West Indies*, edited by Charles U. Clark. Washington D.C.: Smithsonian Institution.

Fernández de Oviedo y Valdés, Gonzalo. 1959. *Natural history of the West Indies*. Edited and translated by Sterling A. Stoudemire. Franklin: Chapel Hill.

García Arévalo, Manuel A. 1978. "Influencias de la dieta Indo-Hispanica en la cerámica Taina." Paper presented at the *7th International Congress for the study of Pre-Columbian Cultures of the Lesser Antilles, 11–16 June, Caracas.*

Goggin, John M. 1960. *Spanish Olive Jar: An Introductory Study*, Publications in Anthropology, No.1. New Haven: Yale University.

Goggin, John M. 1968. *Spanish majolica in the New World: types of the sixteenth to eighteenth centuries*. New Haven: Yale University.

Gullick, Charles J.M.R. 1985. *Myths of a minority: the changing traditions of the Vincentian Caribs, Vol. 30*. Assen: Van Gorcum Ltd.

Hardy, Mereditch D. 2008. "Saladoid Economy and Complexity on the Arawakan Frontier." Unpublished PhD diss., Florida State University.

Hauser, Mark W. and Jerome Handler. 2009. "Change in Small Scale Pottery Manufacture in Antigua, West Indies." *African Diaspora Archaeology Newsletter* 12 (4): 1–15.

Hoff, Berend J. 1994. "Island Carib, an Arawakan language which incorporated a lexical register of Cariban origin, used to address men." *Mixed languages* 15: 161–168.

Hoff, Berend J. 1995. "Language contact, war, and Amerindian historical tradition: The special case of the Island Carib." In *Wolves from the sea: Readings in the Anthropology of the Native Caribbean*, edited by Neil L. Whitehead, 37–60. Leiden: Kitlv Press.

Hofman, Corinne L. 1995. "Inferring inter-island relationships from ceramic style: A view from the Leeward Islands." In *Proceedings of the 15th Congress of the*

International Association for Caribbean Archaeology, edited by Ricardo E. Alegría and Miguel Rodríguez, 233–242. San Juan: Centro de Estudios Avanzados de Puerto Rico y el Caribe.

Hofman, Corinne L. 2013. "The Post-Saladoid in the Lesser Antilles (A.D. 600/800–1492)." In *The Oxford Handbook of Caribbean Archaeology*, edited by William F. Keegan, Corinne L. Hofman, and Reniel Rodríguez Ramos, 205–220. New York: Oxford University Press.

Hofman, Corinne L. 2016. Fieldwork Report La Poterie, Grenada. Unpublished manuscript on file at Faculty of Archaeology, Leiden University.

Hofman, Corinne L. and Alistair J. Bright. 2004. "From Suazoid to folk pottery: pottery manufacturing traditions in a changing social and cultural environment on St. Lucia." *New West Indian Guide/Nieuwe West-Indische Gids* 78 (1/2): 73–104.

Hofman, Corinne L. and Alistair J. Bright. 2010. "Towards a Pan-Caribbean Perspective of Pre-Colonial Mobility and Exchange: Preface to a Special Volume of the Journal of Caribbean Archaeology." *Journal of Caribbean Archaeology, Special Publication* 3: 1–3.

Hofman, Corinne L. and Menno L.P. Hoogland 2011. "Unravelling the Multi-Scale Networks of Mobility and Exchange in the Pre-Colonial Circum-Caribbean." In *Communities in Contact: Essays in Archaeology, Ethnohistory and Ethnography of the Amerindian Circum-Caribbean*, edited by Corinne L. Hofman and Anne van Duijvenbode, 14–44. Leiden: Sidestone Press.

Hofman, Corinne L. and Menno L.P. Hoogland. 2012. "Caribbean encounters: rescue excavations at the early colonial Island Carib site of Argyle, St. Vincent." *Analecta Praehistorica Leidensia:* 63–76.

Hofman, Corinne L. and Menno L.P. Hoogland. 2018. "Arqueología y patrimonio de los Kalinago en las islas de San Vincente y Granada." In *IDe La Desaparición A La Permanencia. Indígenas e indios en la reinvención del Caribe*, edited by Roberto Valcárcel Rojas and Jorge Ulloa Hung, 227–246. Santo Domingo: INTEC.

Hofman, Corinne L., Arie Boomert, Alistair J. Bright, Menno L.P. Hoogland, Sebastiaan Knippenberg, and Alice V.M. Samson. 2011. "Ties with the Homelands: Archipelagic Interaction and the Enduring Role of the South and Cental American Mainlands in the Pre-Columbian Lesser Antilles." In *Islands at the Crossroads: Migration, Seafaring and Interaction in the Caribbean*, edited by L. Antonio Curet and Mark W. Hauser, 73–86. Tuscaloosa: University of Alabama Press

Hofman, Corinne L., Alistair J. Bright, Menno L.P. Hoogland, and William F. Keegan. 2008. "Attractive ideas, desirable goods: examining the Late Ceramic Age relationships between Greater and Lesser Antillean societies." *The Journal of Island and Coastal Archaeology*, *3* (1): 17–34.

Hofman, Corinne L., André Delpuech, Menno L.P. Hoogland, and Maaike S. de Waal. 2004. "Late Ceramic Age survey of the northeastern Islands of the Guadeloupean

Archipelago: Grande-Terre, La Désirade and Petite-Terre." In *Late Ceramic Age Societies in the Eastern Caribbean*, edited by André Delpuech and Corinne L. Hofman, 159–182. Oxford: British Archaeological Reports, International Series 1273.

Hofman, Corinne L., Menno L.P. Hoogland, and Benoit Roux. 2015. "Reconstruire le táboüi, le manna et les pratiques funéraires au village caraïbe d'Argyle, Saint-Vincent." In *Á la recherche du Caraïbe perdu: Les populations amérindiennes des Petites Antilles de l'époque précolombienne à la période coloniale*, edited by Bernard Grunberg, 41–50. Paris: L'Harmattan.

Hofman, Corinne L., Angus A.A. Mol, Menno L.P. Hoogland, and Roberto Valcárcel Rojas. 2014. "Stages of Encounters: Migration, mobility and interaction in the pre-colonial and early colonial Caribbean." *World archaeology* 46 (4): 590–609.

Hofman, Corinne L., Jorge Ulloa Hung, Eduardo Herrera Malatesta, Joseph S. Jean, and Menno L.P. Hoogland. 2018. "Indigenous Caribbean perspectives. Archaeologies and legacies of the first colonized region in the New World?" *Antiquity* 92 (361): 200–216.

Holdren, Ann C. 1998. "Raiders and traders: Caraïbe social and political networks at the time of European contact and colonization in the Eastern Caribbean." PhD diss., University of California.

Honychurch, Lennox. 2000. *Carib to Creole: A History of Contact and Culture Exchange*. Roseau: Dominica Institute.

Hoogland, Menno L.P. and Corinne L. Hofman. 2013. "From Corpse Taphonomy to Mortuary behavior in the Caribbean." In *The Oxford Handbook of Caribbean Archaeology* William F. Keegan, Corinne L. Hofman, and Reniel Rodríguez Ramos, 452–469. New York: Oxford University Press.

Hulme, Peter. 1986. *Colonial Encounters. Europe and the native Caribbean, 1492–1797*. London and New-York: Methuen.

Hulme, Peter and Neil L. Whitehead eds. 1992. *Wild Majesty: Encounters with Caribs from Columbus to the present day – an anthology*. Oxford: Clarendon Press.

Keegan, William F., and Corinne L. Hofman. 2017. *The Caribbean Before Columbus*. New York: Oxford University Press.

Kirby, I.A. Earle. 1973. "The Cayo Pottery of St. Vincent: a Pre-Calivigny Series." In *Proceedings of the 5th International Congress for the Study of Pre-Columbian Cultures in the Lesser Antilles*, 61–64. Antigua: The Antigua Archaeological Society.

Labat, Jean B. (1722) 2005. *Voyages aux isles de l'Amérique (Antilles), 1693–1705*. Paris: L'Harmattan.

Laet, Johannes, de. (1630) 1988. *Nieuvve wereldt ofte beschrijvinghe van VVest-Indien* (Vol. 1). Leyden: bij de Elzeviers.

Las Casas, Bartolomé, de. 1927. *Historia de los Indios* (3 vols.). Madrid: Fondo de Cultura Económica.

Lenik, Stephan. 2012. "Carib as a Colonial Category: Comparing Ethnohistoric and Archaeological Evidence from Dominica, West Indies." *Ethnohistory* 59 (1): 79–107.

Lorimer, Joyce. 1978. "The English contraband tobacco trade in Trinidad and Guiana, 1590–1617." *The Westward Enterprise: English Activities in Ireland, the Atlantic and America, 1460–1650*, edited by Kenneth R. Andrews, Nicholas P. Canny, Paul Edward Hedley Hair, and David B. Quinn, 124–150. Liverpool: Liverpool University Press.

Lorimer, Joyce. 2006. *Sir Walter Ralegh's Discoverie of Guiana*. Works issued by the Hakluyt Society, Third Series, No. 15. London: Ashgate.

Marken, Mitchell W. 1994. *Pottery from Spanish shipwrecks, 1500–1800*. Gainesville: University Press of Florida.

Martin, John A. 2013. *Island Caribs and French Settlers in Grenada, 1498–1763*. St. George's: Grenada National Museum Press.

Moreau, Jean-Pierre. 1992. *Les Petites Antilles de Christophe Colomb à Richelieu: 1493–1635*. Paris: Éditions Karthala.

Nicholl, John. 1605. *An hourglass of Indian news*. London: Printed for Nathaniell Butter.

Oliver, José R. 2009. *Caciques and Cemi Idols. The Web Spun by Taíno Rulers Between Hispaniola and Puerto Rico*. Tuscaloosa: University of Alabama Press.

Oviedo y Banos, Don Jose de. 1987. *The Conquest and Settlement of Venezuela*. Translated by Jeannette Johnson Varner. Berkeley: University of California Press.

Palacio, Joseph O. 2005. *The Garifuna, A Nation Across Borders: Essays in Social Anthropology*. Belize: Cubola Books.

Pelleprat, Pierre. 1965. *Relato de las misiones de los Padres de la Compañía de Jesús en las islas y tierra firme de América Meridional*. Alicante: Biblioteca Virtual Miguel de Cervantes.

Pesoutova, Jana and Corinne L. Hofman. 2016. "La contribución indígena a la biografía del paisaje cultural de la República Dominicana. Una revisión preliminar." In *Indígenas e Indios en el Caribe Presencia, legado y estudio*, edited by Jorge Ulloa Hung and Roberto Valcárcel Rojas, 115–150. Santo Domingo: INTEC.

Petitjean Roget, Henry. 2001/2002. *"De Baloue à Cariacou": Facettes de l'Art Amérindien Ancien des Petites Antilles – Catalogue des Pièces Exposées*. Fort-de-France: Ecomusée de Martinique.

Petitjean Roget, Henry. 2015. *Les Tainos, Les Callinas Des Antilles*. Guadeloupe: Association Internationale d'Archeologie de la Caraïbe.

Pinchon, Robert Père. 1961. "Description de l'Isle de Saint-Vincent. Manuscrit anonyme du début du XVIIIème Siècle." *Annales des Antilles* 9: 35–81.

Reid, Basil A. 2009. *Myths and Realities of Caribbean History*. Tuscaloosa: University of Alabama Press.

Richard, Gérard. 2002. "Capesterre Belle Eau: Arrière plage de Roseau." In *Bilan Scientifique de la Guadeloupe,* edited by the Ministry of Culture and Communication, Paris.

Rochefort, César. 1658. *Histoire naturelle et morale des îles Antilles de I' Amérique.* Rotterdam: Arnold Lucas.

Rostain, Stéphen. 2009. Between Orinoco and Amazon: The Ceramic Age in the Guianas." In *Anthropologies of Guayana: Cultural Spaces in Northeastern Amazonia,* edited by Neil L. Whitehead and Stephanie W. Alemán, 36–54. Tucson: Arizona University Press.

Rouse, Irving. 1992. *The Tainos: Rise and Decline of the People who Greeted Columbus.* New Haven: Yale University Press.

Scholten, Frits T. 1993. *Dutch Majolica & Delftware, 1550–1700; the Edwin van Drecht Collection.* Amsterdam: Van Drecht.

Shafie, Termeh, David Schoch, Jimmy L.J.A. Mans, Corinne L. Hofman, and Ulrik Brandes. 2017. "Hypergraph Representations: A study of Carib attacks on Colonial Forces, 1509–1700." *Journal of Historical Network Research* 1: 3–17.

Sued Badillo, Jalil, ed. 2003. *General history of the Caribbean, Vol. 1, Autochthonous societies.* Paris and Oxford: UNESCO and Macmillan Caribbean.

Taylor, Douglas R. and Berend J. Hoff. 1980. "The Linguistic Repertory of the Island-Carib in the Seventeenth Century: The Men's Language: A Carib Pidgin?" *International Journal of American Linguistics* 46 (4): 301–312.

Twinn, Paul. 2006. "Land ownership and the construction of Carib identity in St.Vincent." In *Indigenous resurgence in the contemporary Caribbean: Amerindian survival and revival,* edited by Maximilian C. Forte, 89–106. New York: Peter Lang Publishing.

Waal, Maaike S., de. 2006. *Pre-Columbian social organisation and interaction interpreted through the study of settlement patterns. An archaeological case-study of the Pointe des Châteaux, La Désirade and Les Îles de la Petite Terre micro-region, Guadeloupe, FWI.* La Désirade and Les îles de la Petite Terre micro-region, Guadeloupe, FWI Leiden: De Waal Publishers.

Watts, David. 1987. *Patterns of development, culture and environmental change since 1492.* Cambridge: Cambridge University Press.

Whitehead, Neil L. 1995. *Wolves from the Sea: Readings in the Anthropology of the Native Caribbean, Vol. 14.* Leiden: KITLV Press.

EPILOGUE

Situating Colonial Interaction and Materials: Scale, Context, Theory

Maxine Oland

As stated in the introduction, the editors of this volume bring together archaeological research from early colonial encounters in the Caribbean and the surrounding mainland. They sought original and fresh field data, and asked scholars to bring a material culture perspective to their interpretations. The result is an impressive body of work that brings into conversation case studies from a diverse set of indigenous cultural traditions.

Although the authors draw from a range of historical data, previously excavated, and recently excavated material, each foregrounds material culture as a way to understand colonial period interactions. How do the artifacts found at early colonial period indigenous sites help us understand the relationships between European, Amerindian, and African peoples? What can these objects tell us about the culture changes that resulted from these interactions? These questions fit into a larger inquiry into the role of material culture in colonial contexts, which is summarized in the introduction.

This epilogue highlights three areas of inquiry that seem critical to the case studies in this volume. What is the scale of contact and material interaction? What are the contexts in which material culture was acquired, created, and deposited? And how do we conceptualize the relationship between foreign or mixed material culture and culture change in a way that honors the indigenous peoples that made, acquired, or deposited these objects? I conclude the epilogue with a short look toward future studies, as inspired by reading these chapters.

1 The Scale of Colonial Interaction

Several authors stress that the material record at colonial period sites is greatly shaped by the scale, type, and frequency of colonial interaction. How long was the period of colonization? Were relationships between colonists and indigenous peoples based in exploitation, or were they on a more equal footing? Authors illustrate how interactions between indigenous and Spanish peoples were shaped by Europeans' economic and political goals. They also point out the longer lasting effects of these interactions on indigenous networks and political systems.

The case for scalar analysis is argued most directly by Fowler and Card, who state that archaeologies of colonialism should address Orser's (2014) "haunts" of modernity (colonialism, mercantilism/capitalism, Eurocentrism, racialization) but that we need to do so at a variety of scales. The authors suggest that we study colonial interaction at the local, regional, and global scales, and consider the length and duration of contact. Their study comparing Ciudad Vieja, the first Spanish *villa* in El Salvador, and Caluco, an indigenous town in the Izalcos region, illustrates the value of a multiscalar analysis, and the differences in indigenous spatial and material use at the two sites.

Mathers takes a macro-regional perspective, and argues that type of contact, frequency, and political histories of indigenous groups are crucial for understanding patterns of consumption of European goods in indigenous communities. His chapter compares the *entradas* of the southeastern and southwestern United States. Whereas the goals of the Spaniards and the materials they carried with them were nearly identical in each region, there have been very few Spanish goods found that were used or modified in indigenous contexts in the Southwest.

Mathers concludes that the explanations for the differences are geographical, historical, climatic, and geopolitical. The Southwest was more sparsely populated, remote, and had questionable economic value to the Spaniards, and was separated from other Spanish territories by hostile hunter-gatherer groups. *Entradas* were also very long, over land, and were later and less frequent than the *entradas* in the southeast. There were fewer indigenous settlements with large food supplies. Mathers also hints at the differences in indigenous political systems. Whereas the southeast was characterized by hierarchical chiefdoms, whose elite leaders relied on exotic foreign goods for prestige (i.e. Beck et al. 2011), the more egalitarian political systems of the Southwest did not depend on an influx of foreign goods.

As illustrated in the next section, the case studies in this volume pay careful attention to the manner and length of colonial exploitation (or lack thereof), and the ways that particular colonial contexts shaped material assemblages. Indigenous residents of a pearl fishery *ranchería* (Antczak et al. this volume), mining town (Ernst and Hofman this volume), or an *encomienda* (Valcárcel Rojas this volume) created different material patterns when relations were structured by domination than when they were structured by trade (Berman and Gnivecki this volume, Hofman et al. this volume, Keehnen this volume, Van den Bel and Collomb this volume). Similarly, residents of a colonial frontier, who experienced occasional visits from priests and encomenderos (Awe and Helmke this volume), interpreted and incorporated European material culture through a much different lens from those in colonial cores (Fowler and Card this volume, Sarcina this volume, Hernández-Sánchez this volume).

2 Contextualizing Colonial Material Culture

Most of the chapters combine an analysis of scale with detailed attention to the context of novel material culture objects and assemblages. How did European goods get into indigenous communities? How were they used, and by whom? What kinds of material changes occurred at Spanish settlements, and what can they tell us about interactions?

The case studies in this book use the careful consideration of documentary sources and archaeological data to inquire into the processes by which European artifacts were incorporated into indigenous life. They pay special attention to the scale and type of contact (how long? under what circumstances?), and to the context in which European materials were acquired and eventually found. Historical documents make explicit the ways Europeans conceptualized their relationships with indigenous peoples, and hint at indigenous motivations. The material record, when considered within the scale and context of the larger site and region, provides a window into indigenous ontologies and agentive choices.

Several chapters deal with situations of contact in which power relations between Europeans and indigenous populations were fairly balanced, or in which indigenous communities did not live under a consistent dominating European presence. For example, for the Lucayans of the Bahamas, there was only sporadic and intermittent contact until 1509, and no direct control of indigenous peoples. Berman and Gnivecki illustrate how European objects at the Long Bay site transformed European objects into indigenous ones. European objects were adopted for their aesthetics (brilliance, smell, sound, color) and their association with distant locales. Berman and Gnivecki argue that European objects were adopted because they were consistent with the Lucayan cosmovision.

Keehnen found that indigenous peoples on Hispaniola obtained European artifacts as gifts, via pilfering, by collecting, and through native exchange. Material patterns illustrate that European objects were used for indigenous purposes. Ritual caches at El Variar contained European materials, and Spanish goods were deposited in indigenous burials at Juan Dolio. At En Bas Saline, most European artifacts and foods were found in elite contexts. Keehnen argues that although European objects did not penetrate much of indigenous life on Hispaniola, their deposition reflects that they were controlled by elites, and played a role in the indigenous political economy of Hispaniola.

Awe and Helmke illustrate how Maya residents of the Belize frontier received European objects from Spaniards as gifts, as payment for services, as rewards for conversion, and in trade. But they also took them by forceful means, or received them indirectly through indigenous trade networks. Although much of the Belize frontier was incorporated into *encomienda* by 1544, the region lacked a consistent Spanish presence. European goods were largely used for Maya purposes, as status symbols for the Maya elite, or as a replacement for native exotics in Maya caches (see also Oland 2014, 2017; Pugh 2009; Pugh et al. 2012). Other European objects, such as axes and machetes, were used for mundane purposes, in addition to their deposition in elite caches.

The case study by Van den Bel and Collomb focuses on the Yao, an indigenous group that positioned itself as a middleman between the Europeans and indigenous populations of Guiana. The Yao were afraid of the Spanish, who were known for their slave raids, but maintained a free zone of trade with the French, British, and Dutch. The Yao traded European goods along their extensive Amerindian networks. Van den Bel and Collomb show that most European objects in Guiana were found in urn burials, and were transformed by Amerindians into symbolic objects of prestige. Although the exchange rate was tilted in the Europeans' favor, power relations between Amerindians and Europeans were otherwise fairly equitable. This lasted until the later seventeenth century, when the Dutch created larger settlements and took Amerindian land for the creation of sugar plantations.

As noted in many of the case studies, Spaniards gave gifts to indigenous leaders and individuals as a strategy to gain their cooperation, particularly when making initial contacts and *entradas*. Valcárcel Rojas points out that the "gift kit" was a product of a mercantile strategy in the Greater Antilles. Between 1503 and 1505, Spanish colonists in Cuba shifted from a mercantile to an imperial strategy of tribute collection. Gift giving went by the wayside, and Spaniards quickly shifted to a system in which they paid for indigenous *encomienda* labor with European goods. Valcárcel Rojas found many objects that would not be found in a traditional gift kit, such as weapons, tools, and elements associated with architecture and horses. These objects were found in indigenous settlements that were refashioned as annexes to Spanish farms, or as mining camps. Unlike the examples from many other chapters, the objects found in the Holguín province were not acquired and used according to indigenous value systems, but in a context of domination.

Henry and Woodward's example from Jamaica shows significant material continuity at the 'Taíno' site of Maima between the precolonial and colonial period. Although there were a few nails, fragments of glass, and pieces of metal found at Maima, these likely date to the initial year of contact between the Spanish and 'Taíno,' which were characterized by more or less equitable trade relations between the two groups. Documentary sources show that subsequent relations were characterized by domination by Spaniards based at Sevilla la Nueva. Henry and Woodward therefore conclude that the lack of European material culture at Maima was due to the conscious rejection of European material culture by the 'Taíno' residents.

This chapter also provides context as to how 'Taíno' ceramics were incorporated at the Spanish settlement of Sevilla la Nueva. Henry and Woodward found that most of the Spanish households used traditional 'Taíno' ceramics, which they see as evidence of 'Taíno' women taken as wives. Only the households of elite Spaniards, found at the governor's fort, had evidence of New Seville Ware. This ceramic ware was a type of colonoware, which combined traditional 'Taíno' ceramic materials and methods, but made into Spanish styles. Unlike other chapters that deal with the blending of indigenous and European ceramic technologies (see below), the careful analysis at Sevilla la Nueva makes it clear that this technological blending took place in a context of domination and European class privilege.

3 Theorizing Material Shifts

The previous sections discuss the way authors use scale and context to deeply understand the relationships that led to particular material patterns. This section examines how authors couch these relationships within a larger theoretical framework. What does it mean when indigenous peoples adopt European technology, religion, or goods? How do we understand the mix of European and indigenous styles and craft techniques within the language of cultural mixing? What does it mean if the material culture remains indigenous, but the overall assemblage changes in response to colonial geopolitics? Finally, what theoretical language do we use for indigenous communities trying to maintain their political, economic, and social systems in spite of the presence of Europeans?

One of the most common forms of "mixed" objects discussed in this volume is pottery that combines both European and indigenous technological and stylistic elements. For example, Hernández Sánchez argues that potters in central

Mexico made choices about whether to adopt, reject, or reinterpret new European technologies. She shows that potters selectively adopted glaze and majolica decoration, but not the potter's wheel. She argues that the choices potters made were not political, but were about maintaining their way of life while at the same time adapting to post-conquest society.

Fowler and Card's study of Ciudad Vieja, the first Spanish *villa* in El Salvador, first uses Bourdieu's structural theory of practice to understand ceramic change. The authors argue that Ciudad Vieja was an "incubator of experimentation and transformation" for the first generation of potters that experimented with new styles. Through a process of creolization the next generation created a more homogeneous style of ceramics, which were Pipil in style, but incorporated Spanish elements.

Sarcina, in contrast, uses the concept of syncretism to understand the indigenous-made pottery with European motifs found at Santa María, Colombia. This pottery was found in an indigenous context, alongside only indigenous materials. Sarcina therefore argues that this pottery was the result of experiments by local potters, who were inspired by the material culture of their masters.

Ernst and Hofman employ the concept of transculturation, which they define as the "creative, ongoing, process of appropriation, revision and survival," to understand the creation of a new pottery style in Hispaniola. Europeans did not enforce a fully Spanish way of life, and contexts of sustained interaction between indigenous, African, and European peoples, such as in mining camps, resulted in new forms of material culture.

Iverson uses Judith Butler's (1990) concept of "resignification" to challenge traditional narratives of religious conversion in her case study from Hidalgo. Iverson illustrates how indigenous religious ontologies shaped the New World church in significant ways, even though they lacked autonomy and power under the domination of the church. The indigenous residents of Tula altered the meanings and values of the church, and subtly changed it, simply by participating in the construction of buildings and the practice of rituals.

Sheptak and Joyce bring up the important point that colonial change in indigenous societies may not be expressed through foreign material culture or styles. Although the material culture at Ticamaya, Honduras, appears to be indigenous, they argue that careful attention to the assemblage reveals changes in practice. Obsidian projectile points, for example, are not in and of themselves hybrid objects, as they were found in earlier deposits at Ticamaya, and have been found in Late Postclassic and Terminal Late Postclassic deposits across the Maya lowlands (i.e. Masson and Peraza Lope 2014, 274; Oland 2013; Simmons 2002). Instead, the increase in indigenous obsidian projectile points at Ticamaya, alongside the addition of new defensive walls (perhaps influenced by the Spaniard turned Maya resistance fighter, Gonzalo Guerrero), represents the ethnogenesis or hybridity of a more militarized culture in response to colonialism.

4 Discussion: Future Challenges and Possibilities

In her chapter about religious conversion, Iverson argues that we need new narratives for understanding the colonial experience of indigenous peoples. The chapters in this volume contribute to that effort, as they contextualize material and cultural changes along the continuum of agentive choice and domination, and within the long-term historical trajectories of indigenous groups. As we move forward, however, we will have to struggle with how to create these new narratives.

What is the balance between focusing on Eurocentrism, as Orser (2012) urges, and understanding colonialism from the perspective of indigenous motivations, worldviews, and long-term histories? Many of the authors in this volume, for example, find it useful to consider the way Amerindian peoples incorporated Europeans and their objects into their own worldviews and value systems, while others are more explicit about the constraints that Europeans placed on indigenous lives and material worlds. There is a tension between these approaches that could be explored further.

How do we intentionally theorize the mixing of European and indigenous material culture elements? In his consideration of hybridity, Liebmann (2013) breaks down the backgrounds and meanings behind various theoretical conceptions of the amalgamation of indigenous and European elements in material culture. I was struck, in this volume, by how many different conceptual terms were used for pottery found with both European and indigenous features. Just as we must take care to consider the power dynamics behind the terms we use for colonial processes (Jordan 2014; Silliman 2005), we must also be aware of the implications of the ways we conceptualize colonial period material culture (Liebman 2013).

Many times throughout this volume I questioned what theoretical constructs are available for understanding the process of consuming and creating objects according to one's own value system and worldview, in spite of the presence of Europeans? Several authors in the book (Keehnen this volume, Antczak et al. this volume, Hofman et al. this volume) employ the term entanglement, seemingly as a metaphor (Silliman 2016) to describe the messy interweaving of multiple ontologies, objects, and cultures that came out of colonial interactions.

Kurt Jordan (2009, 2014), building upon the work of Rani Alexander (1998), argues that cultural entanglement should be used to conceive of interactions in which indigenous peoples held power equal to, or greater than, European colonists. As many authors point out, the earliest colonial period interactions in the circum-Caribbean region were often characterized by equitable power relations, although these quickly shifted to more exploitative relations as the colonists' goals changed from mercantile to imperial. It may be useful to explore this term further as a way to conceptualize both colonial period relations in this region, and the material worlds that resulted from these relations.

One of the most exciting trends in this volume is the move away from binary models that position indigenous/colonized peoples in opposition to Europeans/colonizers. The reality was almost always far more complicated, and included competition between European powers, the personal goals of individual European conquerors, multiple and often competing indigenous groups and individuals, and African slaves and their descendants.

Ernst and Hofman, for example, see the mix of African, Spanish, and indigenous influence in the pottery of Hispaniola. Van den Bel and Collomb's chapter recognizes that European trade in Guiana was shaped by indigenous trade networks, as much as by the competition between the Spanish and other European powers in Guiana. Mathers' chapter brings attention to the native allies brought from Mexico on Southwest *entradas*, some of whom stayed in Zuni communities. The fact that colonial Mexican pottery, made by indigenous potters, was modified and used in Hawikku graves complicates the colonizer/colonized binary in exciting ways.

The chapter by Hofman, Hoogland, Boomert, and Martin shows how the southern Lesser Antilles was a space contested by both Amerindian groups and multiple European powers. Their work is meaningful for contemporary *Kalinago* peoples whose origins have been contested, and suggests the potential and need for more collaboration between contemporary populations and archaeologists studying colonialism in this region. Amerindian groups can challenge narratives of erasure through the creation of archaeologically based heritage sites.

The chapters in this volume point us forward toward the creation of new narratives that do not deny the horrific realities of colonial loss, but also honor the creativity and adaptability of indigenous and enslaved peoples. At the same time, they break down simplistic and binary definitions of colonialism. The diversity of concepts and theoretical constructs in the volume are in part tied to the tremendous diversity present in the contexts, scales, and situations of colonial interaction in this region. Future scholarship will include conceptual and theoretical models that challenge the limiting binary of colonizer and colonized/European and indigenous, paying increased attention to the role that enslaved and free Africans, their descendants, and indigenous allies of Spanish *conquistadores*, played in colonial interactions. It will also include efforts to use the results of archaeological studies for meaningful heritage work with contemporary descendants of indigenous and African peoples.

References

Alexander, Rani. 1998. "Afterword: Toward an Archaeological Theory of Culture Contact." In *Studies in Culture Contact,* edited by John G. Cusick, 467–795. Center for Archaeological Investigations Occasional Paper No. 25. Carbondale: Southern Illinois University.

Beck, Robin A. Jr., Christopher B. Rodning, and David G. Moore. 2011. "Limiting Resistance: Juan Pardo and the Shrinking of Spanish La Florida, 1566–1568." In *Enduring Conquests: Rethinking the Archaeology of Resistance to Spanish Colonialism in the Americas,* edited by Matthew Liebmann and Melissa Murphy, 19–39. Santa Fe: School for Advanced Research Press.

Butler, Judith. 1990. *Gender Trouble: Feminism and the Subversion of Identity.* New York: Routledge.

Jordan, Kurt A. 2009. "Colonies, Colonialism, and Cultural Entanglement: The Archaeology of Postcolumbian Intercultural Relations." In *International Handbook of Historical Archaeology,* edited by David Gaimster and Teresita Majewski, 31–49. New York: Springer.

Jordan, Kurt A. 2014. "Pruning Colonialism: Vantage Point, Local Political Economy, and Cultural Entanglement in the Archaeology of Post-1415 Indigenous Peoples." In *Rethinking Colonial Pasts through Archaeology,* edited by Neal Ferris, Rodney Harrison, and Michael V. Wilcox, 103–120. Oxford: Oxford University Press.

Liebmann, Matthew. 2013. "Parsing Hybridity: Archaeologies of Amalgamation in Seventeenth-Century New Mexico." In *The Archaeology of Hybrid Material Culture,* edited by Jeb J. Card, 25–49. Center for Archaeological Investigations Occasional Paper No. 39. SIU-Carbondale: Center for Archaeological Investigations.

Masson, Marilyn A. and Carlos Peraza Lope. 2014. *Kukulcan's Realm: Urban Life at Ancient Mayapán.* Boulder: University Press of Colorado.

Oland, Maxine. 2013. "The 15th–17th Century Lithic Economy at Progresso Lagoon, Belize." *Lithic Technology* 38 (2): 81–96.

Oland, Maxine. 2014. "'With the Gifts and Good Treatment that He Gave Them': Elite Maya Adoption of Spanish Material Culture at Progresso Lagoon, Belize." *International Journal of Historical Archaeology* 18 (4): 643–667.

Oland, Maxine. 2017. "The Olive Jar in the Shrine: Situating Spanish Objects within a 15th–17th Century Maya Worldview." In *Indigenous People and Foreign Things:*

Archaeologies of Consumption in Native America, edited by Craig N. Cipolla, 127–142. Tuscon: University of Arizona Press.

Orser Jr, Charles E. 2012. "An Archaeology of Eurocentrism." *American Antiquity* 77 (4): 737–755.

Orser Jr, Charles E. 2014. *A Primer on Modern-World Archaeology*. Clinton Corners: Eliot Werner Publications.

Pugh, Timothy W. 2009. "Contagion and Alterity: Kowoj Maya Appropriations of European Objects." *American Anthropologist* 111 (3): 373–386.

Pugh, Timothy W., Jose Romulo Sanchez, and Yuko Shiratori. 2012. "Contact and Missionization at Tayasal, Peten, Guatemala." *Journal of Field Archaeology* 37 (1): 3–19.

Silliman, Stephen W. 2005. "Culture Contact or Colonialism? Challenges in the Archaeology of Native North America." *American Antiquity* 70 (1): 55–74.

Silliman, Stephen W. 2016. "Entanglements of Colonial Encounters and Social Inequality: Disentangling the Archaeology of Colonialism and Indigeneity." In *Archaeology of Entanglement*, edited by Lindsay Der and Francesca Fernandini, 31–48. London and New York: Routledge.

Simmons, Scott E. 2002. "Late Postclassic-Spanish Colonial Period Stone Tool Technology in the Southern Maya Lowland Area: The View from Lamanai and Tipu, Belize." *Lithic Technology* 27 (1): 47–72.

Index

Printed in the United States
By Bookmasters